Sustainable Plant Nutrition in a Changing World

This book series aims to provide new insight and improve understanding of plant nutrition, a crucial component in plant production and food security. Plant nutrition encompasses the study of plant growth and metabolism, external supply of necessary chemical compounds and the biochemistry of various nutrients in plants. This field of study has grown in light of recent scientific advances and climate change and has expanded to include the impact of climate change, pollution, nanomaterials, biofortification and genomics on the nutrition of plants.

The component volumes are written by leading scholars from around the globe and contain the latest research and applications. Each volume is fully illustrated with in-depth references for further reading. This series will be of interest to all scientists, academics, students and professionals in the plant sciences and agriculture, as well as biotechnology, ecology and biochemistry.

Sheikh Adil Edrisi • Pradeep Kumar Dubey
Cherukumalli Srinivasa Rao • Himlal Baral
Rajiv Kumar Chaturvedi
Purushothaman Chirakkuzhyil Abhilash
Editors

Sustainable Plant Nutrition and Soil Carbon Sequestration

Springer

Editors
Sheikh Adil Edrisi
Thapar School of Liberal Arts and Sciences
Thapar Institute of Engineering and Technology
Patiala, Punjab, India

Cherukumalli Srinivasa Rao
National Academy Agricultural Research Management
ICAR
Hyderabad, Telangana, India

Rajiv Kumar Chaturvedi
Department of Humanities and Social Sciences
Birla Institute of Technology and Science
Goa, India

Pradeep Kumar Dubey
Department of Environment & Sustainable Development
Banaras Hindu University
Varanasi, Uttar Pradesh, India

Himlal Baral
Centre for International Forestry Research
Bogor, Indonesia

Purushothaman Chirakkuzhyil Abhilash
Department of Environment & Sustainable Development
Banaras Hindu University
Varanasi, Uttar Pradesh, India

Editor-in-Chief
Hassan El-Ramady

ISSN 2662-2394 ISSN 2662-2408 (electronic)
Sustainable Plant Nutrition in a Changing World
ISBN 978-3-031-53589-5 ISBN 978-3-031-53590-1 (eBook)
https://doi.org/10.1007/978-3-031-53590-1

This Springer imprint is published by the registered company Springer Nature Switzerland AG
The registered company address is: Gewerbestrasse 11, 6330 Cham, Switzerland

Preface

Terrestrial plant systems are an integral part of Earth's land resources. Resources are mutually connected via the nutrient exchange phenomena. Thus, plant nutrition plays a crucial role in managing soil fertility and land productivity. Soil organic carbon (SOC) is one of the critical indicators for assessing the viability of land, and hence soil carbon sequestration, which is a burgeoning issue regarding changing climatic conditions. It has been consistently proclaimed that planting trees vigorously will help counter the challenges in sequestering carbon dioxide (CO_2) in terrestrial landscapes. However, it has been observed that climate change is also affecting the health of plants and trees, making them less efficient at capturing CO_2 in the aboveground and belowground parts. Hence, the usual planting of trees will not suffice the requirements in order to achieve our global commitments like the Bonn Challenge and associated UN Sustainable Development Goals (UN-SDGs). In this backdrop, there is a need to develop advanced strategies that include modeling approaches, climate-adaptive agroforestry practices, and agronomic approaches, viz., utilization of biodegradable waste, biochar technologies, biofertilizers, novel microorganisms, etc., that can support sustainable plant nutrition and carbon sequestration.

In this context, the present book, *Sustainable Plant Nutrition and Soil Carbon Sequestration*, aims to provide an essential linkage between plant nutrition and carbon sequestration and their sustainable management strategies that lead to multidimensional benefits for environmental sustainability. The primary purpose of this book is to explore the nexus between carbon sequestration and plant growth, its role in maintaining ecosystem services, and the modeling of aspects of soil carbon and nutrient dynamics. Furthermore, it aims to address the growing challenges of greenhouse gas emissions and ecological perturbations, unraveling the potential of degraded lands for food, fuel, and nutrition security and accounting for meeting various UN-SDGs, including net-zero targets. Moreover, the prospects and policy implications of the different strategic measures have been discussed. This edited book volume has primarily tried to broaden readers' understanding of various dimensions regarding environmental sustainability and climate justice.

The book is divided into four sections. The first section discusses the overview of plant nutrition and soil carbon sequestration that deals with the fundamental processes and mechanisms involved, different approaches to sequestration, and crucial discussion toward enhancing nutrient use efficiency (NUE) for achieving net-zero emissions under sustainable agricultural practices. The second section deals with the modeling perspectives of soil carbon and nutrient dynamics, covering different insights into the dynamism of carbon sequestration among different ecosystems. The discussion on modeling plant growth, nutrition, and the dynamics of SOC, along with the dedicated chapter on remote sensing and modeling SOC sequestration, built fundamental perspectives. In contrast, the chapter on carbon dynamics and sequestration under past and future climate scenarios via Earth system model simulations brought insightful perspectives. The third section demonstrates various strategies for enhancing plant nutrition and carbon sequestration, and there are crucial discussions about agroforestry practices in a separate chapter. The dedicated chapter on agronomic practices for improving the NUE attracted particular attention. The chapter on recycling biodegradable waste in agriculture brought insightful perspectives toward influencing soil quality and plant nutrition. The different prospects of exploiting the potentials of biochar technology, biofertilizers, and novel microorganisms for agricultural sustainability have been dealt with under individual chapters. Finally, yet importantly, the fourth section is inclined toward the concluding chapters and is dedicated to paving the way for future avenues and critical challenges while attaining environmental sustainability. These discussions were divided into chapters that focus on identifying priority areas for cropland restoration via agroforestry and soil carbon enrichment, challenges and opportunities for soil carbon sequestration, and the global and regional sustainability challenges for plant and soil nutrient management.

Thus, we shall be pleased if this book can be a valuable piece of information envisioned for enriching the insights of undergraduate and postgraduate students, scientists, social scientists, decision-makers, and interested ones. The topics showcased in this volume shape the significant contributions of soil and plant scientists, agricultural scientists, ecologists, and biogeochemists, whose research is mentioned throughout this piece. As editors of this book, we are grateful to all those contributors who are part of this book directly or indirectly for their interests, outstanding contributions, and consistent cooperation that ultimately brought this piece to fruition. We all are indebted to our guides, teachers, seniors, students, well-wishers, and beloved family. Without their never-ending support, encouragement, and motivation, this colossal task would have never been completed.

Patiala, Punjab, India — Sheikh Adil Edrisi
Varanasi, Uttar Pradesh, India — Pradeep Kumar Dubey
Hyderabad, Telangana, India — Cherukumalli Srinivasa Rao
Bogor, Indonesia — Himlal Baral
Sancoale, Goa, India — Rajiv Kumar Chaturvedi
Varanasi, Uttar Pradesh, India — Purushothaman Chirakkuzhyil Abhilash

Acknowledgments

The experience of editing this book comprising the holistic perspectives and sections is enriched by worthy inputs from the contributors and reviewers. An edited book of this level does not find its purview without the indebted support and patience of many enthusiastic personalities. We are earnestly grateful to our students and colleagues who assisted us in this voyage to bring this book to the spotlight. We are sincerely gratified by the stupendous impression, elite guidance, and unwavering encouragement from Prof. Padmakumar Nair, Prof. Rattan Lal, Prof. Panjab Singh, Prof. Kaustuv Roy, Prof. Ajit Sinha, Prof. Santha Kumari, Prof. Vinay Kumar, Prof. N. Tejo Prakash, Prof. Rajeev Mehta, Prof. Anoop Verma, Prof. B.D. Singh, Prof. R.K. Mall, Prof. H.B. Singh, Prof. Amitava Rakshit, Dr. J.P. Verma, and Dr. Shalini Dhyani. We feel glad to wish our heartfelt thanks to Dr. Rama Kant Dubey, Dr. Vishal Tripathi, Dr. Durgesh Kumar Jaiswal, Dr. Anuj Shukla, and Dr. Jay Prakash Mishra for their consistent support and boosting morale during the discourse of this book preparation and compilation.

Sheikh Adil Edrisi is grateful to Science and Engineering Research Board (SERB) for the Start-up Research Grant and Council of Scientific & Industrial Research-National Botanical Research Institute-International Society for Environmental Botanists (CSIR-NBRI-ISEB) for the Young Scientist Medal in Environmental Science. Pradeep Kumar Dubey is indebted to the Society for Plant Research (SPR), India, and the United Nations Intergovernmental Panel on Biodiversity and Ecosystem Services (IPBES) for the Fellow Awards. Special gratitude is also due to Prof. Panjab Singh, the President of the National Academy of Agricultural Sciences (NAAS), for his continuous motivation and encouragement. We also thank the DST-Mahamana Centre for logistics support for Excellence in Climate Change Research (MCCECR).

We would like to express our sincere gratification to Prof. Hassan El-Ramady, Eric Stannard, Kate Lazaro, Natalie Munoz, Shina Harshavardhan, Shanthini Kamaraj, and the production team members from Springer for their swift support, guidance, and noteworthy cooperation throughout the procedure of publishing this worthy piece. Last but not least, we should thank our family, immediate and extended, who always encouraged us to continue the massive task. Despite the best

About the Editors

Sheikh Adil Edrisi is an Assistant Professor at the Thapar School of Liberal Arts and Sciences at Thapar Institute of Engineering & Technology (TIET), Patiala, India. He earned his MPhil and PhD from Banaras Hindu University in Varanasi, India. He is researching to restore degraded lands for multipurpose environmental benefits, exploring the role of GIS to assess land productivity, bioremediation of contaminated and degraded lands for efficient bioenergy production, biodiversity conservation and restoration by developing an apposite policy framework for attaining key UN Sustainable Development Goals (UN-SDGs). He has received the 2022 Prof. Subodh Bhatnagar Innovation Award from the Society for Plant Research, India, the 2021 Early Career Start-up Research Grant from SERB, Govt. of India, the 2019 Green Talents Award from the Federal Ministry of Education and Research (BMBF, Germany) in Germany, and The National Academy of Sciences, India (NASI)-Springer Award (2018) from NASI, India. Additionally, he holds editorial roles in reputed journals in the area of land degradation, restoration, and sustainable production, such as *Energy, Ecology and Environment*, *Anthropocene Science* (Springer), *Restoration Ecology*, *Land Degradation and Development* (Wiley), and *PLoS-One* (PLoS). Dr. Adil is also the "initial member" of Global Soil Biodiversity Initiatives and is currently an active International Union for Conservation of Nature (IUCN)- Commission on Ecosystem Management (CEM), Commission on Environmental, Economic and Social Policy (CEESP), and Commission on Education and Communication (CEC) commission member to resolve pressing global challenges of land restoration and sustainable development.

Pradeep Kumar Dubey is a Program Manager within the Sustainable Landscape and Restoration program at World Resource Institute India (WRI India). His primary area of expertise lies in agroecosystem management following a bottom-up approach, and he currently focuses on landscape restoration in Maharashtra, India. Pradeep holds an MSc-PhD degree in Environmental Science from Banaras Hindu University, Varanasi, India. He has undertaken substantial agricultural surveys and field-level experimental validation, proposing various ecological solutions for agri-environmental sustainability, particularly the land-soil management within the

Vindhyan dryland and Gangetic Plain in eastern UP, India. Dr. Dubey is selected as IPBES fellow for Nexus assessment, fellow of the Society for Plant Research in India, and recipient of the IUCN-CEM Agroecosystems Special Recognition Award and Young Scientists Award from the Society for Science of Climate Change and Environmental Sustainability. He serves as a member of IUCN commissions (CEM, CEC, CEESP), Global Land Program (GLP), and Global Soil Biodiversity Initiative (GSBI) and is invited as a resource person, expert, or speaker in various state, national, and international forums. He published research, reviews, opinions, editorial pieces, book chapters, and a lead-authored book. He has been actively involved in editorial and reviewer roles for esteemed international journals.

Cherukumalli Srinivasa Rao acquired his PhD in Soil Science and Agricultural Chemistry from Indian Agricultural Research Institute (IARI), New Delhi, and his Post-Doctoral (Fellow) from Tel-Aviv University, Tel-Aviv, Israel. Dr. Rao worked in the Indian Council of Agricultural Research (ICAR) system in different capacities, from Scientist to Director. Currently, he is serving as Director of ICAR-National Academy of Agricultural Research Management, Hyderabad, since 2017. He also served the International Crop Research Institute for Semi-Arid Tropics (ICRISAT), Patancheru, Hyderabad, for three years. He is the recipient of 40 national and international awards, including the Rafi Ahmed Kidwai Award for Outstanding Research in Agricultural Sciences in 2019. He contributed to the United Nations Framework Convention on Climate Change (UNFCCC) global climate change negotiator in the agriculture sector from the Government of India. Dr. Rao is a Fellow of the Indian National Science Academy, National Academy of Sciences, and National Academy of Agricultural Sciences and a Member of various national and international committees globally. Dr. Rao is specialized in Climate Change, Soil Carbon Sequestration, Carbon Balance, Dryland Production Systems, and Climate Policy. Dr. Rao is credited with 340 research papers, 50 books authored/edited, 14 copyrights, and 21 policy/strategy/status papers. He visited more than 15 countries for scientific meetings. His Google Scholar total citations = 9151 and h-index = 46.

Himlal Baral is a Senior Forest and Landscape Restoration (FLAR) scientist with CIFOR-ICRAF's Climate Change, Energy, and Low Carbon Development team. Based in Bali, Indonesia, Dr. Baral coordinates CIFOR's work on FLAR, planted forests, and bioenergy in Asia and the Pacific, specifically, (1) promoting FLAR for multiple ecosystem goods and services (EGS); (2) quantifying and valuing EGS under different landscape management scenarios; (3) investigating and developing locally appropriate landscape restoration models; and (4) contributing to land-resource planning for food, biomaterial, and energy security. Prior to joining The Center for International Forestry Research- International Centre for Research in Agroforestry (CIFOR-ICRAF), he worked in the forestry and natural resources management field for over 25 years in both the private and public sectors in Asia and the Pacific. Dr. Baral has a PhD in Land and Environment and two Master's degrees in Forest Science and Social Science. Dr. Baral also serves as a Fellow of the

Melbourne School of Land and Environment and of the Global Evergreening Alliance and leads IUCN-CEM's specialist group on forest ecosystems.

Rajiv Kumar Chaturvedi is currently an Associate Professor at BITS Pilani, Goa campus. He obtained his PhD from the Indian Institute of Science (IISc) in Ecology from the Centre for Ecological Sciences and a Masters in Geography from the Delhi School of Economics, Delhi University. Dr. Chaturvedi has been a National Environmental Sciences Fellow at IISc and has contributed to more than 40 peer-reviewed articles and two books in the areas related to environment and climate change. Dr. Chaturvedi has worked extensively with communities, Government officials, NGOs, and UN organizations. Dr. Chaturvedi is currently the co-chair of IUCN's forest ecosystems' specialist group. He is currently involved with India's national greenhouse gas (GHG) inventory program and is listed as a UN expert on GHG inventory. He has been a lead author for the regional assessment reports of the UNEP (United Nations Environment Programme) and IPBES (Intergovernmental Panel on Biodiversity and Ecosystem Services).

Purushothaman Chirakkuzhyil Abhilash is a Senior Assistant Professor of Sustainability Science in the Institute of Environment and Sustainable Development (IESD) at Banaras Hindu University, Varanasi, India, and Lead of the Agroecosystem Specialist Group of the IUCN-Commission on Ecosystem Management. He is a Fellow of the National Academy of Agricultural Sciences, India. His research interest lies in sustainable biomass production from marginal and degraded lands for supporting a biobased economy, restoring degraded lands for regaining ecosystem services, land system management, sustainable utilization of agrobiodiversity, nature-based solutions, and ecosystem-based adaptations for climate-resilient and planet-healthy food production, and sustainable agriscape management for food and nutritional security. He is particularly interested in sustainability analysis, system sustainability, sustainability indicators, circular economy principles, policy realignment, and the localization of UN SDGs for sustainable development. He is sitting on the editorial board of prestigious journals in Ecology/Environment/Sustainability from leading international publishers and also serving as a subject expert for UN-IPBES, IRP-UNEP, United Nations Development Programme-Intergovernmental Science-Policy Platform on Biodiversity and Ecosystem Services (UNDP-IPBES) Network, Intergovernmental Panel on Climate Change (IPCC), United Nations Convention to Combat Desertification (UNCCD), Asia-Pacific Network (APN), GLP, and IUCN Commissions (CEM, CEC, CEESP, and IUCN-Species Survival Commission (SSC)) for fostering global sustainability.

Abbreviations and Acronyms

ACZ	Agro-climatic zones
AFSs	Agroforestry systems
AMF	Arbuscular mycorrhizal fungi
APSIM	Agricultural production system simulator
AVHRR	Advanced very-high-resolution radiometer
BNF	Biological nitrogen fixation
CCS	Carbon capture and storage
CEC	Cation exchange capacity
CFC	Chlorofluorocarbon
CL-CL	Cropland-remaining-cropland
CMIP5	Coupled models inter-comparison project phase 5
CSMs	Crop simulation models
DCP	Dicalcium phosphate
DOC	Dissolved organic carbon
DPM	Decomposable plant material
DR	Distillation residues
DSSAT	Decision Support System for agro-technology transfer
EMF	Ectomycorrhizal fungi
EOC	Readily oxidized organic carbon
ESA	European Space Agency
FR	Fertilizer response
FYM	Farmyard manure
GEM	Genetically engineered microorganisms
GHGs	Greenhouse gases
GIS	Geographic information systems
GSOCseq	Global SOC sequestration potential map
GWP	Global warming potential
GWRK	Geographically weighted regression kriging
HCFC	Hydrochlorofluorocarbon
HOM	Humified organic matter
INM	Integrated nutrient management

InSAR	Interferometric SAR
IOM	Inert organic matter
IPM	Integrated pest management
$KMnO_4$-C	Permanganate oxidizable carbon
LISS III	Linear imaging self-scanning sensor 3
LULC	Land use land cover
LULCC	Land use land cover changes
MAT	Mean annual temperature
MBC	Microbial biomass carbon
MIPV	Microbe induced plant volatile
MLR	Multiple linear regression
NASA	National Aeronautics and Space Administration
NCBI	National Centre for Biotechnology Information
NDVI	Normalized difference vegetation index
NFs	Nano fertilizers
NIR	Near infrared
NPP	Net primary productivity
NPs	Nanoparticles
NUE	Nutrient/nitrogen use efficiency
OK	Ordinary kriging
OLI	Operational land imager
OTUs	Operational taxonomic units
PCR	Polymerase chain reaction
PEP	Phospho-enol pyruvate
PGPR	Plant growth promoting rhizobacteria
RDF	Recommended dose of fertilizer
REDD	Reducing emissions from deforestation and forest degradation
RK	Regression kriging
RMPs	Recommended management practices
ROTHC	Rothamsted carbon
RPM	Resistant plant material
RS	Remote sensing
RuBisCO	Ribulose 1,5-bisphosphate carboxylase oxygenase
RuBP	Ribulose bis-phosphate
SAR	Synthetic aperture radar
SOC	Soil organic carbon
SOM	Soil organic matter
TNN	Traditional neural network
UNFCCC	United Nations Framework Convention on Climate Change
UN-SDGs	United Nations Sustainable Development Goals
USGS	United states geological survey
VNIR	Visible, near-infrared
VNIR-SWIR	Visible-near infrared-shortwave infrared
VOCs	Volatile organic compounds

Part I
Overview: Plant Nutrition and Soil Carbon Sequestration

Chapter 1
Plant Soil Nutrition and Carbon Sequestration

P. Anitha, Alaknanda J. Adur, M. Manjushree, K. Narendra, S. Srinivasan, R. Murali, and D. Raajasubramanian

Abstract Leaching of nutrients is a notable problem in the agricultural sector. Conventional agricultural practices and intensive farming lead to the loss of essential nutrients and entrapped carbon from the soil, making soil flaccid and unfit for further crop cultivation. Retaining the nutrients in soil has become a major challenge. A judicious irrigation method is highly required depending on the land topography and soil quality. Soil dynamics reveal that soil is highly reactive and is a cradle of energy for microflora and fauna. It facilitates carbon sequestration through plants, adding carbon to the soil reservoir, which is available in organic and inorganic forms. Soil carbon sequestration improves food and nutrition stability, increases water cycling, and improves overall productivity, thereby strengthening nutrient recycling and increasing ecosystem services. Soil can sequester three times

P. Anitha
Department of Botany, BMS College for Women, Bangalore, India

A. J. Adur
Department of Environmental Science, Surana College, Peenya, Bengaluru, India

M. Manjushree
Rotary Educational Society, Mandya, Karnataka, India

K. Narendra
Department of Botany, Annamalai University, Chidambaram, Tamil Nadu, India

S. Srinivasan · R. Murali
Department of Biochemistry and Biotechnology, Faculty of Science, Annamalai University, Annamalainagar, Tamilnadu, India

Postgraduate and Research Department of Biochemistry, Government Arts College for Women, Krishnagiri, Tamil Nadu, India

D. Raajasubramanian (✉)
Department of Botany, Faculty of Science, Annamalai University, Annamalainagar, Tamilnadu, India

Department of Botany, Thiru. A. Govindasamy Government Arts College, Tindivanam, Tamil Nadu, India
e-mail: raajasubramanianbotany@gmail.com

S. A. Edrisi et al. (eds.), *Sustainable Plant Nutrition and Soil Carbon Sequestration*, Sustainable Plant Nutrition in a Changing World,
https://doi.org/10.1007/978-3-031-53590-1_1

more carbon than the atmosphere, acting as a sink for carbon dioxide (CO_2). Nutrients like nitrate, phosphate, and organic carbon can be retained by adding slowly decomposing sclerenchymatous organic matter like tamarind shells. It is rich in calcium and helps exchange cations and essential trace elements, making it available for plants. Therefore, the present chapter aims to converse the critical account of plant nutrition, soil nutrition, and carbon sequestration. It will also highlight the potential application of tamarind shells and the allied mechanisms facilitating the enhancement of plant soil nutrition and sequestering atmospheric carbon.

Keywords Soil nutrition · Plant nutrition · Nutrient retention · Carbon sequestration · Sustainable development

1.1 Introduction

Intensive agriculture and inappropriate land management, including unscientific irrigation practices, have stimulated increased soil runoff, soil erosion, salinity, and ultimately the loss of soil nutrition. Urbanization and improper land use patterns have contributed to desertification, depletion of organic content, and loss of biodiversity and indigenous species. IPBES (2019) states that more than 66% of biodiversity loss worldwide is associated with greenhouse gas (GHG) emissions, land use change, and food consumption. Moreover, the countries with higher income per capita possess a biodiversity loss of 50% globally. The fundamental reason is the higher dependencies and consumption of these nations on non-food goods and services (IPBES 2019). The biodiversity loss leading to ecosystem degradation and vice-versa is affecting the local flora and fauna, inducing habitat fragmentation, which has aggravated the problem and added to the infertility of the soil, resulting in land degradation. IPCC (2019) and IPBES (2018, 2019) have emphasized the impacts of land degradation, which negatively affects the well-being of more than 3.2 billion people worldwide.

According to an estimate by IFAD (2010) and IPBES (2019), around 12 million hectares of land are lost annually due to land degradation, resulting in a loss of US$42 billion annually. Furthermore, the dryland areas contributing to around 41% of the land area at the global scale are at the juncture of desertification and are habitat to 2.1 billion people. Around 44% of the global agricultural systems fall in such regions and sustain 50% of livestock worldwide (MEA 2005; IPBES 2019). Globally, increased mismanaged agricultural practices have resulted in Soil Organic Carbon (SOC) losses of up to 50% (Benbi and Brar 2009). Loss of ecosystem diversity leads to reduced functioning of nutrient cycling due to the unavailability and lack of interaction between the biotic and abiotic communities.

Microbial diversity in the plant and soil is also lost due to the extensive and continuous use of chemical pesticides affecting the biogeochemical cycles. Such chemical pollutants also stimulate direct mortality, weaken the health and development of individuals, and alter community structure (Muturi et al. 2017; IPBES 2019). Therefore, soil restoration through natural nutrient cycling must be prioritized and addressed with prime importance. The dynamic nutrient cycling is being altered due to the abovementioned issues, and its restoration through natural processes is still a major challenge to humanity. Climatologists predict that precipitation events will become less frequent but more severe, leading to calamitous flooding, soil degradation, and acute food scarcity (IPCC, 2019). More appropriate research attention could be paid to augmenting soil organic matter (SOM) content to improve soil water-holding capacity.

Carbon capture, coupled with various scientific and conventional methods, can address the challenges of soil nourishment and restoration, which need to be focused. Carbon sequestration in soils has also been promoted to combat the increasing CO_2 levels in the atmosphere, one of the most vital causes of global warming and climate change (Baveye et al. 2020). In this milieu, the present chapter aimed to ideate the vital discussions on plant nutrition, soil nutrition, and carbon sequestration. It will also highlight the importance of holistic approaches to retain the nutrients and carbon in the soil via amendment-based applications, for instance, tamarind shell as part of strategic agronomic practices, and discuss its benefits in maintaining biogeochemical cycling.

1.2 Plant and Soil Nutrition

Plant and soil nutrition are connected under a myriad of nexus mediated by various phenomena through the regulation of soil and plant microbial communities (Tripathi et al. 2017). Therefore, the plant roots explore the nutrition in the soil by continuously reaching the zones of higher nutrients to meet their nutritional requirements. Nutrient uptake occurs at the differentiated roots, where the root hairs and epidermis facilitate nutrient surface adsorption. Besides, certain other nutrient transfers also occur through the root tip (Robe and Barberon 2023). Considering the soil nutrition that usually comes from biogeochemical cycles and decomposition, fundamentally mediated by the saprophytes through mineralization processes. These processes eventually enable soil nourishment and trigger plant growth and development. The plant requires seventeen essential nutrients absorbed from air, water, and soil (Shaikh and Ali 2021). The macronutrients like nitrogen, potassium, phosphorous, sulfur, calcium, and magnesium, and micronutrients such as manganese, iron, copper, boron, chlorine, molybdenum, nickel, and zinc are absorbed from the soil. Other elements, oxygen, hydrogen, and carbon, are absorbed from the water and air. Besides, some non-essential elements, i.e., selenium, cobalt, silicon, and sodium, are conducive to plant growth (Vatansever et al. 2017). Trace elements or metalloids, including arsenic, aluminum, chromium, lead, and cadmium, have the intense potential to intoxicate the plants (Ghori 2019).

The soil texture influences nutrient absorption (Table 1.1) (Heber and Ampt 1911). Along with the soil, nutrients, soil moisture content, or soil water plays a critical role in the mobility of soil nutrients. Soil moisture analysis and study have become vital in soil nutrition enrichment. Excess of soil moisture will lead to

Table 1.1 Importance of macro and micronutrients and their symptoms of deficiency in plants

S. No.	Nutrient element	The role played in plant	Deficiency symptoms
1.	Nitrogen	The key constituent of proteins, nucleic acids, vitamins, hormones and chlorophyll It is needed in all parts of the plant but is essential in meristematic tissues and metabolically active tissues Soil fertility is maintained by soil-bound humus	Yellowing of leaves due to reduction in chlorophyll content—Chlorosis Reduction in protein content leads to stunted growth of the plant Dormancy of the lateral buds Delay in flowering
2.	Phosphorus	Constituent of cell membranes, nucleic acids, coenzymes, and ATP Mandatory for phosphorylation reactions Influences cell division, stimulating root and plant growth	Stunted growth of plants Premature fall of leaves
3.	Potassium	Required mainly in growing points like in meristematic tissues, buds, leaves, and root tips Maintaining anion-cation stability in cells Maintenance of turgidity of cells Implicated in protein synthesis Opening and closing of stomata Activation of enzymes	Shoot growth is restricted Chlorosis Acute deficiency may cause dieback of shoots
4.	Calcium	Required by meristematic and differentiating tissues during cell division for the synthesis of the cell wall, as it is the chief constituent of middle lamella in the form of calcium pectate Formation of mitotic spindles Normal functioning of cell membranes Activates many enzymes Essential for fat metabolism, carbohydrate metabolism, and nitrogen assimilation	Deficiency symptoms are mainly seen in young leaves and the growing points because calcium is a structural component and, hence, is not easily released or transported to younger parts from the older parts of the plant Necrosis mainly of tissues of leaves
5.	Magnesium	Activates enzymes of respiration and photosynthesis Synthesis of DNA and RNA Constituent of the ring structure of chlorophyll is hence required for the formation of the pigment Maintains ribosome structure by combining the two subunits of ribosomes	As magnesium is mobilized easily, its deficiency symptoms are seen in older leaves firs Chlorosis and necrosis
6.	Sulfur	Forms indispensable part of amino acids cysteine and methionine The main constituent of coenzyme A, vitamins like Thiamine (vitamin B1), Biotin, and Ferredoxin. Constituent of volatile oils.	Chlorosis, mainly of young leaves, as sulfur is also an immobile element Delay in flowering Fruit formation is suppressed

(continued)

Table 1.1 (continued)

S. No.	Nutrient element	The role played in plant	Deficiency symptoms
7.	Iron	Significant constituent of proteins concerned with electron transfer like Ferredoxin and Cytochromes (It is reversibly oxidized to Fe^{2+} from Fe^{3+}) Activates enzymes like catalase and aconitase Essential for the formation of chlorophyll	Chlorosis of leaves
8.	Manganese	Activator of many enzymes involved in photosynthesis, respiration, and nitrogen metabolism Required for splitting of water to liberate oxygen during photosynthesis	Chlorosis of leaves Retarded growth in plants
9.	Zinc	Activates many enzymes like carboxylases Needed in the synthesis of auxin Required in RNA synthesis, nitrogen metabolism, and formation of seed	Chlorosis of leaves Plants show resetting of leaves due to shortened internodes Necrotic spots on leaves
10.	Copper	Essential for overall metabolism Related with certain enzymes involved in redox reactions (it is reversibly oxidized to Cu^{+} from Cu^{2+}) Acts as a catalyst in plant respiration	Vegetative and reproductive growth is reduced Reduction in seed production and fruit size
11.	Boron	Compulsory for uptake and consumption of calcium Membrane functioning Germination of pollen Cell elongation and differentiation Carbohydrate translocation	Stunted growth Apical meristem blackens and dies
12.	Molybdenum	Component of enzymes, nitrogenase, and nitrate reductase, both of which are required in nitrogen metabolism Required for nodule formation in legumes	Chlorosis Delay in flowering
13.	Chlorine	Helps in determining the solute concentration and the anion-cation balance in cells (Along with Na^{+} and K^{+}) Essential in photosynthesis for water-splitting reaction, leading to the release of oxygen	Wilting of leaves Chlorosis and necrosis in leaves

waterlogging and soil erosion. In contrast, a shortfall of water content will lead to the drying up of soil and less mobility of soil nutrients, leading to degradation. The pH of the soil is equally essential in soil nutrition as it affects the plant's availability and uptake of nutrients via the pH-driven enzymatic pathways (Edrisi et al. 2019). The pragmatic soil pH-biogeochemistry associations endow insight into potential applications for increased crop yields through nutrient recycling and availability, facilitating crop development (Neina 2019).

Saprophytic organisms may be microscopic or macroscopic, but they tend to break both macro and micro bio-molecules into their micro-elemental forms. These elements form vital nutrients for the growth of the plants, the producers of the ecosystem. Protein, the building block of a living organism, is made up of amino acids, carbohydrates help in signal transmission, and it is composed of saccharides, lipids constitute glycerol, and fatty acids maintain the integrity of the structure, minerals, and other nutrients play a momentous role in maintaining integrity and functioning of living organisms. Crops and plant and animal debris are broken down to release the elemental forms by saprophytes belonging to the kingdom Monera to Mycota. Most of the organisms belonging to this kingdom are decomposers and participate in the degradation and decomposition of larger biomolecules into their elemental forms (Fig. 1.1). Biomolecules or organic molecules contain carbon atoms in their structure, to which carbon atoms may be attached acids, bases, ions, metals, nonmetals, inorganic ion at various positions based on the type of molecules and its function.

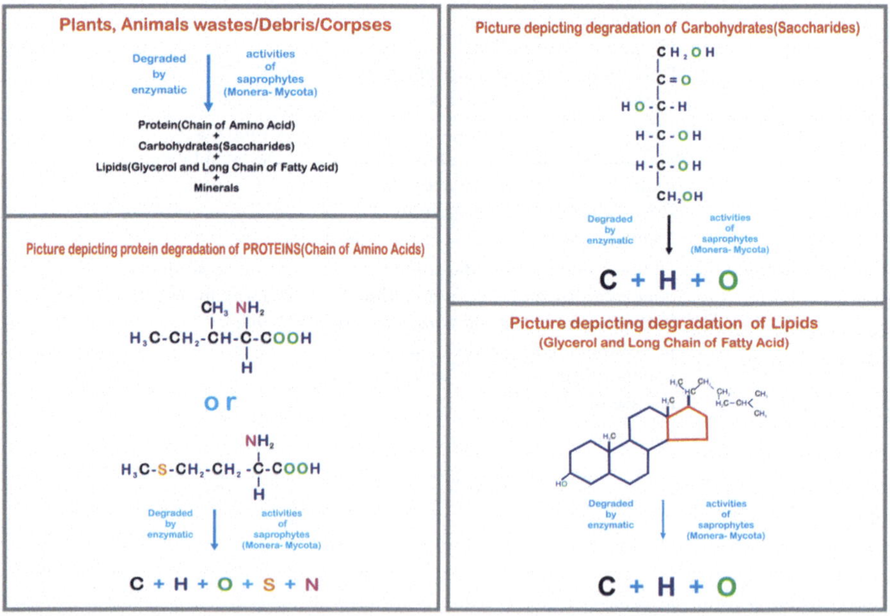

Fig. 1.1 Saprophytic decomposition of complex biomolecules (Adopted and modified from Aro et al. (2005), Douglas (2013), Johansen (2016))

When decomposers break down biomolecules by enzymatic activities, they release various micro substances like Fe, Mg, P, Cu, Ca, Mn, etc., that are essential and play an imperative role in maintaining the vitalities of plants-the producers of the ecosystem, consumers, and the health of the ecosystem. The microflora and fauna play a significant role and keep nutrient flow uninterrupted in various biogeochemical cycles. Therefore, various waste products, including agricultural waste products, animal wastes, and others, are mineralized by the action of microbial processes.

1.2.1 Agricultural Waste to Enhance Soil Nutrition

Agriculture plays a vital role in soil nutrition and carbon sequestration. Scientific approaches, including agronomic practices, intercropping, and various other agriculture methods, can be helpful in carbon sequestration and enhance soil nutrition. India generates around 0.35 billion tons of farming waste. The agricultural waste includes crop residues, unused and unprocessed crops, damaged crops, and spillover during packaging and cleaning, transportation, etc. The unused biomass from farm waste can be utilized as a good carbon and nitrogen source to enrich the soil. In most European countries, the dry biomass is left in the farms for drying (Corato 2020).

In India, farm residues are set on fire to reduce the volume of the waste generated as disposal incurs cost, time, and space occupancy. Instead, this farm waste can be shredded into small pieces and can be used as cattle feed partly and partly spread across the field. The shredded farm waste can decompress naturally by maintaining the moisture content and microbial culture. The moisture content and microbial culture synergistically enhance the decomposition rate, thus increasing the soil's humus and organic carbon content (Akshaya et al. 2023). The crop residues are nothing but plant material.

The plant cell wall comprises polysaccharides such as hemicelluloses, cellulose and pectin, and phenolic polymer lignin (Nagappa et al. 2023). Filamentous fungal enzymes play a vital role in the degradation of cellulose and biopolymers. Various fungal species can be utilized to degrade the plant's cellulosic matter (Aro et al. 2005). However, not ample data is available on the same, but the research reveals a rapid degradation of carbohydrate structures. The degradation pathways are identical to those of decaying plant material degraded and decomposed under analogous conditions like the natural environment, implying that the disparity in recalcitrant carbon from different sources after decomposition is negligible (Spence 2011). However, research in genetic engineering has conferred the early degradation of lignocellulosic material, leading to the release of more simple compounds like sugars and lignin-based alcohols, namely coniferyl alcohol, sinapyl alcohol, and p-coumaric alcohol.

The soil microfauna (invertebrates) mechanically disrupt polysaccharides, rendering them accessible to enzymes from different gut bacteria that catalyze the chemical degradation of cellulose, hemicellulose, and xylan (Johansen 2016; Douglas 2013). Photochemical factors combined with microbial consortium increase the rate of decomposition. Solar radiation contributes to decomposition, releasing simple compounds (Rossel et al. 2013). Thus, agricultural waste can be converted into compostable matter using conventional and biotechnological tools to enhance soil fertility through natural degradation.

1.2.2 Tamarind Shell as a Nutrient Enhancer

The Tamarind shell, a non-food product, is considered an agricultural waste. This agricultural waste is directly lit due to the space occupancy for storing the waste matter. The Tamarind shell possesses numerous phytochemicals (secondary metabolites) like alkaloids, phenols, glycosides, terpenoids, flavonoids, and tannins (Gomathi et al. 2017). Tamarind shell (carbonized and activated) is a great toxic adsorbent used in industries to adsorb lead and cobalt from wastewater (Sahu et al. 2016; Lopez et al. 2017). As an alternative to the currently employed activated carbon method, Radhakrishnan et al. (2018) used the ethylenediamine hydroxypropyl tamarind fruit shell as an adsorbent to extract Eriochrome black-T from an aqueous solution. Tamarind shells are a potential source for the generation of nanosheets. As tamarind shell particles are harder, they can be explored in the making of polymer composites or are used as an additive in polymer composites due to their tensile strength.

The tamarind shell may not be available throughout the year, which limits its availability as a raw material in industrial applications. As the fruits ripen and are deshelled once a year, their availability may be limited. Most of the small farmers use it immediately in burning as an alternative to wood. Hence, it is not stocked. Due to a lack of awareness and limited demand in the open market, the tamarind shell is kept away from the regular commodity. Hence, the farmers dispose of it immediately by burning it or using it as a source of fuel. This is because the storage is bulky and voluminous, thus occupying the place. Tamarind shells are naturally calcium-rich compounds. Combining secondary metabolites and calcium from tamarind shells effectively treats alkaline and acidic soil. Tamarind shell provides the scope for cation exchange, making essential nutrients available for the plants. Commonly, plants grow luxuriously at neutral to slightly acidic pH. If the soil is too acidic, the calcium from the tamarind shell buffers the acidic condition.

$$2H^+ + CaCO_3 \rightarrow Ca^{2+} + CO_2 + H_2O$$

Further, this cation exchange (Ca^{2+}) in soil provides microelements to plants, which are taken up by the roots, while H^+ is exchanged for the metal ions.

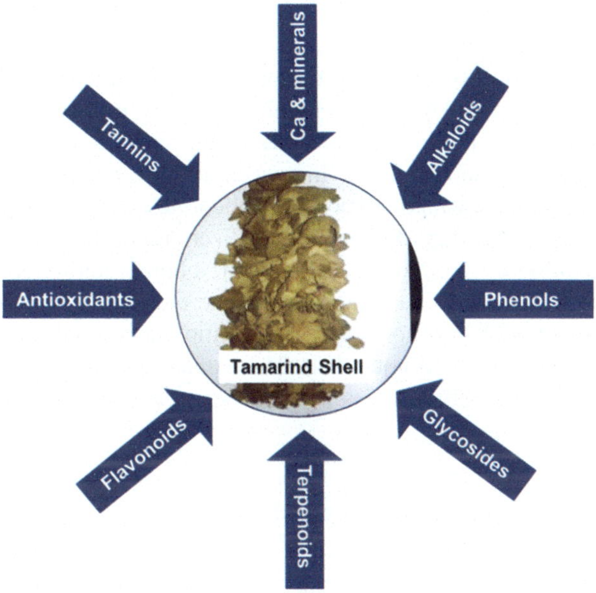

Fig. 1.2 Composition of Tamarind shell. (Adopted from Gomathi et al. (2017), Radhakrishnan et al. (2018))

$$Ca^{2+} + 2CO_2 + 2H_2O \rightarrow 2H^+ + Ca(HCO_3)_2$$

In drought-prone regions, the soil might turn too alkaline due to low rainfall. Alkaline soil may be acidified due to acidic content like tannic acid and natural sulfur content, which is oxidized by soil bacteria-mediated reactions to H_2SO_4. Thus, mixed Tamarind shell powder with soil can be a natural soil enrichment to improve soil fertility (Fig. 1.2). Due to the higher nutrient status of the Tamarind shell, which includes tannins, antioxidants, flavonoids, terpenoids, glycosides, phenolic compounds, alkaloids, and various other macro- and micro-nutrients and minerals, it has a potential for a promising additive under the agricultural practices. Moreover, there is a need for insights for its site-based application depending on the type of requirements for plant growth and development for enriching the nutrient status in the plants and soil. Further, the local people and other stakeholders (particularly the end-users like farmers and the suppliers) must develop a cohort for supplying these kinds of substrates and getting knowledge of specific requirements for its application.

The flavonoids in tamarind shells act as potential antioxidants reacting with free radicals, which can be explored in treating various disorders and as antioxidant supplements in food additives. (Li et al. 2023). Increased amounts of free radicals have a negative impact on the soil biota, which primarily consists of microbial populations. The outcomes of scientific studies suggest that a mixture of environmental factors, such as chemical pollution, including soil contamination, as well as physical factors, like ionizing radiations and ultraviolet radiation, air pollution, and fires, may cause free radical concentrations in cells to rise. Free radicals are actively involved in the process of delignification, which is achieved by fungal enzymes. It encompasses oxidation by hydroxyl and perhaps other free radicals, along with enzymatic hydrolysis. In the process of humification, polyphenols like lignin are depolymerized by fungi, yielding quinones. Reactive free radicals must be removed in order to inhibit the oxidative chain reaction and create the stable semiquinone radicals found in SOM. In scavenging the reactive free radicals, the scavenging compounds, which are probably phenolics, act as antioxidants. According to scientists, the buildup of chemical fertilizer and pesticide residues in soil may cause the creation of free radicals; as a result, a higher antioxidant level must be guaranteed to prevent their harmful effects. The tamarind shell, being identified with ample amounts of flavonoids and other antioxidants, will nullify the oxidative reactions and keep the soil nourished. This could be one of the potent low-cost solutions to treat the contaminated soil recovery process, as soil is a non-renewable resource.

1.3 Carbon Sequestration

United States Geological Survey (USGS) defines carbon sequestration as capturing, accumulating, and storing atmospheric carbon dioxide (CO_2). Terrestrial and aquatic ecosystems are the sinks for CO_2. Carbon sequestration aims to reduce atmospheric carbon released due to anthropogenic activities. Carbon sequestration removes CO_2 from atmospheric air to combat climate change and global warming. Sequestration involves two methods: geologic and biological sequestration. Biologic sequestration involves edaphic factors influencing the carbon cycle. A carbon-neutral economy can be achieved by carbon sequestration using abiotic or biotic, natural or engineered methods. Abiotic sequestration is based on physical, chemical, or engineering techniques without involving biotic communities like plants, animals, and microbes. Biotic sequestration involves natural biological nitrogen fixation and reduced use of synthetic fertilizers and pesticides. It also involves enhancing the humification process, deep-rooted plant growth, and formation of carbonated through natural processes (Lal 2008). Pedogenesis or soil formation process involves carbonation as one of the factors in the chemical weathering process. Under the carbon cycle, atmospheric CO_2 easily combines with rock and soil matter to form carbonic acid. It later combines calcium and magnesium to form carbonates and bicarbonates (Fig. 1.3).

$$CO_2 + H_2O \rightarrow H_2CO_3$$

$$Ca(OH)_2 + CO_2 \rightarrow CaCO_3 + H_2O$$

$$CaCO_3 + H_2O + CO_2 \rightarrow Ca(HCO_3)_2$$

In agricultural systems, how the soil resource is dealt with largely determines how much and how long carbon is stored. Numerous agricultural techniques can improve

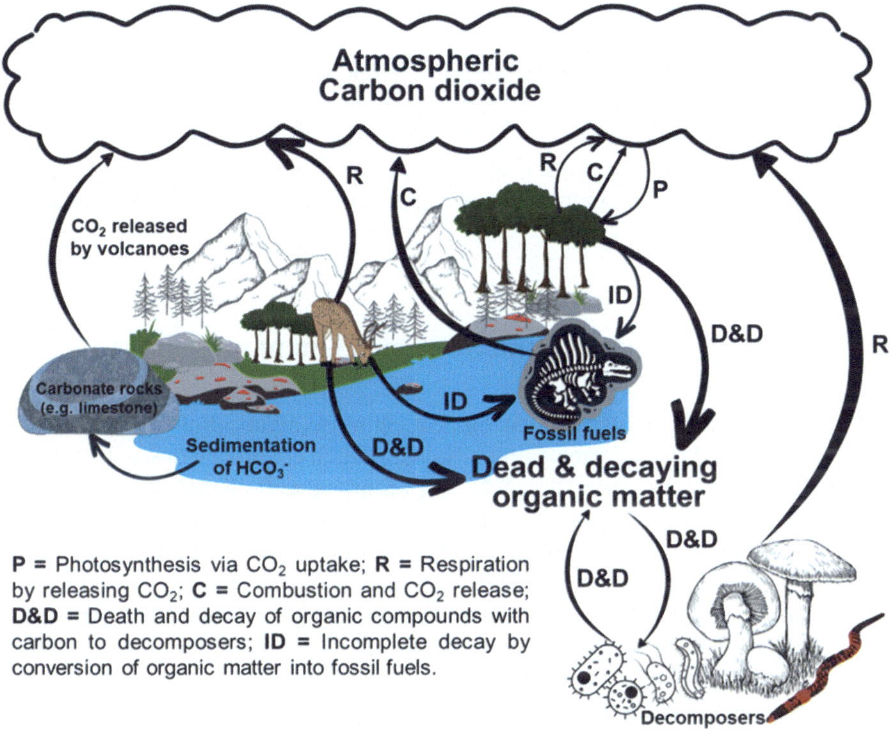

Fig. 1.3 Carbon cycle and the possible natural pathways of carbon sequestration. (Adapted and modified from Edrisi et al. (2022), Chapman and Reiss (1999))

carbon storage. Conservation tillage, cover cropping, crop rotation, and increasing the vegetation in semi-arid regions with stress-tolerant tree species increase the green cover in arid and semi-arid regions. Woody plants can store more carbon in the form of tissues and, if not burnt, remain arrested in the wood. The wood that is converted into long-lived wood can act as a carbon sink. Sustainable agroforestry methods can be deployed to wastelands to develop a carbon sink zone in the localized regions. Proper land use management and sustained practice of agriculture can retain the soil texture, promoting the growth of younger plants, which can capture CO_2 that is more atmospheric than the older forest. Appropriate landscaping, keeping the soil hydrated, and maintaining the environmental determinants of soil like appropriate PH, temperature, moisture, and oxygen balances the carbon cycle. Minimal utilization of chemical fertilizers and promotion of organic, traditional farming will retain natural humification. The non-productive wasteland can be well treated with tamarind shells to enhance the process of humification and improve the soil texture.

It is significant to remember that even when carbon is locked up in organic matter, SOM continues to serve as a food source for various microbial communities in the soil and can mineralize into nutrients that are readily available to plants. Up to

80% of the newly generated organic matter may be devoured in a few years, depending on crop rotation and the quantity of carbon pedaled each year. However, employing carbon-sequestering sustainable farming techniques can sustain or even increase SOM over time. A soil that has more carbon is more biologically active, offers more agricultural nutrients, is more resistant to weathering, can hold more water, and is better able to withstand erosion. Conservation of soil will lead to increased plant production and, in turn, more carbon capture from the environment. (Ramya et al. 2023).

1.4 Waste Management and Soil Enrichment

Natural plant residues (complex structures of cellulose, lignocellulose, pectin, etc.) produce high SOM in the form of simple sugars, which are broken down finally into organic and inorganic carbon (Fig. 1.4). Inappropriate waste handling and ill practices in waste management are detrimental to human and soil health. Waste and waste-derived materials have a high potential to help increase soil fertility and crop production. Waste treated scientifically produces usable products. Solid waste treated aerobically gives compost, while anaerobically gives biofuel. Thus, the partially decomposed sludge matter can also be used as a soil enricher as it contains many soil micro and macro elements.

The waste like wood compost, bio-solids, log yard waste, and wood ash with a controlled amount of fertilizer will enhance the decomposition (Ramya et al. 2023). It can be used in various proportions and in various combinations to reclaim the landmine soil and industrial food waste may be a potential soil enrichment in the

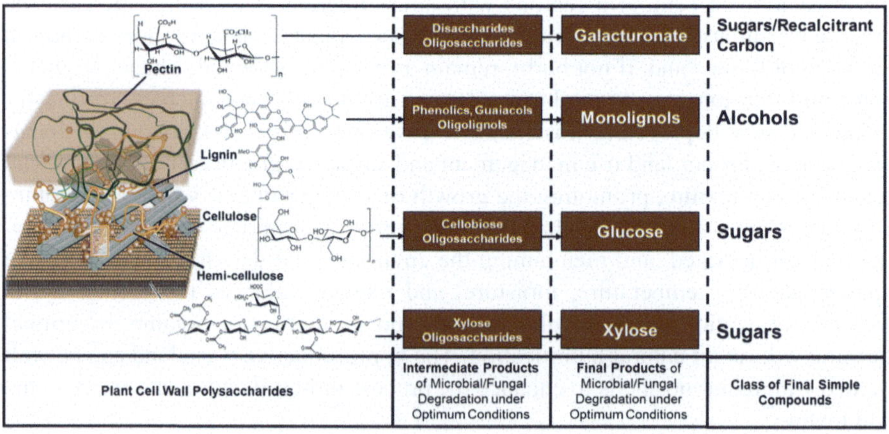

Fig. 1.4 Decomposition of complex plant cell wall materials (mostly plant residues) to the simple compounds for making it readily available for soil system and nutrient enrichment (Adapted from Nagappa et al. (2023))

near future. Residual waste like potato peel is effectively converted to bioethanol, and residues obtained from fermentation in combination with a consortium of microbial cultures lead to higher NPK values in soil (Chintagunta et al. 2016). Food waste compost could be a viable substitute for chemical fertilizers to enhance soil microbial populations and enzyme actions and promote soil nutrient biology (Lee et al. 2004). Lakhdar et al. (2011) revealed through their studies that the amount of nitrogen, phosphorus, calcium, potassium, and carbon in the soil increased significantly after Municipal Solid Waste compost was applied, which resulted in a higher uptake of NPK and a significant increase in plant biomass as well. Thus, waste treated appropriately and amended judiciously can add to soil nutrients.

1.4.1 Soil Enrichment and Carbon Sequestration

The carbon-to-nitrogen ratio (C:N) is essential in deciding soil health. The C:N ratio promotes microbial activity, adding nutritional value and enhancing plant growth. The nitrogen component is added through agricultural waste, and long-term nitrate content in the soil induces carbon uptake from the atmosphere. The studies by Lu et al. 2021 revealed that long-standing Nitrogen inputs raised soil Carbon stocks by 7–21%, primarily due to reduced Carbon production fluxes and physical protection mechanisms without altering the chemical makeup of organic matter. It was discovered that excessive Nitrogen deposition could dramatically raise soil Carbon in Nitrogen-rich tropical forests using more than ten years of continuous Nitrogen addition. Hence, nitrogen-rich agricultural waste can raise the carbon stocks in the soil, thus enhancing carbon sequestration (Lu et al. 2021). The soil's organic content is highly affected by microbial diversity. Higher soil water content, total nitrogen, organic carbon, and an appropriate pH value considerably improve the carbon source's metabolic function and the diversity of the soil microbes under subsoil tillage. The microbial community and carbon sequestration in farming soil change due to subsurface tillage and complete straw returning, which will help boost soil carbon sequestration and give a practical basis for alleviating the problem of soil degradation (Hao et al. 2019). Agriculture waste or crop residues enrich SOM by inoculating the consortia of microorganisms, which promotes carbon mineralization and SOM formation and improves the soil's biological activity. In excess of 70% of soil bacterial taxa have enzymes that can metabolize inorganic substrates from the environment (Bay et al. 2021). Thus, the microbial population is at bay with increased organic content. Agricultural waste regulates soil nutrition and acts as a soil conditioner. The microbial diversity and high nitrate content helps to control the atmospheric carbon content and thus help in carbon sequestration.

1.4.2 Sustaining Agroecosystem for Improving Soil Health

The inclusion of trees in the agricultural mono-cropping system will boost the organic content of the soil ecosystem. Since soil has a greater capacity to store carbon than vegetation by 1.5–3 times (Stefano and Jacobson (2018), but there are other factors that influence carbon sequestration, topography, past and present land use, type of cropping system, other climatic factors, etc. As agroforestry is looked upon as one of the mitigation measures to curb climate change, it might as well help in improving soil health by restoring and promoting the soil microflora and fauna. Studies have revealed that the agroforestry pattern of cultivation has contributed towards retaining the nutrition in the soil and thus improving the overall soil health. Green leaves of *Sesbania grandiflora* and *Moringa oleifera's* nutrient leaching demonstrated excellent results as organic sources of nutrients for crops in agroforestry (Limon et al. 2016). Crop yields increase because of the decomposition and mineralization of litter that delivers ample nutrient stock to agroforestry systems. Despite being the main avenue for nutrient input, litter decomposition can be supplemented by nutrient leaching from leaves and nutrient-rich rains.

1.5 Conclusions

Rapid urbanization, changing lifestyles, and modern agricultural practices have diversified waste generation with minimal importance to the changing environment. Food security, put at the apex of the pyramid, has brought enormous changes in the agricultural sector. The waste generated from agricultural and allied industries should be given prime importance. The unprofessional conduct and mishandling of waste have led to land degradation, loss of soil nutrition, and loss of carbon sequestration by natural means. Hence, this article emphasizes the agglomeration of soil nutrition through carbon sequestration and restoring soil nutrients by utilizing waste products. Tamarind shell and agricultural waste, often charred to reduce the volume, can be aptly used as soil enrichers. Both acidic and alkaline soil can be buffered by the shell's phytochemicals and calcium, which helps keep the soil's macro- and micronutrients intact. Agricultural waste helps increase the NPK and other micronutrient values that are critical for plant growth and development. Plant wastes improve the physical structure and texture of the soil, thereby increasing the organic, inorganic, and water-holding capacity of soil. Degraded soil can be reclaimed by the application of such agricultural waste along with microbial inoculum, which is cost-effective and also an effective waste management strategy that helps naturally in carbon sequestration. Appropriate agricultural waste management strategies can reduce the carbon footprint, enhancing the carbon sink through implementing a scientific approach in agricultural practices and waste management systems. Advantageously, the techniques with a scientific approach can build a sustainable environment and work towards achieving UN sustainable development goal number 15 for combating desertification and reverse land degradation to prevent loss of biodiversity, thus creating harmony for all life forms.

References

Akshaya V, Akila I, Murali R, Raajasubramanian D, Kuppan N, Srinivasan S (2023) Rice straw biomass and agricultural residues as strategic bioenergy: effects on the environment and economy path with new directions. In: Ramanujam PK, Parameswaran B, Bharathiraja B, Magesh A (eds) Bioenergy. Energy, environment, and sustainability. Springer, Singapore. https://doi.org/10.1007/978-981-99-3002-9_9

Aro N, Pakula T, Penttila M (2005) Transcriptional regulation of plant cell wall degradation by filamentous fungi. FEMS Microbiol Rev 29(4):719–739. https://doi.org/10.1016/j.femsre.2004.11.006

Baveye PC, Schnee LS, Boivin P et al (2020) Soil organic matter research and climate change: merely re-storing carbon versus restoring soil functions. Front Environ Sci 8:1–8. https://doi.org/10.3389/fenvs.2020.579904

Bay SK, Dong X, Bradley JA et al (2021) Trace gas oxidizers are widespread and active members of soil microbial communities. Nat Microbiol 6:246–256. https://doi.org/10.1038/s41564-020-00811-w

Benbi DK, Brar JS (2009) A 25-year record of carbon sequestration and soil properties in intensive agriculture. Agronomy for Sustainable Development, Springer Verlag/EDP. Sciences/INRA 29(2):257–265. hal-00886492

Chapman JL, Reiss MJ (1999) Ecology: principles and applications. Cambridge University Press

Chintagunta AD, Jacob S, Banerjee R (2016) Integrated bioethanol and biomanure production from potato waste. Waste Manag 49:320–325. https://doi.org/10.1016/j.wasman.2015.08.010

Corato UD (2020) Agricultural waste recycling in horticultural intensive farming systems by on-farm composting and compost-based tea application improves soil quality and plant health: a review under the perspective of a circular economy. Sci Total Environ. https://doi.org/10.1016/j.scitotenv.2020.139840

De Stefano A, Jacobson MG (2018) Soil carbon sequestration in agroforestry systems: a meta-analysis. Agrofor Syst 92:285–299

Douglas AE (2013) Microbial brokers of insect-plant interactions revisited. J Chem Ecol 39(7):952–961

Edrisi SA, Tripathi V, Abhilash PC (2019) Performance analysis and soil quality indexing for Dalbergia sissoo Roxb. Grown in marginal and degraded land of eastern Uttar Pradesh, India. Land 8(4):63

Edrisi SA, Tripathi V, Dubey PK, Abhilash PC (2022) Carbon sequestration and harnessing biomaterials from terrestrial plantations for mitigating climate change impacts. In: Biomass, biofuels, biochemicals. Elsevier, pp 299–313

Ghori N-HG, Hayat T, Imadi MQ, Gul SR, Altay A, Ozturk V, M. (2019) Heavy metal stress and responses in plants. Int J Environ Sci Te 16

Gomathi AC, Rajarathinam SX, Sadiq AM (2017) Phyto chemical screening of aqueous extract of tamarind (Tamarindusindica L.). Shell 7(11):65–70

Hao M, Hu H, Liu Z, Dong Q, Sun K, Feng Y, Ning T (2019) Shifts in microbial community and carbon sequestration in farmland soil under long-term conservation tillage and straw returning. Appl Soil Ecol 136:43–54

Heber GW, Ampt G (1911) Studies on soil Phyics. J Agric Sci 4(1):1–24. https://doi.org/10.1017/S0021859600001441

IFAD (2010) Desertification. International Fund for Agricultural Development. https://www.ifad.org/documents/38714170/39150184/Desertification+factsheet_e.pdf/40c0689e-a726-42ed-91c4-a5f79273ccd8
IPBES (2018) In: Montanarella L, Scholes R, Brainich A (eds) The IPBES assessment report on land degradation and restoration. https://doi.org/10.5281/zenodo.3237392
IPBES (2019) In: Brondizio ES, Settele J, Díaz S, Ngo HT (eds) Global assessment report on biodiversity and ecosystem services of the Intergovernmental Science-Policy Platform on Biodiversity and Ecosystem Services. IPBES secretariat, Bonn, Germany, 1148 pages. https://doi.org/10.5281/zenodo.3831673
IPCC (2019) Summary for policymakers. In: Shukla PR, Skea J, Buendia EC, Masson-Delmotte V, Pörtner H-O, Roberts DC, Zhai P, Slade R, Connors S, van Diemen R, Ferrat M, Haughey E, Luz S, Neogi S, Pathak M, Petzold J, Pereira JP, Vyas P, Huntley E, Kissick K, Belkacemi M, Malley J (eds) Climate Change and Land: an IPCC special report on climate change, desertification, land degradation, sustainable land management, food security, and greenhouse gas fluxes in terrestrial ecosystems. In press
Johansen KS (2016) Lytic polysaccharide monooxygenases: the microbial power tool for lignocellulose degradation. Trends Plant Sci 21(11):926–936
Lakhdar A, Falleh H, Ouni Y et al (2011) Municipal solid waste compost application improves productivity, polyphenol content, and antioxidant capacity of Mesembryanthemumedule. J Hazard Mater 191(1–3):373–379
Lal R (2008) Carbon sequestration. Philos Trans R Soc B363(1492):815–830. https://doi.org/10.1098/rstb.2007.2185
Lee JJ, Park RD, Kim YW et al (2004) Effect of food waste compost on microbial population, soil enzyme activity and lettuce growth. Bioresour Technol 93(1):21–28
Li W, Huang R, Han S, Li X, Gong H, Zhang Q, Yan C, Li Y, He R (2023) Potential of tamarind shell extract against oxidative stress in vivo and in vitro. Molecules 28(4):1885. https://doi.org/10.3390/molecules28041885. PMID: 36838870; PMCID: PMC9961368
Limon SH, Hossain M, Spiecker H (2016) Nutrients leaching from green leaves of three potential agroforestry tree species. Agrofor Syst. https://doi.org/10.1007/s10457-016-9996-x
López-González H, Serrano-Gómez J, Olguín MT, Hernández-López J, Bulbulian S (2017) Removal of Co by carbonaceous material obtained through solution combustion of tamarind shell. Int J Phytoremediation 19(12):1126–1133
Lu X, Vitousek PM, Mao Q, Gilliam FS, Luo Y, Turner BL, Mo J (2021) Nitrogen deposition accelerates soil carbon sequestration in tropical forests. Proc Natl Acad Sci 118(16):e2020790118
MEA (2005) Millennium ecosystem assessment. Ecosystems & human well-being: synthesis. Island Press, Washington, DC
Muturi EJ, Donthu RK, Fields CJ, Moise IK, Kim CH (2017) Effect of pesticides on microbial communities in container aquatic habitats. Sci Rep. https://doi.org/10.1038/srep44565
Nagappa N, Jyothi SG, Anitha P, Manjushree M, Narendra K, Murali R, Kanagalakshimi A, Raajasubramanian D, Srinivasan S (2023) Recent perception into the extraction of nanocellulose: cross-talk between natural resources and progressive applications. Handbook of Natural Polymers 1:97–122. https://doi.org/10.1016/B978-0-323-99853-6.00024-3
Neina D (2019) The role of soil pH in plant nutrition and soil remediation. Appl Environ Soil Sci. https://doi.org/10.1155/2019/5794869
Radhakrishnan PG, Varghese SP, Das B C (2018) Application of ethylenediamine hydroxypropyl tamarind fruit shell as adsorbent to remove Eriochrome black T from aqueous solutions–kinetic and equilibrium studies. Sep Sci Technol 53(3):417–438
Ramya R, Adur Alaknanda J, Raajasubramanian D, Srinivasan S, Narendra K, Manjushree M (2023) Weed—an alternate energy source. In: Ramanujam PK, Parameswaran B, Bharathiraja B, Magesh A (eds) Bioenergy. energy, environment, and sustainability. Springer, Singapore. https://doi.org/10.1007/978-981-99-3002-9_10
Robe K, Barberon M (2023) Nutrient carriers at the heart of plant nutrition and sensing. Curr Opin Plant Biol 74:102376

Rossel PE, Vahatalo AV, Witt M et al (2013) Molecular composition of dissolved organic matter from a wetland plant (Juncus effusus) after photochemical and microbial decomposition (1.25 yr): common features with deep sea dissolved organic matter. Org Geochem 60:62–71

Sahu JN, Acharya J, Sahoo BK, Meikap BC (2016) Optimization of lead (II) sorption potential using developed activated carbon from tamarind wood with chemical activation by zinc chloride. Desalin Water Treat 57(5):2006–2017. https://doi.org/10.1080/19443994.2014.979446

Shaikh Z, Ali A (2021) Nitrate assimilation: cross-talk between plants and soil. In: Frontiers in plant-soil interaction. Academic, pp 135–151

Spence A, Simpson AJ, Mcnally DJ et al (2011) The degradation characteristics of microbial biomass in soil. Geochim Cosmochim Acta 75(10):2571–2581

Tripathi V, Edrisi SA, Chen B, Gupta VK, Vilu R, Gathergood N, Abhilash PC (2017) Biotechnological advances for restoring degraded land for sustainable development. Trends Biotechnol 35(9):847–859

Vatansever R, Ozyigit II, Filiz E (2017) Essential and beneficial trace elements in plants, and their transport in roots: a review. Appl Biochem Biotechnol 181:464–482. https://doi.org/10.1007/s12010-016-2224-3

Chapter 2
Carbon Sequestration via Plant Nutrition Management: Processes and Mechanisms

Priyanka Yadav, Anjali Malik, Shruti Jha, Sandeep Singh, and Amit Berwal

Abstract Global climate change is a major impact of global warming, a critical environmental issue affecting human life and other organisms on the earth. It is a gradual increase in the overall temperature of the earth's atmosphere through GHGs or other pollutants. CO_2 is one of the main GHGs causing global warming and accelerating climate change. Nowadays, the concentration of CO_2 in the atmosphere is continuously increasing due to natural phenomena like forest fires, the decomposition of organic matter, volcanic eruptions, and anthropogenic activities such as burning fossil fuels and deforestation. Carbon sequestration is one of the best methods to reduce the amount of CO_2 in the atmosphere. It removes and stores carbon (mainly CO_2) from the atmosphere into a carbon sink through physical or biological processes. The storage of CO_2 in a carbon sink for a long period is managed in such a way that it remains safe and not released back into the atmosphere. Carbon sequestration is done by natural (ocean, soil, forest, etc.) and artificial methods (depleted oil reserves, un-mineable mines, etc.). The terrestrial ecosystem sequestered the carbon of nearly 1.2 billion t C yr^{-1}. Therefore, this chapter will highlight the different processes of carbon sequestration and their mechanisms. Understanding the core mechanisms will help increase the carbon content in soil and improve soil structure, which can further contribute to enhanced crop productivity.

Keywords Carbon dioxide · Carbon sequestration · Soil · Climate change · Carbon · Plants · Greenhouse gases

P. Yadav · A. Malik · S. Jha · S. Singh
Department of Biochemistry, Maharshi Dayanand University, Rohtak, Haryana, India

A. Berwal (✉)
Department of Botany, Maharshi Dayanand University, Rohtak, Haryana, India
e-mail: amitberwalamii@gmail.com

S. A. Edrisi et al. (eds.), *Sustainable Plant Nutrition and Soil Carbon Sequestration*, Sustainable Plant Nutrition in a Changing World,
https://doi.org/10.1007/978-3-031-53590-1_2

2.1 Introduction

Carbon The ever-growing population on earth necessitates escalated food production. Soil is targeted to ensure nutrition in the crop-produce to meet the demand of the burgeoning population (Edrisi et al. 2021a). Soil nutrition can only be achievable when the soil is utilized sustainably via organic methods rather than exploited through unconventional practices (Verma and Edrisi 2022). Carbon (C) is an essential element found in all living organisms and maintains life on Earth. Naturally, it occurs in two forms, i.e., organic and inorganic (Gomez et al. 2022). In ecosystems, it is found in various forms, such as plant biomass, SOM, CO_2 dissolved in seawater and in the atmosphere, or carbon exists as gases in the atmosphere, dissolved ions in the hydrosphere, and in solids as a constituent of organic matter. It enters into the atmosphere from both natural (decomposition of plants and animals, volcanoes, forest fires, respiration, methane by ruminants) and anthropogenic sources (vehicles, power plants, etc.) (Panchal et al. 2022). Carbon absorbed by plants (123 Gt C yr^{-1}) is accumulated as biomass, but respiration and decaying biomass in the ecosystem lead to 60 Gt C yr^{-1} being liberated back into the atmosphere (Energy, genomics 2005). Generally, soil accumulates carbon as SOM. Therefore, soil plays a significant role in maintaining a balanced global carbon cycle (Edrisi et al. 2022a; Elbasiouny et al. 2022). The total carbon accumulation in SOM is regulated by adding carbon from the dead plant material and carbon losses from respiration, decomposition, and natural and human soil disturbance. SOC brings in the carbon cycle via the soil, plants, ocean, and the atmosphere. Most soil carbon comes from plants, while some are from mineral sources. Plants capture CO_2 from the atmosphere and accumulate carbon in the soils through roots, root exudates, seeds, and plant residues at various soil profile horizons (Meena et al. 2020) (Fig. 2.1).

During growth, plants liberate CO_2, water, and energy through respiration. As plants grow and die, they give up carbon-based organic compounds in the soil. During favorable situations, soil microorganisms such as earthworms and insects decompose these compounds and assimilate some carbon into novel chemical compounds while the rest of the carbon respires into the atmosphere as CO_2. Moreover, carbon accumulates in stems, branches, and roots (Nowak and Crane 2002) and deposits as carbon-rich organic compounds such as cellulose, hemicelluloses, lignin, starches, lipids, and waxes (Unwin and Kriedemann 2000) (Fig. 1.4). Photosynthetic products used by other living beings cycle carbon between the atmosphere into forests, soils, and oceans, while human energy consumption cycles carbon from fossil fuel pools to the atmosphere. The amount of carbon accumulated in the soil depends on the balance between the amount of carbon taken up by the earth and the amount of carbon back out of the earth as carbon-based respiration gases resulting from microbial activity. Due to the shifting process, continuous carbon supply and stable organization are necessary to retain the carbon pool in the soil in a balanced condition (Kell 2012).

For continuity of life, carbon must be recycled. This cycle remained in balance for thousands of years, and the CO_2 concentration in the atmosphere remained

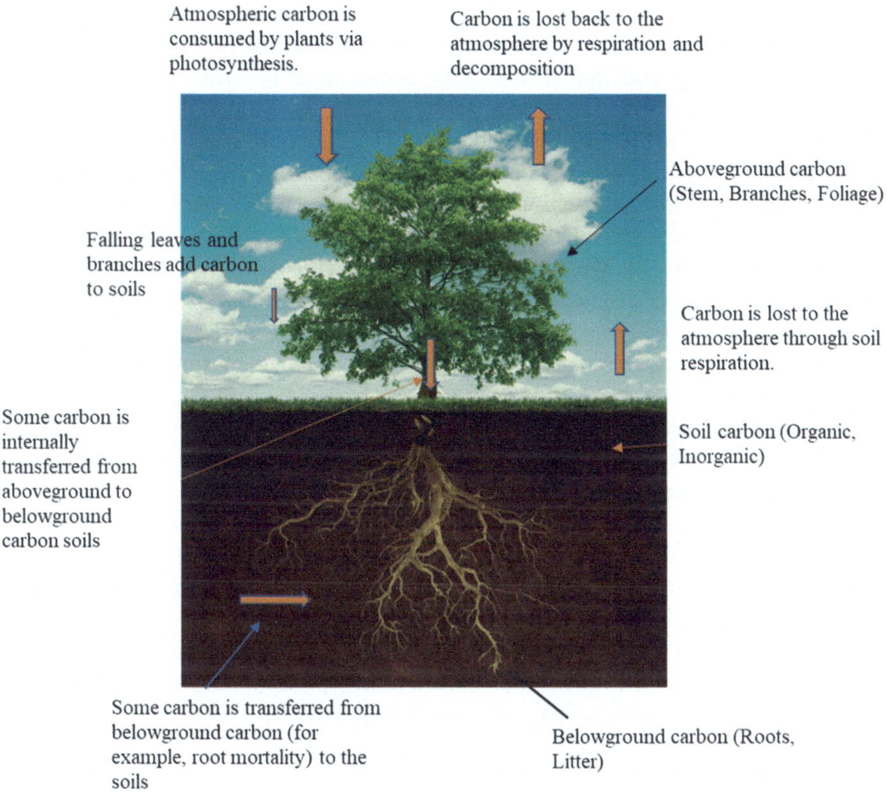

Fig. 2.1 Different processes by which plants facilitate the capturing and losing of carbon (Meena et al. 2020)

uniform. But in the last hundred years or more, human activities have created a disturbance in the global carbon cycle by soil degradation due to land-use change, combustion of fossil fuels, etc., which trigger CO_2 emission into the atmosphere. CO_2 accumulates in the atmosphere at about 3.4 Pg yr^{-1} due to fossil fuel combustion and land use change. However, continued farming increases the decomposition of SOM and gives rise to the depletion of 20–67% of soil carbon (Wei et al. 2015; Lal 2001). The Intergovernmental Panel on Climate Change (IPCC), in its second assessment report, evaluates that it may be possible to eliminate 40–80 Pg of carbon in the upcoming hundred years by sequestering it in farmland soils (Cole et al. 1995). Therefore, carbon sequestration is necessary to maintain the carbon balance in the atmosphere.

GHGs play an essential role in maintaining the global energy balance; a minor shift in their assembly in the atmosphere can alter the climatic conditions on Earth, the melting of polar ice, rising sea levels, and ultimate weather conditions. Most GHGs like methane (CH_4), nitrous oxide (N_2O), chlorofluorocarbons, and carbon dioxide (CO_2) are giving rise to global climate change, but CO_2 contributes to a greater extent. The capability of these gases to cause climate change depends on their global warming potential (GWP). CO_2 is examined to have a GWP of 1, CH_4 has a 100-year GWP of 28, and N_2O has the highest 100-year GWP of 265 (IPCC 2014). CH_4 is liberated from soils by decomposing organic matter in an anaerobic environment. GHGs are emitted continuously into the atmosphere due to anthropogenic activities like the combustion of fossil fuels (which meet 85% of the world's energy demand) (Herzog et al. 1997; DOE 2008; Lal 2008a, b). Overall, agriculture also produces N_2O (70%) (World Bank 2015a) and CH_4 (40%) (World Bank 2015b). The emissions of these GHGs are considered to create warming of about 1°C above pre-industrial levels, and this is likely to reach 1.5°C (considered as dangerous climate change) (Lewis 2016) between 2030 and 2052 if they continue at the current rate (IPCC 2018). CO_2 composition in the atmosphere is 0.03%. The amount of CO_2 in the atmosphere has increased by more than 30% since the beginning of the industrial age (1850). This increase has been caused mainly by burning fossil fuels, land use change, and deforestation (IPCC 2014).

It is also released naturally through the decomposition of plants and animals. It is a very effective GHG that absorbs infrared radiations, which are emitted from the earth's surface. As CO_2 concentrations rise in the atmosphere, more infrared radiation is retained, and the average temperature of the earth's lower atmosphere increases. It has been observed that the atmospheric CO_2 concentration was 280 ppm in 1850 (Etheridge et al. 1996) and enhanced to 411 ppm in the year 2019 (NOAA 2019), equivalent to the mean temperature rise of Earth 01°C during the twentieth century (IPCC 2018). Soil acts as both a source and sink for GHG. As a source, soil releases GHGs into the atmosphere, where they catch thermal radiation that increases the greenhouse effect and gives rise to global warming. In 2018, the built environment was responsible for 40% of global GHG emissions (IEA and UNEP, 2018). The IEA study demonstrates that CO_2 emissions will be enhanced by 130% by 2050 in the absence of new policies (IEA 2008).

Three main strategies of GHG mitigation in agriculture are (1) developing management practices to enhance the assimilation of atmospheric CO_2 by plants, (2) managing the transfer of carbon from the plants/animal's biomass into the soil, and (3) changing the cycling of SOC to increase its residence time. Three effective strategies could be used for reducing the carbon concentration in the atmosphere: (a) reductions in CO_2 emissions in the global atmosphere, (b) carbon capture and long-term storage, and (c) developing alternatives for carbon-based fuel. According to Schrag (2007), CO_2 emissions can be decreased by decreasing global energy use, developing low-carbon fuel, and sequestration of CO_2 via photosynthesis and other practices.

Theoretically, it assumes that elevated atmospheric CO_2 concentration would increase plant photosynthesis and, finally, crop yield, but the prolonged outcome is unpredictable and implies the worst consequence on the food web, diminished nutritional values, declined nitrogen content of plants, etc. (Reddy et al. 2023). Increased levels of CO_2 in ocean water since the beginning of the Industrial Revolution caused an increase in ocean acidification by 30% (formation of carbonic acid) (IOCCP 2009). The high water acidity diminishes the growth of marine organisms.

Climate change has been occurring continuously since the earth came into existence. It is a universal environmental and societal issue faced by humanity nowadays (IPCC 2018). It is one of the most important challenges of the twenty-first century. It affects health, livelihoods, food security, water supply, human security, and economic growth, which are projected to increase with global warming of 1.5°C and increase further with the warming of 2°C (IPCC 2018). The effects associated with climate change are global warming, a rise in sea levels, hurricanes, droughts, loss of biodiversity, heat waves, and floods, which have become a global threat. The IPCC, set up by the WMO (World Meteorological Organization) and UNEP (United Nations Environmental Programme), produced its first scientific assessment of these concerns (IPCC 1990). According to the IPCC, nations may need to remove between 100 billion and 1 trillion tons of CO_2 from the atmosphere this century to prevent the terrible effects of climate change. The '4 per 1,000' initiative launched by the French Minister of Agriculture at the UN Climate Change Conference (COP21) in Paris (2015) aims to increase SOC stocks by 0.4 percent per year through optimized land and soil management in agricultural soils (Rumpel et al. 2018).

Global warming is the regular elevation in the overall temperatures of the earth's atmosphere near-surface air and oceans. It is now universally recognized that global warming is the most pressing environmental issue affecting human lives worldwide. It is firmly associated with increased GHG emissions such as CO_2, CH_4, and N_2O, contributing to 410 (parts per million), 1866 (parts per billion), and 332 (parts per billion) (IPCC 2023). According to the Environmental Protection Agency, an increase in average temperature can lengthen the growing season in regions with relatively cool spring and fall seasons, negatively affect crops in areas where summer heat already limits production, increase soil evaporation rates, and increase the chances of severe droughts.

2.2 Carbon Sequestration and Its Various Methods

As global warming increases and society continues to release GHGs, enhanced carbon sequestration could be needed to stabilize the atmospheric carbon stock to maintain the CO_2 concentration below a threshold level (Lal 2008a, b; Zeng 2008). Article 5 of the Paris Agreement encourages parties to conserve and enhance carbon sinks, including forests (United Nations 2015), while the Bonn Challenge (bonnchallenge.org), initiated by IUCN and the Government of Germany in 2011, has gathered 62 commitments by governments and other organizations to restore over 350 million hectares of degraded and deforested land by 2030 to provide carbon sequestration and other benefits. Carbon sequestration is defined as the long-term storage of carbon in the form of CO_2 in the soil, geologic formations, and the ocean from

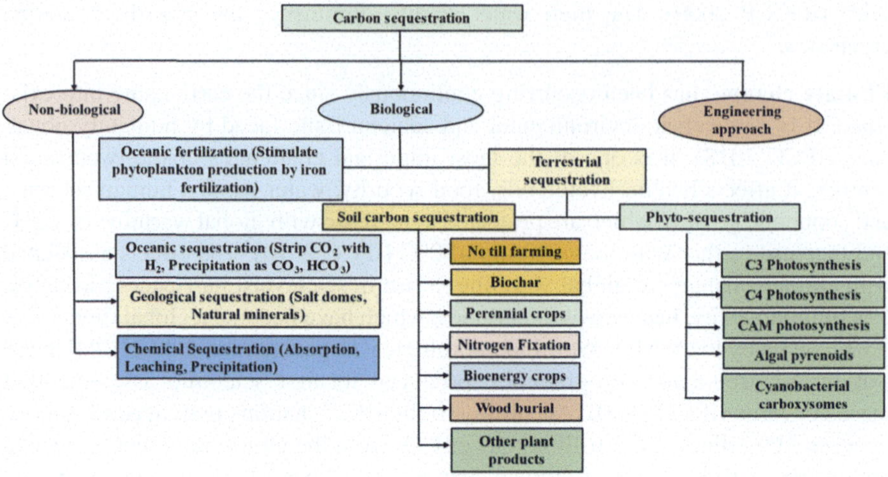

Fig. 2.2 Different ways of carbon sequestration in the different ecosystems (Nogia et al. 2016)

the atmosphere. According to the United Nations Framework on Climate Change, it is a mechanism of capturing CO_2 from the atmosphere that accumulates into a carbon stock.

Agriculture covers a bigger part of the world's land area (approx. 35%) than any other anthropogenic activity and releases a considerable amount of GHGs into the atmosphere (Salinger 2007). Nevertheless, it can also reduce GHG emissions through photosynthesis, the organic carbon cycle, and by increasing the weathering of alkaline-earth silicates (mineral carbonation) (Hartmann et al. 2013; Zomer et al. 2017). The potential of agricultural soil to sequester carbon depends on various factors such as climate, soil type, type of crop, and management techniques. Therefore, some agriculture practices that enhance the carbon capture and accumulation in the soil involve farmyard manure and inorganic fertilizers, conservation tillage, carbon capture, utilization and sequestration, crop residue management, cover cropping, rotational grazing, perennial cropping systems, and other engineered approaches. Besides, there are other ways through which the carbon is sequestered in the different ecosystems (terrestrial and aquatic) via biotic and abiotic means, as depicted in Fig. 2.2.

2.2.1 *Abiotic Sequestration*

In this method, living beings (plants and microorganisms) are not involved. It involves physical-chemical reactions and engineering techniques, for example, mineral carbonation, scrubbing, etc. (Kang et al. 2005). It also includes secondary carbonates formation, leaching, and transfer of these compounds in rivers (Lal and Kimble 2000; Raymond and Cole 2003). Cations produced by the weathering of silicate minerals also participate in carbon sequestration via metal-organic complex formation (Singh et al. 2018).

Ocean Carbon Sequestration In this process, oceans consume, emit, and accumulate large amounts of CO_2 from the atmosphere by increasing the efficiency of ocean biological systems through iron fertilization and inoculating CO_2 into the deep ocean. Iron deposition prompts phytoplankton assembly, which absorbs more CO_2 and increases photosynthesis. Ocean sequestration stores 50 times more inorganic carbon than the atmosphere. Therefore, it acts as a sink of atmospheric carbon (Raven and Falkowski 1999). It can be of two types: biotic (fertilization methods) and abiotic (direct injection of CO_2 into the ocean) (Lal 2008a, b). Dispersal of CO_2 between ocean and atmosphere depends on multiple agents such as pH, temperature, and water salinity; e.g., elevated temperature and salinity negatively affect CO_2 dissolution in water (Raven and Falkowski 1999). Phytoplankton photosynthesis fixes about 45 Pg C yr^{-1} (Falkowski et al. 2000). The organic matter produced in this process is accumulated at the sea floor and thus sequesters carbon (Rivkin and Legendre 2001). A high concentration of essential nutrients increases phytoplankton's growth and biomass, finally increasing CO_2 fixation (by photosynthesis) (Yang et al. 2008).

Direct Inoculation of CO_2 into the Ocean CO_2 inoculation causes acidity in water (because it reduces pH), which during a long period causes toxicity and negatively affects sea microorganisms (Auerbach et al. 1997).

Geologic Carbon Sequestration This method involves absorption, dissolution, transmission, and inoculation of CO_2 into deep geological strata such as oil and gas reservoirs, coal seams, and deep saline aquifers (Baines and Worden 2004; Gale 2004; Klara et al. 2003). It lowers the emission of GHGs and is also helpful in the oil recovery process. Saline aquifers are permeable rocks saturated with salt water. CO_2 inoculated in them reacts with dissolved minerals and is stored on porous sediments (Bergman and Winter 1995). In General, the principal considerations about geological sequestration are the responsibility of accumulating a large amount of CO_2 in geological strata and their cost (Kintisch 2007; Schrag 2007).

Chemical Carbon Sequestration This method, also known as the scrubbing process, involves chemical reactions that convert atmospheric CO_2 into stable compounds (amine, carbonates, etc.). For example, the chemical weathering of rocks by mineral carbonation (CO_2 present in rainwater reacts with different metal oxides that exist on rocks and results in the generation of mineral carbonates). These carbonates persist in solid rocks, so CO_2 is stored for a longer duration (Lal 2008a, b; Kojima et al. 1997).

$$CaSiO_3 + CO_2 \rightarrow CaCO_3 + SiO_2$$

2.2.2 *Biotic Sequestration*

It is a natural process of carbon sequestration in which CO_2 is removed from the atmosphere by photosynthesis (Lal 2008a, b). It relates to living biomass and soil biodiversity (plants, fauna, and microorganisms). During photosynthesis, plants absorb CO_2 from leaves and convert it into O_2, carbohydrates (required for plant growth), and water in the presence of sunlight. The rate of photosynthetic CO_2 assimilation depends on soil fertility, climate, etc.

2.2.3 *Terrestrial Carbon Sequestration*

The idea of terrestrial carbon sequestration was first given by Dyson in 1977. Atmospheric CO2 is consumed by land vegetation via photosynthesis and accumulated as carbon in soil and biomass (tree trunks, branches, foliage, roots, etc.) in the process. This method is also known as a win-win or no-regrets strategy (Lal et al. 2003). It stores carbon mainly through photosynthesis and lives and dead organic matter. Thus, it acts as a significant carbon stock (Lal 2008a, b). It includes the accumulation of CO_2 in plants (Phyto-sequestration) and soil, both above-ground and below-ground biomass (Post et al. 2009). It provides numerous advantages like increased soil and water quality, upgrading degraded land, enhancing plant biomass, and also helping in mitigating global climate change.

No-till farming Generally, tillage farming is preferred by farmers, which causes soil disturbance (by soil respiration, decomposition, etc.). This technique enhances soil aggregate movement and decreases the carbon pool (Paustian et al. 2000; McBratney and Stockmann 2011). Therefore, carbon sequestration is used in soil. No-till farming techniques are favored nowadays. It minimizes soil disturbance because the soil is not merged accurately and persists extra fertile because of the presence of previous crop residues, which increase the carbon content of the soil. This technique decreases soil erosion and increases infiltration, water-holding capacity, and soil aggregate formation (McBratney and Stockmann 2011). It stores carbon up to 0.16–0.49 Mg C ha^{-1} yr^{-1}.

Perennial crops This farming system dominated by annual plants enhances soil carbon by minimizing soil disturbance. These crops have expanded root systems, which produce more belowground biomass and store a large amount of carbon (Lal 2008a, b; Cox et al. 2006).

Biochar It is also called black carbon and is resistant to microbial decomposition. Chemically, it is more stable due to polycyclic aromatic rings (Ameloot et al. 2013). Biochar is usually created in the atmosphere when the plant biomass is heated at a low temperature (300–650°C) in the presence of low oxygen. It increases soil pH, plant production, nutrient availability, water-holding capacity, maintains cation exchange capacity, and finally increases soil fertility (Hecht 1982). It also reduces the release of GHGs into the atmosphere and stores carbon for a long period. Biochar obtained from the transformation of algae is more useful than any other biomass because of its high nitrogen and phosphorus content, which promotes soil fertility and decreases the demand for fertilizers on agricultural soil (Bird et al. 2011). The highest carbon contents of biochar were acquired at 400°C for Mulberry and at 450°C for Eucalyptus. Different reports evaluate the general carbon sequestration ability of biochar to be 1.6 ± 5.14 Mg C ha^{-1} yr^{-1} (Paustian et al. 2016; Smith et al. 2016; Majumder et al. 2019).

Nitrogen fixation Nitrogen fixing plants have a greater capability of accumulating soil carbon (Binkly and Giardina 1998). It also increases litter fall and mycorrhizal association, which finally enhances soil carbon.

Bioenergy crops A plant species that can be used as a sustainable energy source for the future is known as a bioenergy crop. These crops are the link between the sink (biomass and SOC) and the source (fossil fuel combustion) (Edrisi et al. 2022b; Edrisi 2019). The net CO_2 release comes from fossil fuel combustion and biomass because aboveground carbon biomass is recycled (Zan et al. 2001). They include sugarcane and oil crops (Jansson et al. 2010). These crops sequester carbon through roots and soil.

Wood burial In this method, dead or live plants are covered in above-ground areas in anaerobic conditions. Zeng (2008) reported that this application could diminish global climate change by reducing atmospheric CO_2.

Different plant products , such as wood, are used to construct buildings, houses, and furniture (Jansson et al. 2010) and sequester carbon for long periods. Phytoliths are also plant products released after burning, digestion, and decaying plant parts. These can be silicates and calcareous deposits.

Soil carbon sequestration In this method, atmospheric CO_2 absorbed by plants is changed into organic carbon via photosynthesis, while a small amount of it goes into the soil through the plant's roots, where it is accumulated in the soil (Jansson et al. 2010). The capability of soil carbon sequestration depends on the climate, temperature, rainfall, clay content, mineralogy, moisture content, and soil texture (Metting et al. 2001).

C3 photosynthesis In this type of photosynthesis, the first formed stable compound is 3-phosphoglyceric acid (3C). So, it is called C3. In them, ribulose 1,5-bisphosphate carboxylase oxygenase (RuBisCO) accepts atmospheric CO_2, adds it with ribulose bis-phosphate (RuBP), and finally converts it into an organic compound.

C4 photosynthesis In this type of photosynthesis, the first stable compound formed is oxalo-acetic acid (4C). Therefore, it is called C4. It mainly operates in monocot plants. In them, atmospheric CO_2 is accepted by the enzyme phosphoenol pyruvate (PEP) carboxylase, which has a high affinity for CO_2 binding. So, during high temperatures and low CO_2 concentrations, photorespiration is absent in them.

CAM photosynthesis This process is known as Crassulacean Acid Metabolism. It mainly occurs in the xerophytes plant. They take CO_2 during the night with the help of the enzyme phosphoenol-pyruvate (PEP) carboxylase. They close their stomata during the daytime or in a stressed condition.

Algal pyrenoids also help in carbon sequestration by absorbing atmospheric CO_2. In them, CO_2 accumulates close to RuBisCO, increasing photosynthetic capacity. The structure where photosynthesis takes place in algae is called pyrenoids.

Cyanobacteria are fast-growing photoautotrophic, unicellular bacteria that absorb atmospheric CO_2 to prepare their food and help in carbon sequestration.

$$6CO_2 + 12H_2O + \text{Light} \rightarrow C_6H_{12}O_6 + 6O_2 + 6H_2O$$

Engineering approaches These techniques involve genetic engineering to enhance carbon sequestration. They include both biotic and abiotic methods.

The main benefits of SOC sequestration to soil quality are (a) enhancement in soil structure, (b) decline in soil erosion, (c) reduction of non-point source pollution, (d) upgrade of plant-available water reserves, (e) expand plant nutrients availability (f) denaturing of pollutants (g) improve the agronomic yield of food security (h) moderation of climate and improve the aesthetic and economic value of the soil. Hence, carbon sequestration ameliorates ecosystem services by improving agronomic yield.

2.2.4 *Strategies for Soil Carbon Sequestration*

The processes affecting soil carbon sequestration are (i) biomass stored in the soil and (ii) stabilization and decomposition of SOM (Sanderman et al. 2010). Increasing biomass production and storing in soils can increase SOC, mainly depending on soil management applications. While decomposition of organic matter occurs through microbial respiration (Chowdhury et al. 2014; Schlesinger 2000), and it depends on agricultural activities (Fuentes et al. 2009; Sarker et al. 2017), environmental factors and soil biota (Jenny 1980). Therefore, various strategies have been introduced that fundamentally focus on the factors that influence soil carbon sequestration, as shown in Fig. 2.3.

2.2.4.1 Adoption of Recommended Management Practices (RMPs) on Agricultural and Forest Soils

It is a win-win approach (Follett 2001; Lal et al. 2003; Lal 2006). It enhances soil and water quality and agronomic yield, reduces pollution by diminishing breakdown and sediment loads, and declines the net rate of CO_2 emission by SOC sequestration. They sequester carbon through aggregation (enhance stable micro-aggregates

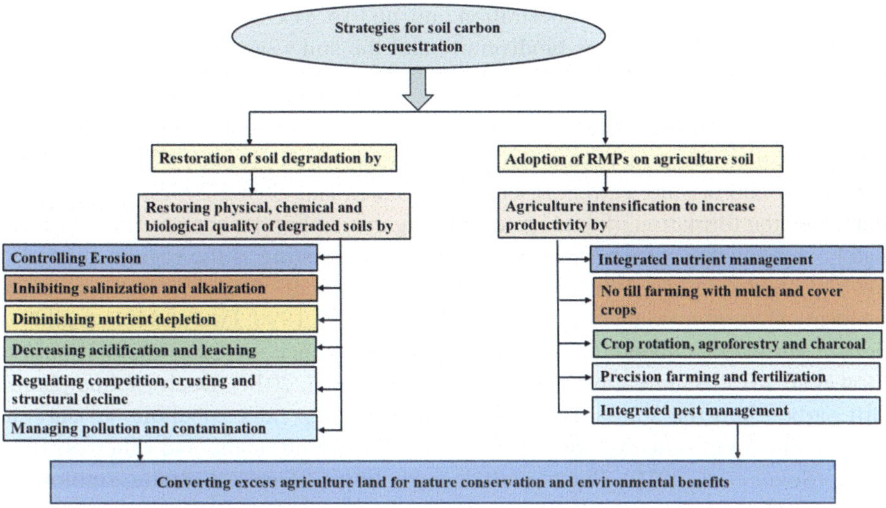

Fig. 2.3 Different management strategies for sequestering the carbon in soil systems (Lal 2008a, b)

via formation of organo-mineral complexes), humification (humic compounds boost SOC), formation of secondary carbonates, and translocation into the subsoil (isolate them from interference by plowing or more agronomic processes and diminish the chance of separated by erosion).

2.2.4.2 Integrated Nutrient Management (INM)

Soil can act as a sink of atmospheric carbon by forming SOM or humus. The productivity of carbon sequestration decreases when carbon and nitrogen are not present in sufficient amounts (Paustian et al. 1997). SOC sequestration depends on the rate and source of nitrogen. Liebig et al. (2002) noticed that high nitrogen rate application amplifies SOC sequestration rates by 1.0–1.4 Mg C ha^{-1} yr^{-1}. A sufficient concentration of nutrients in the soil increases the biomass output during high CO_2 concentrations (Van Kessel et al. 2000). The outcome of fertilizer on the SOC pool depends upon the quantity of biomass of carbon assembled, which comes back to the soil and its humification. Nutrient management is necessary for the humification process (plant residues by soil organisms). Inorganic fertilizers (N, P, K, S) amplify carbon humification in residues and facilitate SOC sequestration (Lal 2003).

Cover crops It is well known that ecosystems with greater biodiversity consume more atmospheric CO_2 and sequester more carbon. These are fast-growing crops like buckwheat, rye, and cowpea. They increase soil productivity (by preventing soil erosion and enhancing nutrients in the soil) and contribute to organic matter (Olson et al. 2014). Cover cropping of olive orchards, vineyards, and almond orchards detected the highest carbon sequestration rate, up to 5.3 t C ha^{-1} yr^{-1}. Growing leguminous cover crops increases biodiversity and the soil's organic pool (Fullen and Auerswald 1998).

Manuring and composting Generally, organic manures and compost boost the SOC stock more than inorganic fertilizers (Gregorich et al. 2001). They enhance soil quality and organic carbon reservoirs (Whalen and Chang 2002). Under dairy manuring, the formation of aggregates and organic carbon is increased, which ultimately stores more carbon (Min et al. 2003). Composting is the regulated breakdown of organic matter. When the compost matures, half of its carbon is present in the form of humic substances (more stable) (Inbar et al. 1990; Post and Kwon 2000). Organic fertilizer is abundant in organic matter and hence increases the SOC stock (Don et al. 2018; Baumgarten et al. 2014; Powlson et al. 2014). Deep-rooting plants shift carbon into the sub-layer of soil (which has a greater capability of carbon sequestration) via root exudates (such as sugars, amino acids, and other organic acids) (Sokol et al. 2019). Plants with deep-rooting systems are alfalfa, sunflower, grass, etc. (Kutschera et al. 2010; Thorup-Kristensen et al. 2020).

Soil biota management Microorganisms ameliorate the physiochemical and biological properties of soil. Mycorrhizae (roots with fungal association) enhance carbon sequestration by forming soil aggregate and aggregates that can retain carbon in the soil. After all, plant roots and mycorrhizal fungi require more carbon; plants capture more atmospheric CO_2 through photosynthesis.

Wood chips Wood is a sustainable energy source used as a fuel to cook food. Bamboo farming helps in carbon sequestration by making greater biomass. That biomass can produce wood chips, pellets, and fuel. Thus, it can sequester carbon of about 1.78 kg of C (Patel et al. 2017).

Crop Residues Agricultural applications like adding crop residues enhance SOM and nutrient contents through integrated nutrient management (Fang et al. 2018). The deterioration of crop residue depends upon its composition. For example, it is hard to degrade a substance rich in lignin content for microbes.

Mulching is a layer of material applied on the soil surface to retain soil moisture and increase soil health and fertility. It plays a vital role in supplying nutrients, the carbon and nitrogen cycle, and the carbon sink. It ameliorates the physiochemical properties of soil and enhances carbon sequestration up to 8–16 Mg ha^{-1} yr^{-1} in agricultural soils.

Biofuel is produced by changing biomass-derived sugars to ethanol and plant-derived oils and fats into bio-diesel. There are ample research that focus on cultivating biomass vis-à-vis biofuel and bioenergy from the marginal and degraded lands (Edrisi 2013, 2019). It is a feasible approach to diminishing the use of fossil fuels and evolving as another energy source (Himmel et al. 2007; Wald 2007). Examples of traditional biofuels are animal dung, crop residues, biomass, wood products, etc. (Himmel et al. 2007; Edrisi et al. 2022b).

Afforestation is the initiation of growing trees (forest) in an area with no former tree cover. It is one of the main methods for carbon sequestration and restoring degraded land (Lamb et al. 2005). The proper establishment of tree species such as *Tamarix* spp. (Athel tree), Eucalyptus spp. (Gum tree), *Casuarina* spp. (Native pine), *Prosopis* spp., (Mesquite) *Azadirachta* (Neem), *Acacia* sp., *Tectona* sp. (Teak), *Dalbergia* sp. (Indian Rosewood), etc., increases soil fertility and sequesters carbon in the soil (Lamb et al. 2005; Edrisi et al. 2020, 2021b, 2023).

Wetlands The U.S. Fish and Wildlife Service describes wetlands as "lands transitional between terrestrial and aquatic systems, where the water table is at or near the land surface or the land is covered by silly water. These special carbon sinks accumulate approximately one-third of the organic soil carbon of the world (Gorham 1991a, b; de Groot et al. 2006). Wetlands also help in mitigating GHG emissions and associated climate change (Sjogersten et al. 2014; Aziz et al. 2016). Wetlands absorb atmospheric CO_2 via photosynthesis and sequester carbon as SOM. Coastal wetlands, like marshes, mangroves, etc., can accumulate more carbon (prevent CH_4 emissions) (Poffenbarger et al. 2011). Wetland soils accumulate 200 times more carbon than the related vegetation (Milne and Brown 1997; Garnett et al. 2001).

2.2.4.3 Soil and Plant Nutrition

Apart from reducing CO_2 concentration in the earth's atmosphere, carbon sequestration also ameliorates land quality and productivity because all carbon sequestration technique improves SOM, which increases soil fertility, structure, aggregate stability, water holding capacity, nutrient availability, and the capacity to diminish toxic elements (Lal 2004). Soil is more important as it provides food, fiber, fuel, fodder, and many other services to the growing population of the world. Along with climate, soil composition (inorganic and organic components) and texture also determine plant distribution and growth factors. Healthy soil improves plant growth by enhancing plant nutrients. Plants obtain most of their carbon compounds (CO_2) from the air but depend on soil for water and minerals. High CO_2 decreases transpiration, enhances water use efficiency, and decreases nutrient absorption from land (Jenny 1980; Kabata-Pendias 2001).

The amount of SOM is the indicator of soil fertility. There are various ways by which the plants uptake the nutrients from the soil mediated by the biotic and abiotic components through various phenomena that include immobilization, assimilation, mineralization, and other mechanisms (Fig. 2.4). It can be provided to the soil naturally (plant and animal residues) or through anthropogenic activity (biowaste, composts, etc.).

Plants contribute to forming stable aggregates (secondary particles formed through combining mineral particles with organic and inorganic substances), which prevent the degradation of SOM. In general, finer-textured soils contain more SOC than coarse-textured soils. SOC stock involves humus and charcoal. It contains decomposed matter of plant and animal residues, material synthesized from the

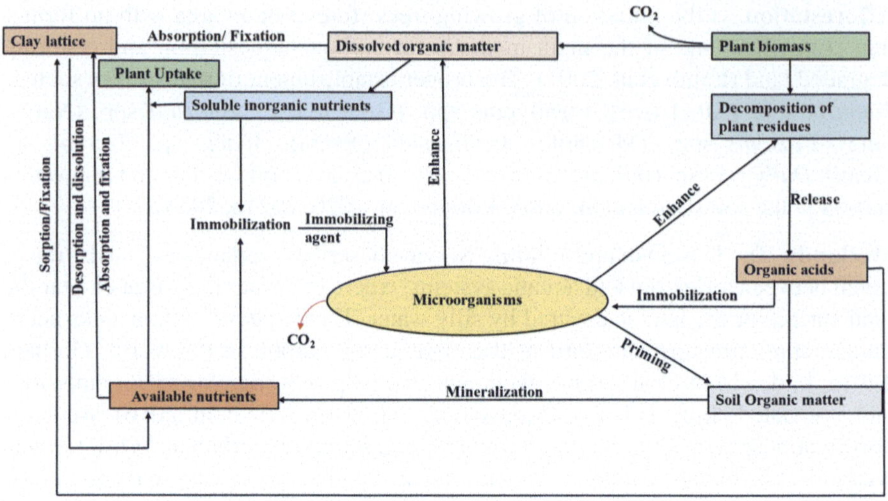

Fig. 2.4 Avenues of nutrient uptake by plants and its release in soil via plant residues (Chowdhury et al. 2021)

breakdown products, and the bodies of microbes. Soil inorganic carbon stock involves elemental carbon and carbonate minerals such as calcite, dolomite, gypsum, and primary and secondary carbonates.

The primary carbonates are obtained from the weathering of parent material, while secondary carbonates are formed by the reaction of atmospheric CO_2 with CaC_2 and MgC_2. Maintaining SOM at an ample quantity prohibits soil degradation (Blume et al. 2016). SOM not only acts as a sink for CO_2 but also improves soil quality (nutrient and their delivery to plants), hydrology and drainage, and other ecosystem functions (Pacala and Socolow 2004; Jiang et al. 2006; Lal 2004). Soil carbon sequestration ameliorates soil quality and enhances yield by amplifying the extent of SOM. Nitrogen (N) fertilizer applications are frequently advised to boost SOM, particularly in areas that have previously seen significant SOM losses due to farming. Thus, soil carbon sequestration means improving soil organic and inorganic carbon, which is achieved by different applications like crop rotation, nutrient management, etc. (Lal 2004). In an agricultural system, this can be achieved by various methods: (i) by reducing soil disturbance and (ii) by enhancing soil microbial diversity.

The accessibility of nutrients to plants in soil depends on soil chemical properties, the position of the ion respective to the root surface, and the length of the pathway through which nutrients move in the soil to reach the root surface. The availability of plant nutrients is also affected by many factors, such as soil pH, soil aggregates (accumulating more carbon), microorganisms, cation exchange capacity, etc. (Mathers 2001). Macronutrients are mostly available at pH 6.5–7.5 (Brady and Weil 2011), while micronutrients are at lower pH (except Mo, Cu, and B). In mineral soils, plant nutrients are mostly available at pH 6.0–7.0, while in organic soils, this range is between 5.0 and 6.0 (Mathers 2001) because at this pH range, microorganisms survive well and they decompose SOM, which increases nutrients available for the plant to uptake. Soil pH influences many biogeochemical processes like soil enzyme activities, nitrification and denitrification, rhizosphere processes, organic amendments, precipitation and mineralization of organic matter, ammonia volatilization, microbial eco-physiological indicators, etc. Plant nutrient uptake also depends on soil cation exchange capacity (CEC). A soil with CEC 10 cmol (+) kg^{-1} is favorable for plant growth (DPI 2002) and regulates plant productivity. Soils with low CEC cause mineral deficiencies such as potassium, magnesium, and other cations (Brown and Lemon 2008). Root exudates and CO_2 produced by root respiration acidify the rhizosphere, making some nutrients available to plants (Kabata-Pendias 2001). Soil moisture and temperature also affect plant nutrient availability and root growth.

2.3 Conclusions and Future Recommendations

Nowadays, GHGs are increasing continuously in the atmosphere, and the threat of global climate change requires our attention. In the upcoming years, determining soil carbon supplies for reducing climate change will be a significant problem. Therefore, carbon sequestration is a better strategy to mitigate atmospheric GHGs. In carbon sequestration, novel techniques increase soil quality, fertility, plant production, and atmospheric CO_2 storage. This method also increases the availability of nutrients to plants. Thus, Carbon Sequestration is a winning method to mitigate GHG emissions and increase crop production for the population. Novel soil and crop management techniques must be used to improve soil carbon storage. In order to reduce CO_2 emissions and increase Carbon Sequestration, producers will likely be willing to implement practices like reduced tillage, crop rotations rather than monocultures, increased cropping frequency instead of bare fallow in arid and semi-arid environments, forage crop cultivation on marginal land, and more effective fertilizer use based on soil test criteria. There has been little study on SOC distribution in aggregates and the intrinsic biophysical properties of aggregates that promote Carbon Sequestration. So, it became unclear how SOC interacts physically and chemically with aggregates to increase soil Carbon Sequestration capacity. Future studies must improve our understanding of the mechanisms behind aggregate dynamics in connection to SOC and their relevance and roles in Carbon Sequestration.

References

Ameloot N, De Neve S, Jegajeevagan K, Yildiz G, Buchan D, Funkuin YN, Prins W, Bouckaert L, Sleutel S (2013) Short-term CO_2 and N_2O emissions and microbial properties of biochar amended Sandy loam soils. Soil Biol Biochem 57:401–410

Auerbach DI, Caulfield JA, Adams EE (1997) Impacts of ocean CO_2 disposal on marine life: I. A toxicological assessment integrating constant concentration laboratory assay data with variable-concentration field exposure. Environ Model Assess 2:333–343

Aziz, A.A., Thomas, S., Dargusch, P. And Phinn, S. (2016). Assessing the potential of REDD+ in a production mangrove forest in Malaysia using stakeholder analysis and ecosystem services mapping. Mar Policy 74, 6–17

Baines SJ, Worden RH (2004) Geological storage of carbon dioxide, vol 233. The Geological Society, London, UK, pp 1–6

Baumgarten A, Geitner C, Haslmayr HP, Zechmeister-Boltenstern S (2014) Der Einfluss Des Klimawandels Auf Die Pedosphäre. In: Österreichischer Sachstandsbericht Klimawandel 2014 (AAR14). Verlag der Österreichischen Akademie der Wissenschaften, Vienna, Austria. ISBN 978-3-7001-7723-4

Bergman PD, Winter EM (1995) Disposal of carbon dioxide in aquifers in the U.S. Energ Convers Manage 36:523–526

Binkly D, Giardina C (1998) Why do tree species affect soils? The warp and woof of trees-soil interactions. Biogeochemistry 42:89–106

Bird MI, Wurster CM, Paula Silva PH, Bass AM, Nys R (2011) Algal biochar—production and properties. Bioresour Technol 102:1886–1891

Blume HP, Brümmer GW, Fleige H, Horn R, Kandeler E, Kögel-Knabner I (2016) Soil organic matter. Scheffer/Schachtschabel Soil Science. Springer, Berlin, Heidelberg

Brady NC, Weil RR (2011) The nature and properties of soils, 13th edn. Collier Macmillan, New York. 50 Saikat Chowdhury et al., ARTICLE IN PRESS

Brown K, Lemon J (2008) Cations and cation exchange capacity University of Western Australia and Department of Agriculture and Food. WA

Chowdhury S, Farrell M, Bolan NS (2014) Priming of soil organic carbon by malic acid addition is differentially affected by nutrient availability. Soil Biol Biochem 77:158–169

Chowdhury, S., Bolan, N., Farrell, M., Sarkar, B., Sarker, J. R., Kirkham, M. B., ... & Kim, G. H. (2021). Role of cultural and nutrient management practices in carbon sequestration in agricultural soil. *Advances in agronomy, 166*, 131–196

Cole CV et al (1995) Agricultural options for mitigation of greenhouse gas emissions. In: Climate change 1995—impacts, adaptations, and mitigation of climate change: scientific-technical analysis. IPCC Working Group II, Cambridge University Press, pp 7450–7771

Cox TS, Glover JD, Tassel DLV, Cox CM, DeHaan LD (2006) Prospects for developing perennial-grain crops. Bioscience 56(8):649–659

de Groot RS, Wilson MA, Boumans RMJ (2006) Valuing wetlands: guidance for valuing the benefits derived from wetland ecosystem services. (No. 3 CBD Technical Series No. 27). Ramsar Convention Secretariat, Gland

DOE (2008) Carbon cycling and biosequestration workshop US Department of Energy. In: Office of Biological and Environmental Research Workshop. March 4–6 and 17–18, 2008

Don A, Flessa H, Marx K (2018) Die 4-Promille-Initiative "Böden Für Ernährungssicherung Und Klima"—WissenschaftlicheBewertung Und DiskussionMöglicherBeiträge in Deutschland. Johann Heinrich von Thünen-Institut: Braunschweig, Germany

DPI (2002) Cation exchange capacity. Department of Primary Industries, Government of New South Wales

Edrisi SA (2013) Sustainable biofuel production from marginal and degraded lands: an ecological and socio-economic approach. Doctoral dissertation, Banaras Hindu University

Edrisi, S. A. (2019). Sustainability analysis of biomass biofuel production from marginal and degraded lands of India

Edrisi SA, El-Keblawy A, Abhilash PC (2020) Sustainability analysis of Prosopis juliflora (Sw.) DC based restoration of degraded land in North India. Land 9(2):59

Edrisi SA, Rakshit A, Dubey PK, Abhilash PC, Singh SK, Patra AK, Pathak H (2021a) Managing soil resources for human health and environmental sustainability. Soil Science: Fundamentals to Recent Advances, pp 3–11

Edrisi SA, Tripathi V, Chaturvedi RK, Dubey DK, Patel G, Abhilash PC (2021b) Saline soil reclamation index as an efficient tool for assessing restoration progress of saline land. Land Degrad Dev 32(1):123–138

Edrisi SA, Tripathi V, Dubey PK, Abhilash PC (2022a) Carbon sequestration and harnessing biomaterials from terrestrial plantations for mitigating climate change impacts. In: Biomass, biofuels, biochemicals. Elsevier, pp 299–313

Edrisi SA, Dubey PK, Chaturvedi RK, Abhilash PC (2022b) Bioenergy crop production potential and carbon mitigation from marginal and degraded lands of India. Renew Energy 192:300–312

Edrisi SA, Singh A, Dubey PK, Abhilash PC (2023) Tectona grandis Lf. mediated restoration of marginal lands in eastern Uttar Pradesh, North India. Land Degrad Dev 34(4):929–942

Elbasiouny H, El-Ramady H, Elbehiry F, Rajput VD, Minkina T, Mandzhieva S (2022) Plant nutrition under climate change and soil carbon sequestration. Sustain For 14(2):914

Energy, Genomics: GTL Roadmap, U.S. Department of Energy (2005)

Etheridge DM, Steele LP, Langenfelds RL, Francey RJ, Barnola JM, Morgan VI (1996) Natural and anthropogenic changes in atmospheric CO over the last 1000 years from air in Antarctic ice and firn. J Geophy Res 101(D2):4115–4128

Falkowski P et al (2000) The global carbon cycle: a test of our knowledge of earth as a system. Science 290:291–296

Fang Y, Nazaries L, Singh BK, Singh PB (2018) Microbial mechanisms of C priming effects revealed during the interaction of crop residue and nutrient inputs in contrasting soils. Glob Chang Biol 24(7):2775–2790

Follett RF (2001) Soil management concepts and carbon sequestration in cropland soils. Soil Till Res 61:77–92

Fuentes M, Govaerts B, Leon DF, Hidalgo C, Sayre KD, Etchevers J, Dendooven L (2009) Fourteen years of applying zero and conventional tillage, crop rotation, and residue management systems and its effect on physical and chemical soil quality. Eur J Agron 30:228–237

Fullen MA, Auerswald K (1998) Effect of grass ley set aside on runoff, erosion and organic matter levels in sandy soil in East Shropshire, U.K. Soil Till. Res. 46:41–49

Gale J (2004) Why do we need to consider the geological storage of CO2? In: Baines SJ, Worden RH (eds) Geological storage of carbon dioxide, vol 233. The Geological Society, London, UK, pp 7–15

Garnett MH, Ineson P, Stevenson AC, Howard DC (2001) Terrestrial organic carbon in a British moorland. Glob Chang Biol 7:375–388

Gomez C, Chevallier T, Moulin P, Bouferra I, Hmaidi K, Arrouays D et al (2022) Prediction of soil organic and inorganic carbon concentrations in Tunisian samples by mid-infrared reflectance spectroscopy using a French national library. Geoderma 375:114469

Gorham E (1991a) Northern peatlands: role in the carbon cycle and probable responses to climate warming. Ecol Appl 1:182–195

Gorham E (1991b) Northern peatlands: role in the carbon cycle and probable responses to climatic warming. Ecol Appl 1:182–195

Gregorich EG, Drury CF, Baldock JA (2001) Changes in soil carbon under long-term maize in monoculture and legume-based rotation. Can J Soil Sci 81:21–31

Hartmann J, West AJ, Renforth P, Köhler P, De La Rocha CL, WolfGladrow DA et al (2013) Enhanced chemical weathering as a geoengineering strategy to reduce atmospheric carbon dioxide, supply nutrients, and mitigate ocean acidification. Rev Geophys 51(2):113–149

Hecht SB (1982) Agroforestry in the Amazon Basin: practice, theory, and limits of promising land use. In: Hecht SB (ed) Amazonia: agriculture and land use research. Proc. Int. Conf, Cali, pp 331–372

Herzog H, Drake E, Adams E (1997) CO_2 capture, reuse, and storage technologies for mitigating global climate change. A White Paper Final Report DE-AF22-96PC01257

Himmel ME, Ding SY, Johnson DK, Adney WS, Nimlos MR, Brady JW, Foust TD (2007) Biomass recalcitrance: engineering plants and enzymes for biofuel production. Science 315:804–807

IEA (2008) Capture and storage: a key carbon abatement option. OECD Publishing

IEA and UN Environment Programme (2018) 2018 global status report: towards a zero-emission. Efficient and Resilient Buildings and Construction Sector

Inbar Y, Chen Y, Hadar Y (1990) Humic substances formed during the composting of organic matter. Soil Sci Soc Am J 54:1316–1323
Intergovernmental Panel on Climate Change (IPCC) (1990) Climate change: the IPCC impacts assessment. In: Tegart WJ, Mc G, Sheldon GW, Griffiths DC (eds) World meteorological organization/United Nations Environment Programme. Australian Government Publishing Service, Canberra
IOCCP (2009) The ocean in a high CO_2 world: ocean acidification. The International Ocean Carbon Coordination Project, Monterey
IPCC (2014) Climate Change 2014: Synthesis, Report Contribution of Working Groups I, II, and III to the Fifth Assessment Report of the Intergovernmental Panel on Climate Change. IPCC, Geneva
IPCC (2018) Strengthening and implementing the global response in global warming of 1.5°C
Jansson C, Wullschleger SD, Kalluri UC, Tuskan GA (2010) Phytosequestration: carbon biosequestration by plants and the prospects of genetic engineering. Bioscience 60(9):685–696
Jenny H (1980) The soil resource. Ecological studies, vol 37. Springer, New York
Jiang HM, Jiang JP, Jia Y, Li FM, Xu JZ (2006) Soil carbon pool and effects of soil fertility in seeded alfalfa fields on the semi-arid Loess Plateau in China. Soil Biol Biochem 38:2350–2358
Kabata-Pendias A (2001) Trace elements in soils and plants. CRC Press, Boca Raton
Kang QJ, Tsimpanogiannis IN, Zhang DX, Lichtner PC (2005) Numerical modeling of pore-scale phenomena during CO_2 sequestration in oceanic sediments. Fuel Process Technol 86:1647–1665
Kell DB (2012) Large-scale sequestration of atmospheric carbon via plant roots in natural and agricultural ecosystems: why and how. Philos Trans R Soc B Biol Sci 367(1595):1589–1597
Kessel VC, Horwath WR, Hartwig U, Harris D, Luscher A (2000) Net soil carbon input under ambient and elevated CO_2 concentrations: isotopic evidence after 4 years. Glob Chang Biol 6:435–444
Kintisch E (2007) Report backs more projects to sequester CO_2 from coal. Science 315:1481
Klara SM, Srivastava RD, McIlvried HG (2003) Integrated collaborative technology development program for CO2 sequestration in geologic formations—United States Department of Energy R&D. Energy Convers Manag 44:2699–2712
Kojima T, Nagamine A, Ueno N, Uemiya S (1997) Absorption and fixation of carbon dioxide by rock weathering. Energy Convers Manag 38:S461–S466
Kutschera L (2010) Wurzelatlas, mitteleuropäischerAckerunkräuter und Kulturpflanzen; Band 1. DLG Verlags GmbH, Frankfurt. ISBN 978-3-7690-0758-8
Lal R (2001) World cropland soils as a source or sink for atmospheric carbon. Adv Agron 71:145–191
Lal R (2003) Offsetting global CO_2 emissions by restoration of degraded soils and intensification of world agriculture and forestry. Land Degrad Develop 14:309–322
Lal R (2004) Soil carbon sequestration impacts on global climate change and food security. Science 304:1623–1627
Lal R (2006) Enhancing crop yields through restoration of soil organic carbon pool in agricultural lands. Land Degrad Dev 17:197–206
Lal R (2008a) Carbon sequestration. Philos Trans R Soc B 363:815–830
Lal R (2008b) Crop residues as soil amendments and feedstock for bioethanol production. Waste Manag 28:747–758
Lal R, Kimble JM (2000) Pedogenic carbonates and the global carbon cycle. In: Lal R, Kimble JM, Eswaran H, Stewart BA (eds) Global climate change and pedogenic carbonates. CRC/ Lewis Publishers, Boca Raton, FL, pp 1–13
Lal R, Follett RF, Kimble JM (2003) Achieving soil carbon sequestration in the United States: a challenge to the policymakers. Soil Sci 168:827–845
Lamb D, Erskine PD, Parrotta JA (2005) Restoration of degraded tropical forest landscape. Science 310:1628–1632
Lewis SL (2016) The Paris agreement has solved a troubling problem. Nature 532:283

Liebig MA, Varvel GE, Doran JW, Wienhold BJ (2002) Crop sequence and nitrogen fertilization effects on soil properties in the western corn belt. Soil Sci Soc Am J 66:596–601

Majumder S, Neogi S, Dutta T, Powel MA, Banik P (2019) The impact of biochar on soil carbon sequestration: meta-analytical approach to evaluating environmental and economic advantages. J Environ Manag 250:109466

Mathers H (2001) Nutrient availability and pH. In: Plant Health Care Workshop-"Out and-About". Wooster

McBratney A, Stockmann U (2011) Some key management options for increasing soil organic carbon. Retrieved from http://ussc.edu.au/s/media/docs/other/keymanagement options primer.pdf

Meena RS, Kumar S, Yadav GS (2020) Soil carbon sequestration in crop production. Nutrient dynamics for sustainable crop production:1–39

Metting FB, Smith JL, Amthor JS, Izaurralde RC (2001) Science needs new technology for increasing soil carbon sequestration. In: Storing carbon in agricultural soils: a multi-purpose environmental strategy. Springer, Dordrecht, pp 11–34

Milne R, Brown TA (1997) Carbon in the vegetation and soils of Great Britain. J Environ Manag 49:413–433

Min DH, Islam KR, Vough LR, Weil RR (2003) Dairy manure effects on soil quality properties and carbon sequestration in alfalfa-orchard grass systems. Commun Soil Sci Plant Anal 34:781–799

NOAA (2019). https://www.esrl.noaa.gov/gmd/ccgg/trends/global.html

Nogia P, Sidhu GK, Mehrotra R, Mehrotra S (2016) Capturing atmospheric carbon: biological and non-biological methods. Int J Low-Carbon Technol 11(2):266–274

Nowak DJ, Crane DE (2002) Carbon storage and sequestration by urban trees in the USA. Environ Pollut 116(3):381–389

Olson K, Ebelhar SA, Lang., J.M. (2014) Long-term effects of cover crops on crop yields, soil organic carbon stocks, and sequestration. Open J Soil Sci 4:284–292

Pacala S, Socolow R (2004) Stabilization wedges: solving the climate problem for the next 50 years with current technologies. Science 305:968–972

Panchal P, Preece C, Peñuelas J, Giri J (2022) Soil carbon sequestration by root exudates. Trends Plant Sci 27:749–757

Patel B, Gami B, Patel P (2017) Carbon sequestration by bamboo farming on marginal land and sustainable use of wood waste for bioenergy: case studies from Abellon clean energy. Springer Singapore, Singapore, pp 451–467

Paustian K, Collins HP, Paul EA (1997) Management controls on soil carbon. In: Paul EA, Paustian K, Elliott T, Cole CV (eds) Soil organic matter in temperate agroecosystems: long-term experiments in North America. CRC Press, Boca Raton, pp 15–49

Paustian K, Six J, Elliott ET, Hunt HW (2000) Management options for reducing CO_2 emissions from agricultural soils. Biogeochemistry 48(1):147–163

Paustian K, Lehmann J, Ogle S, Reay D, Robertson GP, Smith P (2016) Climate-Smart Soils. Nature 532:49–57

Poffenbarger HJ, Needelman BA, Megonigal JP (2011) Salinity influence on methane emissions from tidal marshes. Wetlands 31:831–842

Post WM, Kwon KC (2000) Soil C sequestration and land-use change: processes and potential. Glob Chang Biol 6:317–327

Post WM, Amonette J, Birdsey RA, Garten CT Jr, Izaurralde DRC, Jardine PM, Metting FB (2009) Terrestrial biological carbon sequestration: science for enhancement and implementation. Oak Ridge National Lab. (ORNL), Oak Ridge

Powlson DS, Stirling CM, Jat ML, Gerard BG, Palm CA, Sanchez PA, Cassman KG (2014) Limited potential of no-till agriculture for climate change mitigation. Nat Clim Chang 4:678–683

Raven JA, Falkowski PG (1999) Oceanic sinks for atmospheric CO_2. Plant Cell Environ 22:741–755

Raymond PA, Cole JJ (2003) Increase in the export of alkalinity from North America's largest river. Science 301:88–91

Reddy SB, Srinivasarao C, Rao PC, Lal R, Rakesh S, Kundu S, Singh RN, Dubey PK, Abhilash PC, Rao KV, Abrol V, Somasundaram J (2023) Greenhouse gas emission and agronomic productivity as influenced by varying levels of N fertilizer and Tanksilt in degraded semi-arid Alfisol of Southern India. Land Degrad Dev 34(4):943–955. https://doi.org/10.1002/ldr.4507
Rivkin RB, Legendre L (2001) Biogenic carbon cycling in the upper ocean: effects of microbial respiration. Science 291:2398–2400
Rumpel C, Amiraslani F, Koutika L-S, Smith P, Whitehead D, Wollenberg E (2018) Put more carbon in soils to meet Paris climate pledges. Nature 564:32–34
Salinger J (2007) Agriculture's influence on climate during the Holocene. Agric Forest Meteorol 142:96–102
Sanderman J, Farquharson R, Baldock J (2010) Soil carbon sequestration potential: a review for Australian agriculture. CSIRO
Sarker JR, Singh BP, He X, Fang Y, Li GD, Collins D, Cowie AL (2017) Tillage and nitrogen fertilization enhanced belowground carbon allocation and plant nitrogen uptake in a semi-arid canola crop–soil system. Sci Rep 7:10726
Schlesinger WH (2000) Carbon sequestration in soils: some cautions amidst optimism. Agric Ecosyst Environ 82:121–127
Schrag DP (2007) Preparing to capture carbon. Science 315:812–813
Singh M, Sarkar B, Sarkar S, Churchman J, Bolan N, Mandal S, Menon M, Purakayastha TJ, Beerling DJ (2018) Stabilization of soil organic carbon as influenced by clay mineralogy. Adv Agron 148:33–84
Sjögersten S, Black CR, Evers S, Hoyos-Santillan J, Wright EL, Turner BL (2014) Tropical wetlands: a missing link in the global carbon cycle? Glob Biogeochem Cycles 28:1371–1386
Smith P (2016) Soil carbon sequestration and biochar as negative emission technologies. Glob Chang Biol 22:1315–1324
Sokol, N.W., Kuebbing, S.E., Karlsen-Ayala, E. And Bradford, M.A. (2019). Evidence for the primacy of living root inputs, not root or shoot litter, in forming soil organic carbon. New Phytol 221, 233–246
Thorup-Kristensen K, Halberg N, Nicolaisen M, Olesen JE, Crews TE, Hinsinger P, Kirkegaard J, Pierret A, Dresbøll DB (2020) Digging deeper for agricultural resources, the value of deep rooting. Trends Plant Science 25:406–417
United Nations (2015) Paris Agreement
Unwin G, Kriedemann P (2000) Principles and processes of carbon sequestration by trees. Forest Protection:1–29
Verma, V., & Edrisi, S. A. (2022). Sustainable management of land resources via the co-designing approach
Wald ML (2007) Is ethanol for the long haul? Sci Am January 296:42–49
Wei X, Shao M, Gale W, Li L (2015) Global pattern of soil carbon losses due to the conversion of forests to agricultural land. Sci Rep 4:4062
Whalen JK, Chang C (2002) Macroaggregate characteristics in cultivated soils after 25 annual manure applications. Soil Sci Soc Am J 50:1637–1647
World Bank (2015a) Agricultural nitrous oxide emissions (% of total). Available at http://data.worldbank.org/indicator/EN.ATM.NOXE.AG.ZS/countries/1W?display=graph (verified 1 Oct 2015)
World Bank. (2015b). Agricultural methane emissions (% of total). Available at http://data.worldbank.org/indicator/EN.ATM.METH.AG.ZS/countries/1W?display=graph (verified 1 Oct 2015)
Yang H, Xu Z, Fan M (2008) Progress in carbon dioxide separation and capture: a review. J Environ Sci 20:14–27
Zan CS, Fyles JM, Girourard P, Samson RA (2001) Carbon sequestration in perennial bioenergy, annual corn, and uncultivated systems in southern Quebec. Agric Ecosyst Environ 86:135–144
Zeng N (2008) Carbon sequestration via wood burial. Carbon Balance Manag 3:1
Zomer RJ, Bossio DA, Sommer R, Verchot LV (2017) Global sequestration potential of increased organic carbon in cropland soils. Scientific reports, vol 7. University of Guelph, Guelph, pp 1–8

Chapter 3
Carbon Sequestration: Different Approaches to Address Climate Change Effects

Pooja Singh, Vikram Kumar, and Asha Sharma

Abstract Global warming is a major environmental issue that gives rise to different types of ecological problems. A number of techniques and plans are adopted to overcome these problems, one of which is carbon sequestration. Carbon sequestration is a phenomenon of capturing carbon safely to reduce its atmospheric concentration and minimize the global warming effect. Apart from concern about atmospheric conditions, it is also a serious topic for the agricultural sector due to the gradual decrease in soil carbon content, which results in a decline in agricultural productivity. Carbon sequestration converts atmospheric carbon into different carbon compounds and ultimately enhances the organic content of soil. Roots have a significant role in the transfer of atmospheric carbon to the soil environment in the form of carbon compounds such as phenolic compounds and organic acid and directly contribute to sustainable agricultural management. The decomposition of plant roots comes up with remarkable amounts of organic carbon in the soil environment. This deposition of organic carbon improves soil pH, the growth of soil microbes, and the movement of soil nutrients into the plants. It forms a new approach in which plant roots transfer organic carbon for the enrichment of soil and fertility. Thus, the goal of this chapter is to emphasize the plant's contribution to the process of carbon sequestration for the maintenance of agricultural lands and reducing global warming.

Keywords Agriculture · Carbon sequestration · Global warming · Organic carbon · Photosynthesis

---

P. Singh · V. Kumar · A. Sharma (✉)
Department of Botany, M.D. University, Rohtak, Haryana, India
e-mail: drasha.botany@mdurohtak.ac.in

S. A. Edrisi et al. (eds.), *Sustainable Plant Nutrition and Soil Carbon Sequestration*, Sustainable Plant Nutrition in a Changing World,
https://doi.org/10.1007/978-3-031-53590-1_3

3.1 Introduction

Global warming is a major environmental issue that gives rise to different types of ecological problems (Edrisi et al. 2022a, b). The increase in global warming is due to the burning of fossil fuels that release different types of GHGs into the atmosphere, i.e., carbon dioxide (CO_2), methane (CH_4), water vapor (H_2O), ozone (O_3) nitrous oxide (N_2O), chlorofluorocarbon (CFC), hydrochlorofluorocarbon (HCFC), and other fluorinated gases that heat the earth's surface and atmosphere (USEPA 2020). The most common effect of global warming is an increase in the average temperature of the Earth and an increase in the effects of climate change, resulting in extreme weather events (IPCC, 2007). According to a report by the U.S. Environmental Protection Agency, CO_2 is the most common GHG present in the atmosphere. The main source of CO_2 released from many countries is the burning of fossil fuels such as coal, fuel materials, and natural gases. Industrialization, transportation, and the generation of energy are all the most prominent sources of GHGs, which release a major portion of the overall GHGs produced (USEPA 2020). This increased carbon content in the environment can be deducted through the carbon sequestration process. Carbon sequestration is the process via which the free carbon present in the atmosphere is fixed and stored in carbon pools for a longer duration for their later use. Photosynthesis is the major carbon sequestration process by which atmospheric CO_2 is converted into organic matter, i.e., biomass. Later, these organic products are released into the soil via different mechanisms, which increases the carbon content of the soil (Fig. 1.3). This increased organic content helps the plant to grow better and increases its productivity. Based on total carbon content, we can categorize global carbon pools into five pools. The first one is the oceanic pool, which contains approximately 38,000 Pg of carbon content, followed by the geologic pool, which has around 5000 Pg of carbon content. The third carbon pool is the soil carbon pool, which is also divided into two sub-pools: SOC and soil inorganic carbon. The total carbon content in soil carbon pools is estimated to be around 2300 Pg up to a depth of 1 m (Eswaran et al. 1995; Batjes 1996). The fourth-largest carbon pool is the atmospheric pool, which has a total carbon content of 760 Pg. The fifth and smallest carbon pool is the biological pool, which has a total carbon content of 560 Pg (Lal 2004). Therefore, carbon sequestration is the process through which atmospheric carbon can be captured and stored in the soil in organic or inorganic form or in the ocean. This carbon form can be utilized by plants to maintain their growth and metabolism. Ultimately, it means that carbon sequestration is an important process that not only helps maintain the level of carbon content in the atmosphere but also helps provide nutrition to plants. In this chapter, we will discuss how different processes of carbon sequestration help control the effects of global warming as well as help maintain the quality of soil. Apart from that, we will also discuss the potential of different carbon pools found on Earth at present and in the near future.

3.2 Causes of Global Warming

The GHGs mentioned in the previous section absorb solar irradiance in the form of heat, and some amount of radiation is released back into space. This phenomenon forms a series of continuously absorbing and releasing back heat in the atmosphere, eventually increasing the temperature of the Earth's atmosphere. This series of cycles is called the greenhouse effect. A substantial rise in the level of GHGs is observed at the global level due to anthropogenic activities, resulting in an enhanced greenhouse effect occurring at the global scale, leading to global warming. According to the United Space Environmental Protection Agency (US-EPA), CO_2 is a major GHG contributing to global warming as compared to CH_4, N_2O, and fluorinated gases because of its highest share in the GHG emissions, as shown in Fig. 3.1. The problem is much worsened due to continuous deforestation, industrialization, and many other anthropogenic activities, resulting in increased global warming effects. Different anthropogenic activities responsible for the increased global warming effects and their consequences have been showcased in Fig. 3.2 (IPCC, 2007; Edrisi et al. 2022b).

3.2.1 Industrialization

Human activities are highly responsible for the emission of CO_2 gas into the atmosphere. From the Industrial Revolution, people started burning fossil fuels like coal, petroleum, and fuel oil for energy on a large scale, which enhanced the CO_2 content in the environment at a high rate. This energy is used for industrial purposes such as the transportation of cars, trains, trucks, and airplanes and for electricity and heat.

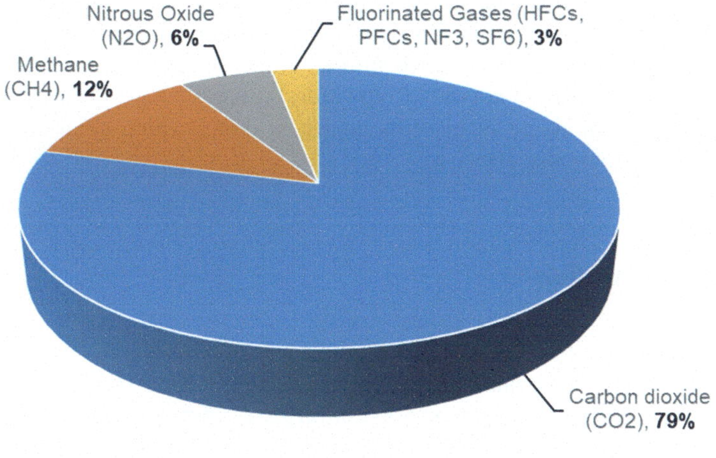

Fig. 3.1 Share of different gases while greenhouse gases (GHGs) emission due to anthropogenic activities from the different sectors in the year 2021. (Data source: www.epa.gov)

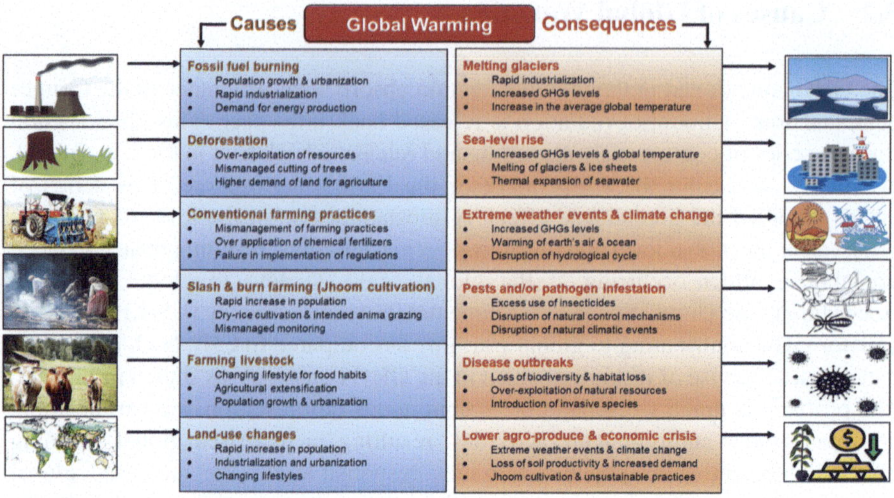

Fig. 3.2 Different human-induced activities causing the global warming effects and their plausible consequences. (Adopted from IPCC (2007), Edrisi et al. (2018, 2019, 2022b))

The rest of the energy is used for oil extraction, agricultural purposes, etc. The energy used in these processes emits other GHGs like chloro-fluoro-carbons (CFCs), nitrous oxide, and methane. However, the level of CFCs in the atmosphere has decreased since their use was banned in 1988 (Mgbemene et al. 2016).

3.2.2 Agricultural Practices

The agricultural activities that give rise to food for humans are another reason for climate change. The application of organic and commercial fertilizers on agricultural land emits a strong GHG called nitrous oxide in the environment. Another GHG, methane, is produced not only from natural sources but also from the burning of fuels, dumping of waste in landfills, and cow dung, which contributes to the increase of the greenhouse effect.

3.2.3 Deforestation

Increased demand for food and other essential needs of human-like dairy cattle feed are the main causes of the large-scale cutting of trees. Forests are also cleared on a large scale to produce paper and wood. According to a report by Scientific American, deforestation is highly responsible for the enhancement of GHGs in the atmosphere. In one year, an aged tree takes up almost 48 pounds of CO_2, and the estimated cost of cutting trees is around 7 billion per year.

3.3 Carbon Sequestration

As we previously discussed, CO_2 is the principal GHG that causes global warming (IPCC, 2007). One of the foremost approaches for declining CO_2 levels in the atmosphere is carbon sequestration, the method of eliminating carbon from the atmosphere or the long-time storage of carbon in plants, soils, oceans, and geological pools. Carbon sequestration occurs naturally and with human actions, which denotes carbon storage, which can form CO_2 gas instantly. It is an area of concern that the increased level of CO_2 gas in the atmosphere results in climate change and global warming. According to the IPCC report (2022), different impacts like drought, food security, fires, biodiversity loss, and temperature fluctuations caused by climate change are already widely spread and severe (Levin et al. 2022).

The rate of carbon sequestration continues to rise due to changes in land use and forestry and the implementation of carbon capture and storage (CCS) technology (Selin 2019). These techniques have become common in the Kyoto Protocol to the United Nations Framework Convention on Climate Change (UNFCCC), the first and largest international agreement to reduce the emission of GHGs that cause global warming. These approaches are used for carbon sequestration through reforestation and afforestation processes. Crop, forest, and vegetative land management were listed in the land-use change and forest activity of carbon sequestration in 2001. This results in awareness of the importance of trees as a means to capture and store atmospheric CO_2 in soils, vegetative lands, and plant biomass products (Malhi et al. 2008). The method of agroforestry became familiar as the carbon sequestration approach, specifically through reforestation and afforestation methods, which have been proven to be the mitigation policies against GHGs under the Kyoto Protocol. So, agroforestry gained increased attention as a carbon sequestration approach in developing and industrial countries (Albrecht and Kandji 2003; Nair and Nair 2003; Makundi and Sathaye 2004; Sharrow and Ismail 2004; Haile et al. 2008; Takimoto et al. 2008).

3.3.1 Soil Organic Carbon

During the photosynthesis process, plants receive CO_2 from the environment and transform it into carbohydrates and oxygen. After photosynthesis, a major amount of CO_2 from the environment remains in the plant until it dies. Reabsorption of CO_2 through photosynthesis occurs at a faster rate, which is out during plant respiration. Green plants add carbon to the soil environment, which has been absorbed in the form of CO_2 from the atmosphere through photosynthesis. Plants produce different carbon compounds like lignin and cellulose. Plants add carbon to the soil environment as SOC in the form of root exudates, root material, decay of leaves (litter), and excreta from animal consumption. Plants in the form of litter material accumulate at the soil surface and produce organic matter through various decomposition processes. Soil accumulated carbon vanishes due to soil erosion. The rest of the carbon in soil is used by the soil microbes such as fauna, decomposing bacteria, and fungi for their nutrition system. Many heterogeneous organic mineral compounds, fungi, soil microbes, and their exudates contribute to the soil masses by binding with organic matter. Soil masses, with multiple organic compounds and other soil carbon pools, produce and collapse uninterruptedly. The rate of collapse and loss of these matters in the environment are very delicate concerning human impact. Most soil microbes use carbon in the form of SOC as a nutritional source for their life and release CO_2 back into the atmosphere (Follett 2001). A report by the U.S. Forest Service (2011) showed that the combustion of forests in the form of fossil fuels contributes to around one-third of the emission of CO_2 gas. Almost all plants, including trees and shrubs, store the carbon and subsequently decline the rate of CO_2 from the environment. Tropical region plants have a great role in mitigating the greenhouse effect as they get enough rays from the sun to photosynthesize more than other regions.

When plants grow up, they store carbon in them by photosynthesis, and when they die, their different organic carbon compounds are distributed in the soil. In appropriate environments, soil microbes decompose these organic compounds and store some carbon to make SOC compounds within them for their life processes, and the rest of the carbon is left in the environment in CO_2 form through respiration or back to the soil by excretion. This continuous series of carbon from plants to soil shows that carbon is changing its form continually in the soil. It also combines with SOC compounds in soil organisms to form various new organic compounds. Some scientists reported the different carbon pools based on the timing of carbon persistence in the soil environment (described later in this chapter).

3.3.2 *Carbon Cycle*

The series of carbon cycling is continuous between various carbon pools as it constantly changes its form to make new organic complex compounds. Due to photosynthesis, plants form carbon-containing compounds, and these compounds are further entered into the carbon cycle after the death of plants. Therefore, the atmospheric CO_2 in the form of carbon enters this cycle through plants into various carbon pools such as soils, oceans, and forests. CO_2 constantly moves from the atmosphere to the soil and oceans and then returns to the atmosphere, forming the carbon cycle. On the other hand, human activities like the burning of fossil fuels add carbon from the forest carbon pool to the atmosphere. These various pools act as a source or sink for the carbon, as carbon is constantly moving between them. Carbon sources are pools that emit more carbon than they store, while carbon sinks store more carbon than they release into the environment. Consideration of the relationship between these sources and sink dynamics and normalizing the dimension of sinks to take out the carbon from the atmosphere and store it is a great need to counter human-induced climate change. The declined rate of SOC affects the fertile rate of soil and reduces the availability of minerals to crop plants (Kane and Solutions 2015).

3.3.3 Carbon Sources and Carbon Sinks

Due to anthropogenic actions, the burning of fossil fuels occurs in environments that emit carbon in long-term storage forms, such as natural gas, petroleum, and coal, and add CO_2 to the atmosphere. Rapid industrialization upsurges the quantity of CO_2 in the atmosphere, but this increment is mostly because of the combustion of fossil fuels. Decomposition and decay of plant parts and excreta of animals also release CO_2 gas naturally (Selin 2019). These are the sources of carbon that release more carbon than they store.

Reservoirs act as carbon sinks, which maintain the amount of carbon and prevent it from entering the atmosphere by storing it (Fig. 1.3). For instance, cutting trees and forests is a source of carbon, which releases CO_2 gas into the atmosphere, but afforestation and reforestation increase the efficiency of carbon sinks, as they store carbon and enhance the carbon sequestration process. Through the process of photosynthesis, carbon is shifted naturally from the atmosphere to terrestrial carbon sinks. Carbon can be stored in the underground part as well as the aerial part of the plants. Land-use change, land use, forestry, revegetation, and other terrestrial processes are helpful in carbon sequestration. These processes help in enhancing plant growth due to the increased level of carbon and nitrogen in the soil. Carbon sequestration in soils and aerial parts of plants can emit carbon back into the atmosphere through changes in climate and land management. Likewise, we can say that the burning of fossil fuel or forests and the breakdown or decay of plant parts by soil microbes are both sources of carbon emission from forests to the outer environment.

The emission of carbon from these combines with oxygen in the atmosphere and forms CO_2 gas. Terrestrial sinks can be a carbon source, too, through these processes of decomposition and burning, and have the ability to emit a large quantity of CO_2 into the atmosphere and oceans. Worldwide, the overall amount of carbon in plants, soil, and litter material is approximately 2200 gigatons, i.e., one gigatons is equal to 1 billion tons. The terrestrial pools alone sequestered carbon by almost 2.6 gigatons every year. A large amount of contribution is shared by oceans in storing carbon under them, around 920 gigatons. In addition, this amount is more than the amount of carbon in an atmosphere, which is about 760 gigatons. Oceans are a great pool of carbon and act as carbon sinks. Out of the total carbon released into the environment by anthropogenic activity, only 45% of carbon remains in the atmosphere, while 30% of carbon content is stored in the ocean. The rest of the carbon content is involved in different terrestrial ecosystems (Selin 2019).

3.3.4 Carbon Pool

Decomposition, which results in the decay of the litter material of the plant, produces GHGs in the environment. Likewise, in deforestation, the carbon present in the plants is released into the atmosphere after decomposition. After that, plants cannot take part in the removal of CO_2 from the atmosphere through photosynthesis, which leads to an increase in the greenhouse effect. Many technologies have developed to decrease the rate of an augmented level of CO_2 in the atmosphere. The rate of emission is 8.6 Pg C year^{-1} annually from energy, industrial area, land-use change, and vegetation. It is a very serious issue of the twenty-first century. The development of low or no-carbon fuel and low-energy use helps reduce carbon emissions. This chapter describes the process and mechanism of carbon sequestration and discusses the biotic and abiotic sequestration technologies. According to a commonly used model, carbon pools are divided into three groups: the fast or labile pool, the slow pool, and the stable pool (Jenkinson and Rayners 1977). The first carbon pool is the fast or labile pool in which soil carbon returns to the atmosphere within several days to several years. In this pool, carbon is formed from newly combined plant deposits and simple organic carbon compounds that are released by plant roots. This pool is frequently used by soil organisms to form CO_2. The second pool, i.e., the slow pool, contained extra-decomposed plant matter or by-products of the first pool and carbon that escaped from the different processes of soil microbes. The time ranges between years to centuries and can be affected by soil management and climate change. The third and last pool is the stable pool, which is highly unaffected by any interference or disturbances and slowly changes its structure. In addition, their meantime ranges between centuries to millennia. This pool adds humus to the environment, which is composed of a group of carbon compounds that are highly resistant to the processes of soil microbes and decomposition (Six et al. 2002). All pools can differ in size in different soils, but the stable pool is almost constant with respect to size, while fast and slow pools are sensitive to the environment.

3.3.5 Carbon Capture and Storage

Many strategies are adapted for carbon sequestration by different scientists and engineers to reduce the level of global warming in the environment. The strategy used for carbon sequestration involves a geoengineering application called carbon capture and storage (CCS). In this application, CO_2 is detached first from the other mixture of gases released by industries and then compressed for transport to a new place separated from the atmosphere for storage for a long time. The place for storage of CO_2 may include large oceans, reservoirs, geographic creations, and sediment rocks, which contain pore spaces filled with water and dissolved salts. Even though, in CCS, CO_2 is directly captured from the release point of the source before it flows out into the environment. Artificial trees and towers are used to sequester CO_2 from the atmosphere. However, many challenges have to be faced in executing the CCS technique on a broad scale. According to IPCC, the CCS technique has increased the cost of electricity production relying on oil, strategies, and place. The storage of carbon and its leakage from the reservoirs is a great concern. However, it is predicted that carbon can be stored for many years by sequestering it and stored in geologic formations (Selin 2019).

3.4 Process and Mechanisms of Carbon Sequestration

The method of transmission and storing of CO_2 from the atmosphere for long-term storage in the carbon pool is known as carbon sequestration. This process is used to diminish the quantity of carbon in the atmosphere. So, we can say that carbon sequestration is a natural or an anthropogenic process. The main aim of human-based activities for carbon sequestration is to create a "Carbon neutral" approach in which the carbon is stabilized in the atmosphere for future perspective, i.e., there is no net addition or gain of CO_2 in the atmospheric pool. This kind of method for carbon sequestration is good, safe, and environmentally friendly, with minimum risk of escape for almost all human activities that release CO_2 (Lal 2008). According to Pacala and Socolow (2004), there are many possibilities to maintain the CO_2 level in the atmosphere to be around 550 ppm by 2050. In these possibilities, three were used in carbon sequestering under terrestrial carbon pools. There are various possibilities for sequestering atmospheric carbon into various carbon pools. The one or combination of these different possibilities for carbon sequestration is essential for future financial growth at a large scale. These possibilities or strategies can be divided into two groups, i.e., abiotic and biotic sequestration (Lal 2008) (Fig. 3.3).

3.4.1 Abiotic Carbon Sequestration

This is a type of carbon sequestration that is established on different engineering strategies and physical or chemical reactions in the absence of lifeforms, i.e., plants and soil microbes. Oceanic and geologic carbon pools have gained significant attention on carbon sequestration through an abiotic approach, as the carbon sink potential of abiotic sequestration is more than that of biotic sequestration (Freund and Ormerod 1997). The technologies for capturing carbon and transferring it by injecting it into global pools are in great progress (Kerr 2001).

Oceanic Injection

For a few decades, CO_2 has been injected into deep oceanic pools, which are highly accepted by some scientists and engineers. Consideration of injecting atmospheric CO_2 into the oceanic pool has been in progress since the late 1970s. CO_2 is injected

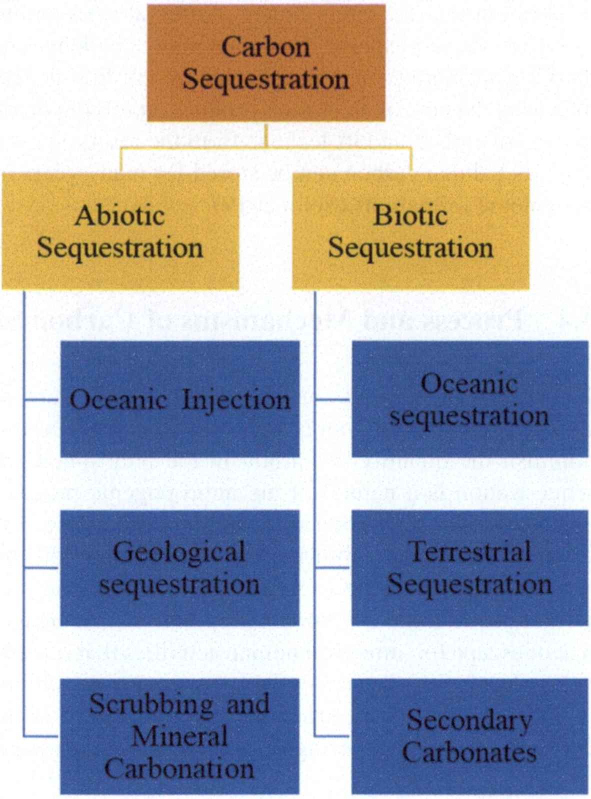

Fig. 3.3 Different strategies of carbon sequestration

deep down into the oceans to stabilize the carbon level and reduce the chance of leakage. Consequently, CO_2 in liquid form is extracted from industrial sources and injected into the oceanic pool. The carbon injection into the oceanic pool can be done through 4 processes: (i) It can be injected around 1000 m depth of the ocean layer, and due to being lighter than water, it makes a droplet plume; (ii) Thicker CO_2 is injected under 500–1000 m depth of the ocean water mixture, and this matter sinks into deep in the ocean; (iii) It can be released with the help of a large tube dragged behind the boat; (iv) it is injected into the last layer of the ocean forming a bottom lake of the ocean. This zone of injecting CO_2 remains stable at around 3000 m depth of the ocean (O'Connor et al. 2001). The carbon sink potential of the oceanic pool is very high compared to that of fossil fuel matter. According to an estimate, the carbon sink potential of the ocean is between 5000 and 10,000 Pg of carbon (Herzog et al. 1997, 2002). On the other hand, a high level of CO_2 injection is harmful to sea organisms (Auerbach et al. 1997; Caulfield et al. 1997; Seibel and Walsh 2001).

Geological Injection It deals with the process of capturing, storing, liquifying, transferring, and injecting CO_2 from industrial sources into deep geological formations. The carbon can be injected into the coal bed, sediment rocks with pore space of water, and dissolved salts and oil mills (Tsang et al. 2002; Klara et al. 2003; Baines and Worden 2004; Gale 2004). Saline aquifers are also sinks of carbon matter, which is present beneath the reservoirs with a separate layer between them. The CO_2, which comes from industrial sources, can be injected into these saline aquifers, where the sequestration of carbon is done by reaction with different salts to produce carbonate compounds. In this method, the CO_2, which is lighter in volume and has low viscosity, is injected into the carbon pool in a critical form. CO_2 forms a gaseous and aqueous phase and makes a multiphase component in the surroundings. The injection of this type of CO_2 into the reservoir's pool shifts the oil or gas from the pool. It could be a financial plan for the recovery of oil in the near future. The consistency of carbon storage in large amounts in geologic formation and its cost are the two main principles of geological sequestration (Kintisch 2007a; Schrag 2007).

Carbonation of Minerals and Scrubbing The transformation of CO_2, released from industrial sources, into calcium carbonate, magnesium carbonate, and many other minerals in the form of purely stable mineral carbonates is observed. It is formed by imitating the natural inorganic chemical conversion of CO_2 (Fan and Park 2004). This process involves two steps: scrubbing and mineral carbonation. Chemical absorption of CO_2 gas using carbonate or amine solvent is mostly used in the process of CCS in scrubbing. Purification of CO_2 is done by absorption column using amine and other solvents such as K_2CO_3, nickel compounds, ceramic, etc. Heating CO_2 with an amine solvent purifies the carbon gas. The CO_2 is further re-precipitated by mineral carbonation, and then these carbonates form stable rock in which carbon is stored for a long time.

3.4.2 Biotic Carbon Sequestration

Biotic components of an ecosystem involve living organisms, including animals, plants, algae, etc. Capturing carbon and its storage from the atmosphere for a longer duration with the help of these biotic components is known as biotic sequestration (Fig. 3.3). This process involves different biological processes that help capture carbon from the environment and store it. It involves different types, which are briefly explained below:

Carbon Sequestration in Oceanic System Oceanic carbon sequestration is one of the major biotic sequestrations in which carbon is fixed and stored in the form of usable carbon via the photosynthesis method. Falkowski et al. (2000) estimated that around 45 Pg C $year^{-1}$ of carbon is fixed by the phytoplankton. The growth of phytoplankton in the sea environment is controlled by the availability of iron, so indirectly, we can say that the oceanic carbon sequestration via phytoplankton is dependent upon the availability of iron (Martin and Fitzwater 1988; Falkowski 1997; Boyd et al. 2004). It is also believed that in some countries, incremental carbon is transferred to other developed nations in the international market.

Carbon Sequestration in the Terrestrial System The second major biotic sequestration component is terrestrial sequestration, in which atmospheric carbon is converted into pedological carbon pools. The major process involved in terrestrial sequestration is photosynthesis, which transfers and stores atmospheric carbon into living or dead organic matter. Lal et al. (2003) believed that it is a no-regret process because it results in several beneficial outcomes, like maintaining soil fertility without any damage to our environment. It provides a number of profits to the ecosystem without disturbing any global stability. Terrestrial sequestration is divided into three different categories, which are briefly described below:

Carbon Sequestration in Forest Ecosystem Forest sequestration is the major terrestrial sequestration in which atmospheric carbon is converted into lignin or other polymeric carbon compounds (Table 3.1). Fan et al. (1998) stated that the rate of carbon sequestration in the forest is around 1.7 ± 0.5 Pg C year^{-1}.

The total amount of carbon that is seized by the plant through the process of photosynthesis in a unit period is demarcated as Terrestrial Net Primary Productivity (NPP) (Pan et al. 2014). At the present level of CO_2 concentration, terrestrial NPP cannot be saturated (Falkowski et al. 2000). A rise in the level of CO_2 in the atmosphere results in a high rate of photosynthesis, which is known as the CO_2 effect. Krishnamurthy and Machavaram (2000) believed that this effect might lead to increased forest sinking. Increased concentration of CO_2 enhances the NPP, but it will be limited by other factors like the limited availability of nitrogen, phosphorus, and water put a limitation on it. Therefore, if we can manage the interaction among carbon, nitrogen, phosphorus, and water, we can indirectly maintain and increase the rate of carbon sequestration under the terrestrial section. Apart from this, planting trees on a large scale and re-establishing already damaged tropical forests can also be a good option to increase the efficiency of terrestrial carbon sequestration (IPCC, 1999; Fang et al. 2001; Lamb et al. 2005). Planting Pinus, Acacia, or Eucalyptus trees is also a good idea to improve the rate of terrestrial carbon sequestration. Along with these, a few other methods are also listed below, which can recover the efficiency of carbon sequestration (Richard and Stokes 2004).

Modifying the process and methods of forest management

- Increasing the period of the forest rotation cycle
- Protection of forest from conversion into bare land or for cultivation purposes
- Use of agroforestry methods

Carbon Sequestration in Wetlands Wetlands are the areas of the land that are saturated with water or covered with water all the time or seasonally. These are rich sources of diversity and constitute a large portion of the terrestrial sequestration process. Due to its high diversity, the soil of wetland areas has approximately 200 times more carbon as compared with associated vegetation (Milne and Brown 1997; Garnett et al. 2001). It is found that after the glaciation period, the rate of carbon accumulation in the wetland is nearly about 0.1 Pg C year^{-1} (Gorham 1991; Kobak et al. 1998). However, due to the usage of peatlands for cultivation purposes after draining, they are converted into a CO_2 pool. A number of wetlands are also used for

Table 3.1 Different types of forest and area-wise carbon stock in India (www.fsi.nic.in)

S. No.	Forest type	Area (in km^2)	Carbon stock in tons
1.	Tropical wet-evergreen forests	20,872	328,070
2.	Tropical semi-evergreen forests	98,039	1,124,429
3.	Tropical moist-deciduous forests	144,747	1,323,014
4.	Littoral and swamp forests	6640	73,566
5.	Tropical dry-deciduous forests	279,507	2,670,344
6.	Tropical thorn forests	13,398	44,846
7.	Tropical dry-evergreen forests	877	8247
8.	Subtropical broadleaved hill forests	19,159	215,638
9.	Subtropical pine forests	19,894	197,363
10.	Subtropical dry-evergreen forests	183	2163
11.	Montane wet-temperate forests	4985	56,116
12.	Himalayan moist-temperate forests	29,216	415,139
13.	Himalayan dry-temperate forests	5183	78,846
14.	Sub-alpine forests	12,924	164,175
15.	Moist alpine forests	1141	11,231
16.	Dry alpine forests	1625	15,731
17.	Plantation	49,883	353,507
Total		708,273	7,082,425

agricultural and forest formation (Armentano and Menges 1986; Paavilainen and Paivanen 1995). So, the re-establishment of the wetland is a good option to enhance the process of terrestrial sequestration and the establishment of the wetland as a carbon pool. According to Dixon and Krankina (1995), wetlands are the major source of terrestrial biological sequestration, and that is why they play a major part in maintaining biotic sequestration.

Soil carbon Sequestration Soil carbon sequestration is the third major part of terrestrial sequestration. It mostly deals with the agricultural sector, which also constitutes a small part of overall biotic sequestration. Different ideas and methods are suggested to enhance the effectiveness of soil sequestration. Restoration of disturbed land is suggested as the most commonly used natural method. Apart from this, the use of biochar is also an option to improve soil carbon sequestration (Fowles 2007). As compared to the geological injection carbon sequestration, in which carbon is directly injected into soil strata at around 2 km, SOC sequestration stores the fixed carbon at around 1 m in depth via the process of humification. According to Lal (2001a, b), a huge loss has occurred in the SOC pool in the last 50 years due to agricultural practices. It is observed that a natural ecosystem contains a higher SOC pool than a managed ecosystem, which leads to the idea that cultivation of a land area causes a decrease in the rate or efficiency of carbon sequestration. This lower rate of carbon sequestration may be due to erosion, leaching, or oxidation processes (Lal 2008). Human activities that influence the organic content of the soil are depicted in Fig. 3.4 (Bot and Benites 2005).

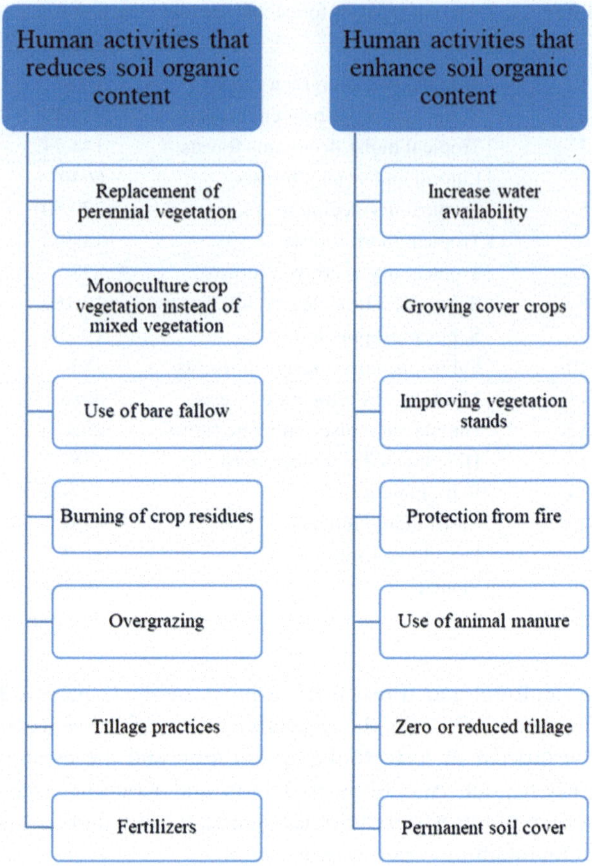

Fig. 3.4 Human activities that influence the organic content of soil. (Source: Bot and Benites (2005))

Paustian et al. (1997) suggested that the process of carbon sequestration will be affected or decreased if the carbon and nitrogen are not in a properly balanced concentration. So, it is observed that the efficiency of soil sequestration can be improved by the balanced usage of carbon and nitrogen (Campbell et al. 1991; Janzen et al. 1998; Halvorson et al. 1999, 2002). It is also reported that the use of nitrogen as a fertilizer in the soil can improve the rate of SOC sequestration as compared to the unfertilized soil (Liebig et al. 2002). As described above, carbon sequestration is the conversion of atmospheric carbon into the stored form for a longer duration in the carbon stock pools. These pools can be the plant biomass involving both aboveground and belowground parts of plants like stems and roots, soil microorganisms, and the formation of manure or other carbon compounds in the soil. It is believed that the presence of trees in the cropland improves the total carbon sequestration of plants (Palm et al. 2004; Haile et al. 2008).

3.5 Potential of Agroforestry Practice in Carbon Sequestration

It can be concluded that the large-scale planting of trees can enhance the efficiency of carbon sequestration of soil. The effectiveness of agroforestry carbon soil sequestration depends on the quality as well as quantity of trees and properties of the soil. Change in these factors results in alteration in the process of carbon sequestration, which can ultimately result in altered climate conditions of that particular region (Schlesinger et al. 1990; Ojima et al. 1999). According to Lal (2001a, b), SOC sequestration is maintained by leaching, humification, and aggregation of secondary particles such as organic and mineral complexes. The lack of proper approaches and methodologies for carbon sequestration is a major challenge to use agroforestry in the sequestration process to its full potential (Ramachandran et al. 2009). Mutuo et al. (2005) suggested that the potential of an agroforestry region to capture carbon is not only dependent on its environment but also on the socio-economic factors of that system. According to them, in humid tropic regions, the carbon sequestration potential of the agroforestry system can be up to 70 Mg ha^{-1} in the top 20 cm of soil. The constituent of agroforestry also affects the carbon sequestration potential of an agroforestry system. Kandji et al. (2006) reported that the agri-silvicultural type of agroforestry system acts as a net sink for GHGs. However, the silvopastoral type of agroforestry system acts as a source for GHGs. Jose (2009) believed that by the year 2040, unproductive croplands and grassland could be converted into agroforestry systems that could have a carbon sequestration potential of approximately 586,000 Mg C $year^{-1}$. Gupta et al. (2019) conducted their experiments on carbon biomass sequestration in relation to tree density. They reported that carbon mass sequestration is relatively high in low-density areas as compared to high-density. The potential of the agroforestry system for trees is assessed as 0.25–76.55 Mg C ha^{-1} $year^{-1}$, and for SOC, it is 3.98 Mg C ha^{-1} $year^{-1}$ in India (Dhyani et al. 2016). However, it varies from region to region, species type, and phases of the agroforestry system in India (Dhyani et al. 2016; Kaul et al. 2010; Gupta et al. 2019). It is estimated that in India, around 8.2% of the total geographical area is covered by agroforestry (Dhyani et al. 2013; Dagar et al. 2014; CAFRI 2015). According to the data from the FSI (Forest Survey of India) (2017), Maharashtra, Gujarat, and Rajasthan are the major states in India for agroforestry areas (Table 3.2).

3.6 Soil Behaviour in Carbon Sequestration

Plant residues are converted into different kinds of carbon complexes by the different microorganisms found in the soil, which is a kind of carbon sequestration because, via this process, C, which was present in the plant parts, is converted and stored in the soil in the form of long durable carbon compounds (Six et al. 2006). Large parts of plant residues are converted into smaller parts with the help of insects

Table 3.2 Carbon retention capacity of agroforestry systems across various eco-regions

Continent	System	Eco-region	Capacity (Mg C ha^{-1})
Africa	Agro-silvicultural	Humid tropical high	29–53
South America		Humid tropical low dry lowlands	39–102 39–195
Southeast Asia		Humid tropical dry lowlands	12–228 68–81
Australia	Silvopastoral	Humid tropical low	28–51
North America		Humid tropical high humid tropical low dry lowlands	133–154 104–198 90–175
Northern Asia		Humid tropical low	15–18

Murthy et al. (2013)

and worms. After that, these small parts can be utilized by the fungi and bacteria for their conversion into metabolized form. During the whole process, different complexes of carbon are synthesized that can be stored in the soil for a longer duration by combining with soil particles (Six et al. 2006; Grandy and Neff 2008; Grandy and Wickings 2010). In this way, it can be concluded that both fungi and bacteria play an essential role in enhancing the efficiency of soil carbon sequestration. Fungi and roots of many plants are present in mycorrhizal association, from which the participants benefit (Govindarajulu et al. 2005). Due to this association with fungi, a significant increase in plant biomass is observed, which results in an increased supply of carbon to the soil via carbon sequestration (Rillig et al. 2001). Wilson et al. (2009) observed that Arbuscular mycorrhizal fungi (AMF) show a positive correlation with the carbon level in soil. Along with fungi, soil bacteria, like nitrifying bacteria, also have a major role in soil carbon sequestration. Lignin is decomposed and converted into a simplified form by actinomycetes. Formations of these simplified substances are required for increasing biomass formation of the plant and for the conversion of unstable atmospheric carbon into well-stored and stable carbon complexes for a longer duration. Proper management of soil microorganisms and minimum tillage practice are excellent strategies to boost the capability of soil for carbon sequestration (Six et al. 2006).

3.7 Carbon Farming

As agricultural products, carbon sequestration can also be operated and exchanged in the international market. Selling and purchasing carbon credits is a common trend in the last few decades (Gressel 2007; McGowan 2007). This topic is closely related to the release of GHGs because carbon sequestration is also concerned with the capturing process of CO_2. According to Kerr (2007) and Kintisch (2007b), the trading of carbon credits is considered an option to control the emission of GHGs. Different policies are formed for carbon trading in the international market, like "Carbon tax" and government-aided subsidies. These systems help maintain the success of environmental goals. Trading of carbon sequestration is already well established in developed countries like Europe, and it is a good option for undeveloped countries, which lack proper methodologies and technologies for the carbon sequestration process. It is a good source of income and helps these undeveloped and developing nations financially. It also helps these countries invest in the agricultural sector, land restoration, and food security (Lal 2008).

3.8 Role of Plant Root

Roots are the major part of any plant, which helps the plant uptake of different nutrients, minerals, and water from the soil. Along with the absorption and transfer of these substances, roots also produce different compounds. These compounds can be accumulated or released in the adjacent sol area, which helps to increase soil fertility. These chemical compounds, water molecules, gases, and other substances released by plants in the soil are collectively termed rhizodeposition (Kuzyakov and Domanski 2000). Compounds released by the plant root also include simple organic compounds like amino acids, sugar molecules, and some enzymes. These products are produced by the plants via the process of photosynthesis, after which photosynthetic products undergo several modifications to be converted into a final product. CO_2 is fixed in the process of photosynthesis, and in another way, we can say that carbon present in the environment is converted into different organic compounds, some of which are released into the soil, which helps to improve soil fertility. Out of the overall carbon assimilated by the cereal plants, about 20–30% is secreted in the soil (Gregory and Atwell 1991). Kuzyakov et al. (1999) suggested that around 30–50% of total carbon assimilated by grassland plants is transferred to soil in the same way as observed in cereal. According to Bowen and Rovira (1999), both kinds of cereal and pasture plants transferred nearly the same amount of carbon into the soil in the same growth duration.

Plants are the main source of the entire organic carbon present in soil. These carbon compounds can be released into the soil from the plant via two primary mechanisms: humification and root exudates. Humification is a process that takes place after the death of a plant, and different microorganisms and fungi are involved in this process. On the other hand, root exudates are the substances that are released by the root in its surrounding soil when it is living. Generally, carbon is transferred to the soil from roots via death, exudates, and respiration. In some cases, it is observed that after the non-apoptotic death of a plant root, all the carbon content of the root is distributed in the soil. In some other cases, plant activity does not cease immediately and slowly degrades into a simpler carbon compound.

3.8.1 Role of Rhizodeposits

As already explained, rhizodeposits are all compounds that are lost by the plant roots, including water molecules, sugar, amino acids, and enzymes. These exudates or rhizodeposits help to maintain soil integrity as well as fertility. Plant roots release carbon compounds in the soil, which maintain and improve the growth of microorganisms present in nearby soil areas. Along with the positive impact of root exudates on microorganism growth, rhizodeposits also include some chemicals that help the plants compete with other plant species and prevent the detrimental activity of harmful microorganisms (Haynes 1990).

3.8.2 Carbon Sequestration and Plant Growth

C stores or accumulated into the soil can play different roles in agriculture, forest management, and restoration of bare or degraded land. It also has a great impact on controlling global warming but is a concern with plant growth; it offers a good chance to maintain the quality of the soil. It is suggested that the global carbon cycle has the capability of restoring or increasing the organic content of soil. It can also help maintain the soil's water-holding capacity and improve soil tillage practice if proper techniques are applied, which ultimately helps sustain agriculture (Kumar et al. 2006). Different compounds are decomposed at different rates, and many factors are involved in deciding the decomposition rate of a compound. Soil chemistry and the type of organism present in soil are the essential features that govern the rate of decomposition of compounds (Nordt et al. 2000).

3.9 Conclusions

Carbon sequestration is a phenomenon of capturing atmospheric carbon via a different process involving both biotic and abiotic components and its storage for a longer duration. Carbon is mainly found in the atmosphere as CO_2, which is a GHG, so carbon sequestration is a beneficial method in terms of global warming. Apart from this, carbon sequestration also helps ensure and improve the quality or fertility of the soil by enhancing its carbon content. This transfer of carbon into soil ultimately helps the plant to grow better. However, due to a lack of advancement and proper methodology, the full mechanism of carbon sequestration via plants is not well understood. However, in the near future, it will be a great option to maintain and improve soil fertility. The studies on carbon sequestration will create a completely new idea that will allow better decisions on fertilization enrichment in the soil through different methods and in climate change. These approaches make the desired tools available for dealing with various problems in the soil and natural environment. This process will be the most viable approach toward global warming and sustainable agriculture and a potential option to ameliorate global change. So, in the near future, carbon sequestration can be proven as the major tool to ameliorate global warming and can also help enhance soil fertility.

References

Albrecht A, Kandji ST (2003) Carbon sequestration in tropical agroforestry systems. Agric Ecosyst Environ 99:15–27

Armentano TV, Menges ES (1986) Patterns of change in the carbon balance of organic wetlands of the temperate zone. J Ecol 74:755–774. https://doi.org/10.2307/2260396

Auerbach DI, Caulfield JA, Adams EE, Herzog HJ (1997) Impacts of ocean CO_2 disposal on marine life: a toxicological assessment integrating constant-concentration laboratory assay data with variable concentration field exposure. Environ Model Assess 2:333–343. https://doi.org/10.1023/A:1019029931755

Baines SJ, Worden RH (eds) (2004) Geological storage of carbon dioxide. The Geological Society, London

Batjes NH (1996) Totalcarbonand N in the soils of the world. Eur J Soil Sci 47:151–163

Bot A, Benites J (2005) The importance of soil organic matter: key to drought-resistant soil and sustained food production, vol No. 80. Food Agric Org

Bowen GD, Rovira AD (1999) The rhizosphere and its management to improve plant growth. Adv Agro 66:1–102

Boyd PW, Law CS, Wong CS, Nojiri Y, Tsuda A, Levasseur M, Takeda S, Rivkin R, Harrison PJ, Strzepek R, Gower J (2004) The decline and fate of an iron-induced subarctic phytoplankton bloom. Nature 428(6982):549–553

CAFRI (2015) Vision 2050. Central Agroforestry Research Institute, Jhansi, India

Campbell CA, Biederbeck VO, Zentner RP, Lafond GP (1991) Effect of crop rotations and cultural practices on soil organic matter, microbial biomass and respiration in a thin, black Chernozem. Can J Soil Sci 71:363–376

Caulfield JA, Adams EE, Auerbach DI, Herzog HJ (1997) Impacts of ocean CO2 disposal on marine life: probalistic plume exposure model with a time-varying dose-responsive analysis. Environ Model Assess 2:345–353. https://doi.org/10.1023/A:1019081915826

Dagar JC, Singh AK, Arunachalam A (2014) Introduction. In: agroforestry Systems in India: Livelihood Security & Ecosystem services (eds.) JC Dagar, AK Singh and a Arunachalam. Springer, India. Adv Agron 10:1–20

Dhyani SK, Handa AK, Uma (2013) Area under agroforestry in India: an assessment for present status and future perspective. Indian J Agrofor 15(1):1–11

Dhyani SK, Ram A, Dev I (2016) Potential of agroforestry systems in carbon sequestration in India. Indian J Agric Sci 86:1103–1112

Dixon RK, Krankina ON (1995) Can the terrestrial biosphere be managed to conserve and sequester carbon? In carbon sequestration in the biosphere. Springer, Berlin, Heidelberg, pp 153–179

Edrisi S A (2019) Sustainability analysis of biomass biofuel production from marginal and degraded lands of India

Edrisi SA, Tripathi V, Abhilash PC (2018) Towards the sustainable restoration of marginal and degraded lands in India. Trop Ecol 59:397–416

Edrisi SA, Tripathi V, Dubey PK, Abhilash PC (2022a) Carbon sequestration and harnessing biomaterials from terrestrial plantations for mitigating climate change impacts. In: Biomass, biofuels, biochemicals. Elsevier, pp 299–313

Edrisi SA, Sarkar P, Son J, Prakash NT, Baral H (2022b) Assessing the realization of global land restoration: a meta-analysis. Anthr Sci 1(1):179–194

Eswaran H, Van den Berg E, Reich P (1995) Global soil carbon resources. In: Lal R, Kimble JM (eds) Soils and global change. E Levine, BA Stewart, pp 27–43

Falkowski PG (1997) Evolution of the nitrogen cycle and its influence on the biological sequestration of CO2 in the ocean. Nature 387:272–275. https://doi.org/10.1038/387272a0

Falkowski P, Scholes RJ, Boyle EE, Canadell J, Canfield D, Elser J, Gruber N, Hibbard K, Högberg P, Linder S, Mackenzie FT (2000) The global carbon cycle: a test of our knowledge of earth as a system. Science 290(5490):291–296

Fan LS, Park A (2004) CO_2 mineral sequestration in a high pressure, high temperature, 3-phase fluidized bed reactor. Can J Chem Eng 81:855–890

Fan S, Gloor M, Mahlman J, Pacala S, Sarmiento J, Takahashi T, Tan P (1998) A large terrestrial carbon sink in North America implied by atmospheric and oceanic carbon dioxide data and models. Science 282:442–446. https://doi.org/10.1126/science.282.5388.442

Fang J, Chen A, Peng C, Zhao S, Ci L (2001) Changes in forest biomass carbon storage in China between 1949 and 1998. Science 292:2320–2316. https://doi.org/10.1126/science.1058629

Follett RF (2001) Soil management concepts and carbon sequestration in cropland soils. Soil Till Res 61(1–2):77–92

Forest Survey of India (2017) India state of Forest report. Ministry of Environment and Forests, Dehradun

Fowles M (2007) Black carbon sequestration as an alternative to bioenergy. Biomass Bioenergy 31:426–432

Freund P, Ormerod WG (1997) Progress towards storage of carbon dioxide. Energy Convers Manag 38:198–205. https://doi.org/10.1016/S0196-8904(96)00269-5

Gale J (2004) Why do we need to consider geological storage of CO2? In: Baines SJ, Worden RH (eds) Geological storage of carbon dioxide, vol 233. The Geological Society, London, pp 7–15

Garnett MH, Ineson P, Stevenson AC, Howard DC (2001) Terrestrial organic carbon in a British moorland. Glob Change Biol 7:375–388. https://doi.org/10.1046/j.1365-2486.2001.00382.x

Gorham E (1991) Northern peatlands: role in the carbon cycle and probable responses to climate warming. Ecol Appl 1:182–195. https://doi.org/10.2307/1941811

Govindarajulu M, Pfeffer PE, Jin H, Abubaker J, Douds DD, Allen JW, Bücking H, Lammers PJ, Shachar-Hill Y (2005) Nitrogen transfer in the arbuscular mycorrhizal symbiosis. Nature 435(7043):819–823

Grandy AS, Neff JC (2008) Molecular carbon dynamics downstream: the biochemical decomposition sequence and its impact on soil organic matter structure and function. Sci Total Environ 404(2–3):297–307

Grandy AS, Wickings K (2010) Biological and biochemical pathways of litter decomposition and soil carbon stabilization. Geochim Cosmochim Acta 74(12):A351–A351

Gregory PJ, Atwell BJ (1991) The fate of carbon in pulse-labelled crops of barley and wheat. Plant Soil 136(2):205–213

Gressel N (2007) From greener production to carbon trading: sustainable energy careers. Science 315:868–869. https://doi.org/10.1126/science.315.5813.868

Gupta DK, Bhatt RK, Keerthika A, Noor Mohammed MB, Shukla AK, Jangid BL (2019) Carbon sequestration potential of Hardwickia binata Roxb. Based agroforestry in hot semi-arid environment of India: an assessment of tree density impact. Curr Sci 116(1):112–116

Haile SG, Nair PR, Nair VD (2008) Carbon storage of different soil-size fractions in Florida silvopastoral systems. J Environ Qual 37:1789–1797

Halvorson AD, Reule CA, Follett RF (1999) Nitrogen fertilization effects on soil carbon and nitrogen in a dryland cropping system. Soil Sci Soc Am J 63:912–917

Halvorson AD, Wienhold BJ, Black AL (2002) Tillage, nitrogen and cropping system effects on soil carbon sequestration. Soil Sci Soc Am J 66:906–912

Haynes RJ (1990) Active ion uptake and maintenance of cation-anion balance: a critical examination of their role in regulating rhizosphere pH. Plant Soil 126(2):247–264

Herzog H, Drake E, Adams EE (1997) CO2 capture, reuse and storage technologies for mitigating climate change. A white paper prepared for USDOE by the MIT Energy Lab

Herzog H, Eliasson B, Kaarstad O (2002) Capturing greenhouse gases. Sci Am 282:72–80
Intergovernmental Panel on Climate Change (2007) Assessment report
IPCC (1999) Land use, land use change and forestry. Intergovernment panel on climate change. Cambridge University Press, Cambridge
IPCC (2007) Climate change 2007: mitigation of climate change. Working Group III contribution to the Intergovernmental Panel on Climate Change, Fourth Assessment Report, Bangkok. http://www.ipcc.ch/ipccreports/index.htm. Accessed: 28 Jan 2008
Janzen HH, Campell CA, Izaurralde RC, Ellert BH, Juma N, McGill W, Zentner RP (1998) Management effects on soil carbon storage on the Canadian prairies. Soil Till Res 47:181–195. https://doi.org/10.1016/S0167-1987(98)00105-6
Jenkinson DS, Rayners JH (1977) The turnover of soil organic matter in some of the Rothamsted classical experiments. Soil Sci 123(5):298–305
Jose S (2009) Agroforestry for ecosystem services and environmental benefits: an overview. Advances in Agroforestry, 1–10. https://doi.org/10.1007/s10457-009-9229-7
Kandji ST, Verchot LV, Mackensen J, Boye A, Van Noordwijk M, Tomich TP, Ong CK, Albrecht A, Palm CA, Garrity DP (2006) Opportunities for linking climate change adaptation and mitigation through agroforestry systems. World Agrofor Future World Agroforesty Centre, Nairobi, pp 113–122
Kane D, Solutions LLC (2015) Carbon sequestration potential on agricultural lands: a review of current science and available practices. Natl Sustain Agric Coalit Breakthr Strateg Solut LLC:1–35
Kaul M, Mohren GMJ, Dadhwal VK (2010) Carbon storage and sequestration potential of selected tree species in India. Mitig Adapt Strat Gl Change 15:489–510
Kerr RA (2001) Bush backs spending for a "global problem". Science 292:1978. https://doi.org/10.1126/science.292.5524.1978
Kerr RA (2007) Scientists tell policy makers we're all warming the world. Science 315:754–757. https://doi.org/10.1126/science.315.5813.754
Kintisch E (2007a) Report backs more projects to sequester CO2 from coal. Science 315:1481. https://doi.org/10.1126/science.315.5818.1481a
Kintisch E (2007b) New congress may be warming up to plans for capping emissions. Science 315:444. https://doi.org/10.1126/science.315.5811.444
Klara SM, Srivastava RD, McElvried HG (2003) Integrated collaborative technology development program for CO2 sequestration in geologic formations. Energy Convers Manag 44:2699–2712. https://doi.org/10.1016/S0196-8904(03)00042-6
Kobak KI, Knodrasheva NY, Turchinovich IE (1998) Changes in carbon pools of peatland and forests in northwestern Russia during the Holocene. Glob Planet Change 16(17):75–84. https://doi.org/10.1016/S0921-8181(98)00011-3
Krishnamurthy RV, Machavaram M (2000) Is there a stable isotope evidence for the fertilization effect? Proc Indian Acad Sci Earth Planet Sci 109:141–144
Kumar R, Pandey S, Pandey A (2006) Plant roots and carbon sequestration. Curr Sci:885–890
Kuzyakov Y, Domanski G (2000) Carbon input by plants into the soil. Review J Plant Nutr Soil Sci 163(4):421–431
Kuzyakov Y, Kretzschmar A, Stahr K (1999) Contribution of Lolium perenne rhizodeposition to carbon turnover of pasture soil. Plant Soil 213(1):127–136
Lal R (2001a) Potential of desertification control to sequester carbon and mitigate the greenhouse effect. Clim Chang 51:35–72
Lal R (2001b) World cropland soils as a source or sink for atmospheric carbon. Adv Agron 71:145–191
Lal R (2004) Agricultural activities and the global carbon cycle. Nutr Cycl Agroecosystems 70(2):103–116
Lal R (2008) Carbon sequestration. Philos Trans R Soc B: Biol Sci 363(1492):815–830
Lal R, Follett RF, Kimble JM (2003) Achieving soil carbon sequestration in the United States: a challenge to the policy makers. Soil Sci 168:827–845. https://doi.org/10.1097/01.ss.0000106407.84926.6b

Lamb D, Erskine PD, Parrotta JA (2005) Restoration of degraded tropical forest landscape. Science 310:1628–1632. https://doi.org/10.1126/science.1111773
Levin K, Boehm S, Carter R (2022) 6 big findings from the IPCC 2022 report on climate impacts, adaptation and vulnerability
Liebig MA, Varvel GE, Doran JW, Wienhold BJ (2002) Crop sequence and nitrogen fertilization effects on soil properties in the western corn belt. Soil Sci Soc Am J 66:596–601
Makundi WR, Sathaye JA (2004) GHG mitigation potential and cost in tropical forestry—relative role for agroforestry. Environ Dev Sustain 6:235–260
Malhi Y, Roberts JT, Betts RA, Killeen TJ, Li W, Nobre CA (2008) Climate change, deforestation, and the fate of the Amazon. Science 319:169–172
Martin JH, Fitzwater SE (1988) Iron deficiency limits of phytoplankton growth in the north-East Pacific sub-arctic. Nature 331:341–343. https://doi.org/10.1038/331341a0
McGowan E (2007) WM joins carbon market model discussion. Waste News 22 January 2007, 23
Mgbemene CA, Nnaji CC, Nwozor C (2016) Industrialization and its backlash: focus on climate change and its consequences. J Environ Sci 9(4):301–316
Milne R, Brown TA (1997) Carbon in the vegetation and soils of Great Britain. J Environ Manag 49:413–433. https://doi.org/10.1006/jema.1995.0118
Murthy IK, Gupta M, Tomar S, Munsi M, Tiwari R, Hegde GT, Ravindranath NH (2013) Carbon sequestration potential of agroforestry systems in India. J Earth Sci Clim Change 4(1):1–7
Mutuo PK, Cadisch G, Albrecht Palm CA, Verchot L (2005) Potential of agroforestry for carbon sequestration and mitigation of greenhouse gas emissions from soils in the tropics. Nutr Cycl Agroecosyst 71:43–54
Nair PR, Nair VD (2003) Carbon storage in North American agroforestry systems. In: Kimble J, Heath LS, Birdsey RA, Lal R (eds) The potential of U.S. Forest soils to sequester makundi carbon and mitigate the greenhouse effect. CRC Press, Boca Raton, pp 333–346
Nordt LC, Wilding LP, Drees LR (2000) Pedogenic carbonate transformations in leaching soil systems: implications for the global carbon cycle. Glob Clim Change Pedogenic Carbonates:43–64
O'Connor WK, Dahlin DC, Nilsen DN, Rush GE, Walters RP, Turner PC (2001) Carbon dioxide sequestration by direct mineral carbonation: results from recent studies and current status. www.netl.doe.gov/publications/proceedings/o1/carbon_seq/bC2.pdf
Ojima DS, Kittel TGF, Rosswall T, Walker BH (1999) Critical issues for understanding global change effects on terrestrial ecosystems. Ecol Appl 1:316–325
Paavilainen E, Paivanen J (1995) Peatland forestry: ecology and principles. Springer, New York
Pacala S, Socolow R (2004) Stabilization wedges: solving the climate problem for the next 50 years with current technologies. Science 305:968–972. https://doi.org/10.1126/science.1100103
Palm CA, Tomich T, van Noordwijk M, Vosti S, Alegre J, Gockowski J, Verchot L (2004) Mitigating GHG emissions in the humid tropics: case studies from the alternatives to slash-and-burn program (ASB). Environ Dev Sust 6:145–162
Pan S, Tian H, Dangal SR, Ouyang Z, Tao B, Ren W, Lu C, Running S (2014) Modeling and monitoring terrestrial primary production in a changing global environment: toward a multiscale synthesis of observation and simulation, vol 2014. Adv Meteorol, pp 1–17
Paustian K, Collins HP, Paul EA (1997) Management controls on soil carbon. In: Paul EA, Paustian K, Elliott T, Cole CV (eds) Soil organic matter in temperate agroecosystems: long-term experiments in North America. CRC Press, Boca Raton, pp 15–49
Ramachandran NPK, Mohan KB, Nair VD (2009) Agroforestry as a strategy for carbon sequestration. J Plant Nutr Soil Sci 172(1):10–23
Richards KR, Stokescarbon (2004) A review of forest carbon sequestration cost studies: a dozen years of research. Clim Chang 63(1):1–48
Rillig MC, Wright SF, Nichols KA, Schmidt WF, Torn MS (2001) Large contribution of arbuscular mycorrhizal fungi to soil carbon pools in tropical forest soils. Plant Soil 233(2):167–177
Schlesinger WH, Reynolds JE, Cunningham GL, Huenneke LF, Jarrell WM, Virginia RA, Whitford WG (1990) Biological feedbacks in global desertification. Science 247:1043–1048

Schrag DP (2007) Preparing to capture carbon. Science 315:812–813. https://doi.org/10.1126/science.1137632

Seibel BA, Walsh PJ (2001) Potential impacts of CO2 injection on deep-sea biota. Science 294:319–320. https://doi.org/10.1126/science.1065301

Selin NE (2019) Carbon sequestration. Encyclopedia Britannica. https://www.britannica.com/technology/carbon-sequestration

Sharrow SH, Ismail S (2004) Carbon and nitrogen storage in agro-forests, tree plantations, and pastures in western Oregon, USA. Agrofor Syst 60:123–130

Six J, Conant RT, Paul EA, Paustian K (2002) Stabilization mechanisms of soil organic matter: implications for C-saturation of soils. Plant Soil 241(2):155–176

Six J, Frey SD, Thiet RK, Batten KM (2006) Bacterial and fungal contributions to carbon sequestration in agroecosystems. Soil Sci Soc Am J 70(2):555

Takimoto A, Nair PR, Alavalapati JRR (2008) Socio-economic potential of carbon sequestration through agroforestry in the West African Sahel. Mit Adapt Strat Glob Change 13:745–761. https://doi.org/10.1007/s11027-007-9140-3

Tsang CF, Benson SM, Kogelski B, Smith RE (2002) Scientific considerations related to regulation development for CO2 sequestration in brine formations. Environ Geol 42:275–281. https://doi.org/10.1007/s00254-001-0497-4

USEPA (2020) Global Greenhouse Gas Emissions Data | Greenhouse Gas (GHG) Emissions | US-EPA. https://www.epa.gov/ghgemissions/global-greenhouse-gas-emissions-data. Accessed 14 Jan 2021

Wilson GW, Rice CW, Rillig MC, Springer A, Hartnett DC (2009) Soil aggregation and carbon sequestration are tightly correlated with the abundance of arbuscular mycorrhizal fungi: results from long-term field experiments. Ecol Lett 12(5):452–461

Chapter 4
Greenhouse Gas Emissions and Plant Soil Nutrition: Towards Nutrient-Use Efficiency for Achieving Net-Zero Emissions

Sheikh Adil Edrisi, Rajan Chaurasiya, and Vivek Verma

Abstract The present scenario of environmental pollution due to the escalated greenhouse gas (GHG) generation has made humanity rethink its societal and economic development. It has been anticipated that the Earth will surpass the global population of 9.2 billion by 2050 and will have more than 11 billion by the end of this century. It will create additional pressure due to increased demand for natural resources, resulting in the escalated emissions of GHGs. Furthermore, the rapidly changing lifestyle of the ever-increasing global population will also have the challenge of getting an optimal supply of food, fuel, fiber, and various other ecosystem services. It may result in the unsystematic implementation of weaker policies that could eventually result in the exploitation of natural resources, including the degradation of land, water, and many other resources. For instance, agricultural production seldom focuses on efficiently utilizing natural resources regarding the supply and uptake of nutrients in the field. Therefore, the present chapter discusses the importance of understanding the cusp of judicious development for resource utilization rather than exploitation. It will also focus on harnessing the enormous potential of understanding the mechanism of plant-soil nutrition and its relation to sequestering the GHG (i.e., CO_2 due to its predominance) from the atmosphere. Through this avenue, there is a window of opportunity to enhance nutrient-use efficiency (NUE) under different systems like agronomic, agroforestry, and forestry systems, which can help increase the sequestration rate and nutrient management. Besides CO_2, it will also highlight the use of essential nutrients like nitrogen, phosphorus, and

S. A. Edrisi (✉) · V. Verma
Thapar School of Liberal Arts and Sciences, Thapar Institute of Engineering and Technology, Patiala, Punjab, India
e-mail: sheikhadil.edrisi@thapar.edu

R. Chaurasiya
Institute of Environment and Sustainable Development, Banaras Hindu University, Varanasi, Uttar Pradesh, India

Department of Botany Shri Murli Manohar Town Post Graduate College, Ballia, Uttar Pradesh, India

S. A. Edrisi et al. (eds.), *Sustainable Plant Nutrition and Soil Carbon Sequestration*, Sustainable Plant Nutrition in a Changing World,
https://doi.org/10.1007/978-3-031-53590-1_4

potassium. Furthermore, focusing on such approaches in different systems will help meet multiple UN-SDGs and achieve ambitious targets towards net-zero emissions.

Keywords Soil nutrition · Plant nutrition · Nutrient retention · Carbon sequestration · Sustainable development

4.1 Introduction

The overall environmental deterioration has also been linked to the exponentially increasing population and changing lifestyles, leading to escalated demand for food, fuel, and fodder, consequently triggering resource exploitation (Edrisi et al. 2023). Extensive agriculture and inappropriate land management, including unscientific irrigation practices, have stimulated increased soil run-off, soil erosion, salinity, and, ultimately, the loss of soil nutrition (Mohanavelu et al. 2021). In addition, such unsustainable agricultural practices and industrialization have also escalated the global levels of greenhouse gases (GHGs), particularly carbon dioxide (CO_2). Agriculture itself depicts around 12% of the global GHG emissions annually (7.1 Gt CO_2 equivalent) (Rosa and Gabrielli 2023). It comprises majorly the CH_4 (54%), N_2O (28%), and CO_2 (18%) liberated by the different sectors of the food systems, including agriculture, as depicted in Fig. 4.1a–c.

Therefore, the manifestation of climate change is fundamentally driven by the agricultural sector as the United Nations- Intergovernmental Science-Policy Platform on Biodiversity and Ecosystem Services (UN-IPBES) classified agriculture as one of the major drivers of environmental and land degradation. Considering the prevailing debates between the prioritization of food production from agricultural land and afforestation for various ecosystem services, agroforestry has been suggested to resolve such conflicts (Edrisi et al. 2018). Moreover, there are multiple impediments during the profuse implementation of such adept practices at a wide scale (Sileshi et al. 2020; Schmidt et al. 2021). It is driven by inappropriate recommendations for optimized agroforestry systems at a site-specific level. Further, the system's inefficiency towards the sustainable utilization of the particular land area is also a limitation in providing requisite services for food, fuel, fodder, fiber, and many others (Edrisi and Abhilash 2021). It may be due to the lack of a cohort between the selected agricultural and forestry systems, which could have supported each other in their flourishment.

As stated in Chap. 1, soil dynamics reveal that soil is highly reactive and is a cradle of energy for microflora and fauna. It facilitates carbon sequestration through plants, adding carbon to the soil reservoir, which is available in organic and inorganic forms. Usually, it has been observed that there is a phenomenon of nutrient leaching from the soil of the agricultural systems under conventional practices. Therefore, it is the need of the hour to enhance nutrient-use efficiency (NUE) by

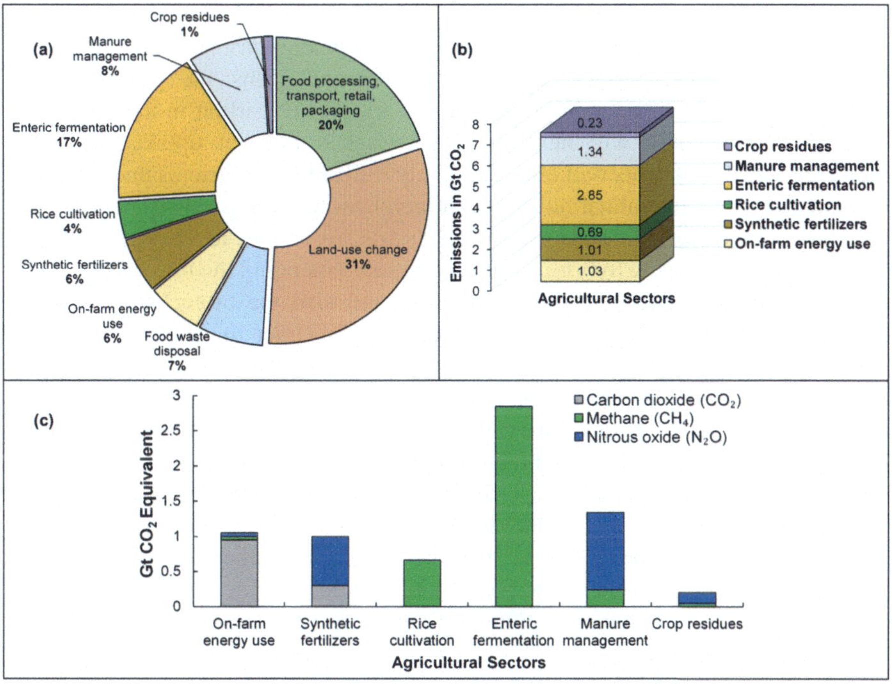

Fig. 4.1 Agricultural GHG emissions in 2020 (**a**) Global GHG emission share by food system sectors (17.3 Gt CO_2 equivalent), (**b**) GHG emissions estimates by different agrarian sectors (7.1 Gt CO_2 equivalent), (**c**) Estimates breakups GHG emissions by agrarian sectors and GHGs (CO_2, CH_4, and N_2O). (Data source: FAO; Available at: http://www.fao.org/faostat)

improving the cohort of plant-soil nutrition. Soil nutrient dynamics can be optimized by improving NUE, predominantly the carbon use and the other macro- and micro-nutrients under the agricultural set-ups, forestry practices, and agroforestry systems. It has been observed that the environmental quality may be improved via an integrated nutrient management (INM) strategy that can include the application of compost and chemical fertilizers (Oyetunji et al. 2022). NUE is generally explained as the capacity of the soil system to provide plant-needed nutrients optimally and appropriately maintain the plant's uptake and remobilization efficiency for optimum productivity (Fixen et al. 2015). It is critically essential to assess and evaluate the productivity and profitability of the arable crop during the discourse of lowering the inputs of chemical fertilizer (Hawkesford et al. 2016; Xin et al. 2017). Therefore, NUE can be defined via multiple avenues that depend on the purpose for which the indicator will be put (Norton et al. 2015). Agronomic efficiency depicts the appropriate performance indicator, particularly in screening more proficient genotypes regarding nutrient uptake from the soil pool. However, these approaches seek no application of the fertilizers to disregard the evaluation of the surplus production induced by the fertilizer addition. Therefore, utilizing the fractional nutrient balance to account for NUE is recommended.

In order to achieve the optimal levels of sequestration rates of carbon, nitrogen, phosphorus, and potassium (depicting the primary macronutrients), there is a need to focus on multiple perspectives through different means and processes. For instance, focusing on enhancing the ability to fix greater carbon in the above and below-ground biomass will help acquire enhanced nutrient uptake efficiency. Acquiring higher nutrients will eventually help utilize those nutrients under various processes, from mineralization to de-mineralization via various phenomena. Therefore, enhancing such ability at the field scale, agronomic levels, phenotypic or genotypic levels by using hybridization techniques, amending the microbial traits in the soil stimulating the plants' performance, and also the biotechnological and molecular tools will help resolve many dimensions under the umbrella of NUE prospects. In addition, the current scenario of ecosystem restoration under the initiatives of the United Nations (UN), global commitments of the Bonn Challenge, UN-Sustainable Development Goals (UN-SDGs), and the latest ambition to achieve Net-Zero emissions have a special focus on preventing degradation of resources and emissions under the unsustainable agricultural practices. Therefore, there is a great challenge to upscale conventional agrarian technologies, particularly in developing and under-developed economies. Hence, incentivization rather than subsidization should be promoted to boost public participation in this venture towards sustainable agricultural practices to enhance plant-soil nutrition for increasing NUE under the Net-Zero commitments.

In this regard, the present chapter discusses the crucial phenomena and sources of different macro- and micro-nutrients in the soil systems. It will be followed by a discussion about the know-how regarding the optimization of NUE along with the understanding of the different indicators for NUE under the different practices. For the sake of the current discussion, the chapter fundamentally focuses on agronomic, agroforestry, and forestry practices for their potential to improve NUE under net-zero targets. Further, this chapter will also elucidate the different challenges in enhancing the NUE with the ambition of attaining net-zero emissions along with suggestive policy measures.

4.2 Sources of Macro- and Micro-Nutrients in Soil and Their Uptake Mechanisms

Multiple factors are involved in the regulation of the plant growth and development. These physiological factors majorly relied on the presence and the abundance of the nutrients. Though there are myriads of elemental nutrients sufficiently available in the soil systems naturally, seventeen elements are considered significantly important during the growth and development of the higher plant species. Amongst seventeen elements, nitrogen, phosphorus, potassium (primary), magnesium, calcium, and sulfur (secondary) are the potential macro-nutrients required at higher levels (>1000 mg kg^{-1}) to maintain plant metabolism. However, zinc, iron, copper,

Table 4.1 Different types of macro- and micro-nutrients and their levels required in the soil for plant metabolism

Type of nutrient	Element	Symbol	Range of levels in the soil (mg kg^{-1})	Range of levels in plants on a dry weight basis (mg kg^{-1})	Adequate levels in plants on a dry weight basis (mg kg^{-1})
Micro-nutrient	Boron	B	2–150 (~15–25 usual average levels)	0.2–800	20
	Chlorine	Cl	~6–15	10–80,000	100
	Cobalt	Co	1–17	0.05–10	0.10
	Copper	Cu	~30	2–50	6
	Iron	Fe	7000–500,000 (~38,000 usual average levels)	20–600	100
	Manganese	Mn	10–3000 (~600–1000 usual average levels)	10–600	50
	Molybdenum	Mo	0.5–5 (~2 usual average levels)	0.10–10	0.10
	Nickel	N	~20–30 usual average levels	0.05–5	0.05
	Silicon	Si	1–45%	0.10%–10%	1000
	Sodium	Na	~ below 40	10–80,000	10
	Zinc	Zn	100–300 (~50 usual average levels)	10–250	20
Macro-nutrient	Calcium	Ca	~14,000	0.10%–6%	5000
	Carbon	C	0.5%–3.0%	–	450,000
	Hydrogen	H	–	–	60,000
	Magnesium	Mg	~5000	0.05%–1%	2000
	Nitrogen	N	15,000 in dry weight of soil. 1–5 mM (NO_3^-), 20–200 µM (NH_4^+)	0.5%–6%	15,000
	Oxygen	O	~50%	–	450,000
	Phosphorus	P	150–800 (Pi), 20%–80% (Po)	0.15%–0.50%	2000
	Potassium	K	3000–100,000 kg K^+/ha of soil (only 1%–2% is available)	0.80%–8%	10,000
	Sulfur	S	30–10,000 (~1000 usual average levels)	0.10%–1.50%	1000

Gowda et al. (2024), Lohry (2007)

manganese, and molybdenum are deemed micronutrients that are needed at lower levels by plants (adequately required ≤150 mg kg^{-1}). Table 4.1 illustrates different types of macro- and micro-nutrients required in the soil for plant metabolism, which needs to be maintained throughout the cohort of agricultural and forestry practices for the expected outcomes.

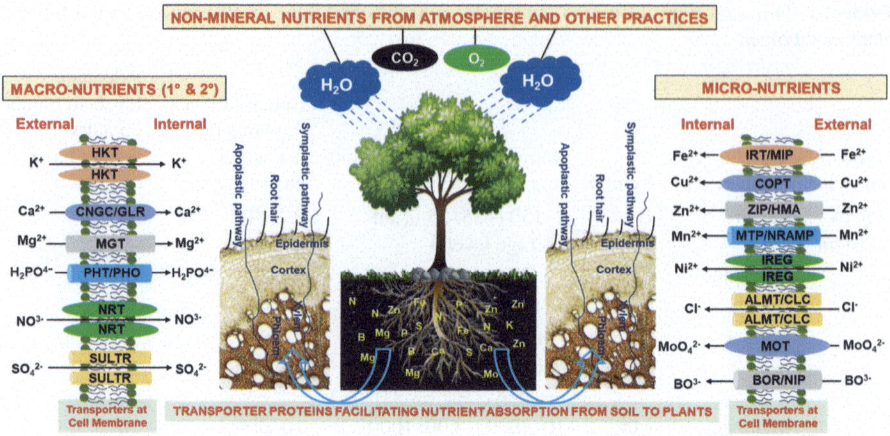

Fig. 4.2 Diagrammatic illustration of different transporter proteins facilitating nutrient absorption (Modified from Gowda et al. (2024), Tripathi et al. (2017)). (Note: *ALMT* Aluminum-activated malate transporters, *BOR* BOR transporter, *CLC* Chloride-transporting protein, *CNGCs* Cyclic nucleotide-gated channels, *COPT* Copper transport protein, *GLRs* Ionotropic glutamate receptors, *HKTs* High-affinity potassium transporters, *HMA* Heavy metal associated transporters, *IREG1* Iron-regulated transporter, *IRT* Iron-regulated transporter, *MGT* Magnesium transporters, *MOT* Molybdate transporter, *NIP* Nodulin 26–like intrinsic protein, *NRT* Nitrate transporter, *PHO* Phosphate transporter, *PHT* High-affinity phosphate transporter, *SULTRs* Sulfate transporters, *ZIP* IRT-like protein, *ZRT* Zinc-regulated transporters)

The significance of micronutrients is generally ignored in agricultural practices, leading to multiple deficiencies in the soil. For example, many soils in India are classified for manganese and zinc deficiencies (Shukla et al. 2012). Out of the seventeen elements that are significantly required for plant metabolism, thirteen are taken up through the plant roots. These uptakes from the soil via plant roots are a complex phenomenon that involves selective transport mechanisms guided by specialized transporters. These nutrients are generally present in the ionic forms in the soil, and their uptake is either selective, differentially accumulative, or depends on inherent genotype (White 2012), which are further classified and controlled by various cellular processes fostered by the membrane transporters.

These transporters not only carry nutrients to the plants from the soil but also shield them from toxic elements like salts and ionic metals (Von Uexküll and Mutert 1995). There are myriads of transporter proteins that facilitate nutrient absorption from the soil to the plants, including Aluminium-activated malate transporters (ALMT), BOR-transporters (BOR), Chloride-transporting protein (CLC), and many others, as depicted in Fig. 4.2. Therefore, understanding these complex mechanisms of nutrient uptake can help to optimize the NUE. Further, the different indicators of NUE can also be emphasized for optimizing plant-soil nutrition as an approach towards the mitigation and adaptation process under climate change scenarios.

4.3 Nutrient-Use Efficiency: Descriptors and an Indicator of Plant Growth

NUE is also called the transfer-to-utilize ratio or the output-to-input ratio. This indicator is derived from the particular nutrient, for example, the total nitrogen in the product transferred or removed from the field (i.e., the harvested crop or livestock product) and the total nitrogen applied to the field, farm, or region. The fertilizer denotes the nitrogen applied, compost, added green manure, animal manure, or other soil amendments. However, atmospheric nitrogen deposition is generally avoided because it is relatively lower than agrarian application. However, it can be counted when required (Norton et al. 2015). Typically, the NUE denotes the ratio of the recovered nutrient in the agricultural product under a specific system boundary compared to the nitrogen levels supplied to the land or soil. Therefore, it can be represented as per the following equation:

$$\text{NUE} = \frac{\sum \textit{output of the produce}}{\sum \textit{inputs supplied to the site}}$$

NUE neither elucidates the mechanisms of the intrinsic nitrogen transformation occurring in the system (i.e., N-nitrification or mineralization) nor is it primarily a direct quantitative indicator of nitrogen loss from the system. This is because the nitrogen has not been removed during the harvest, and it might be retained in the soil system. Hence, soil nitrogen stocks usually remain unchanged over a prolonged period or change to lower levels, so the NUE remains low under such circumstances. Therefore, it is a consistent indirect indicator of significant nitrogen flux in the environment and soil systems (Norton et al. 2015). Similarly, the nutrient parameters such as phosphorus, potassium, and others also reflect relatable indices depending on the flux stimulation and involved cyclic processes.

Therefore, NUE in plants depicts the overall function of soil's ability to supply appropriate amounts of nutrients and the potential of the plant to attain the nutrients and transport them in roots, shoots, and other plant parts through the remobilization (Congreves et al. 2021). There are several environmental and biotic factors or indicators that influence the plant's NUE. The environmental factors include the temperature, precipitation, and solar irradiance, whereas the biotic factors are the plant's response to insects, pests, diseases, root microbes, and allelopathy (Baligar and Fageria 1997; Blair 1993; Duncan 1994; Baligar et al. 2001).

There are different descriptors through which the NUE is assessed under various practices such as agronomic, agroforestry, and forestry practices (Oyetunji et al. 2022; Randhawa et al. 2021; Dobermann 2007; Fageria and Baligar 2003). These efficiencies usually depict the adeptness of the applied fertilizers (Cassman et al. 2002), majorly for analyzing the response of a short-term crop to a particular nutrient. The different descriptors are generally agronomic efficiency (AE), apparent recovery efficiency (ARE), apparent recovery efficiency by difference (RE), physiological efficiency (PE), utilization efficiency (UE), agro-physiological efficiency

(APE), internal utilization efficiency (INE), partial nutrient balance (PNB), and partial factor productivity (PFP) (Oyetunji et al. 2022; Randhawa et al. 2021; Dobermann 2007; Fageria and Baligar 2003). Their expressions and significance are elucidated below.

$$\text{Agronomic efficiency}\left(\mathbf{AE}\right) = \frac{\left(G_f - G_u\right)}{N_a} = \text{kg}\,\text{kg}^{-1}$$

Significance: It is utilized to estimate the alterations and improvements in the productivity of crops based on the applied nutrient (i.e., alterations in productivity with each unit of additionally applied fertilizer). This efficiency is influenced by the practice of fertilizer application and its influence on Physiological efficiency (PE) and Apparent Recovery Efficiency by Difference (RE).

$$\text{Apparent recovery efficiency}\left(\mathbf{ARE}\right) = \frac{\left(N_f - N_u\right)\times 100}{N_a} = \%$$

Significance: It is a measure of assessing the change ratio in the ability of the crop to uptake nutrients as compared to the amount of nutrients applied. It relies on the genetic factors of the crop, soil adeptness, and the different management practices adopted during the fertilizer application.

$$\text{Apparent Recovery Efficiency by Difference}\left(\mathbf{RE}\right) = \frac{\left(U_f - U_u\right)}{F} = \text{kg}\cdot\text{kg}^{-1}\text{ha}$$

Significance: It assesses the nutrients taken up by the plants in comparison to plant demands and capacity for the residual nutrients loss from the soil. It is influenced by the genetic factors of the crop, soil adeptness, and the different management practices adopted during the fertilizer application.

$$\text{Physiological efficiency}\left(\mathbf{PE}\right) = \frac{\left(Y_f - Y_u\right)}{\left(N_f - N_u\right)} = \text{kg}\,\text{kg}^{-1}$$

Significance: It estimates the plant's potential to transform nutrients from fertilizer into grain yields that are viable economically. This measure is influenced by the genetic factors of the crop and the practices adopted during the fertilizer application.

$$\text{Utilization efficiency}\left(\mathbf{EU}\right) = \text{PE} \times \text{ARE} = \text{kg}\,\text{kg}^{-1}$$

Significance: It is a measure of the assessing the association of a plant's potential to transform nutrients from fertilizer with the ability of the nutrient uptake by the crop as compared to the amount of the nutrients applied (i.e., the association of physiological efficiency with the apparent recovery efficiency by difference).

$$\text{Agro} - \text{physiological efficiency}\left(\mathbf{APE}\right) = \frac{\left(G_f - G_u\right)}{\left(N_f - N_u\right)} = \text{kg kg}^{-1}$$

Significance: It is used to estimate the alterations and improvements in the productivity of crops based on the plant's potential to transform nutrients from fertilizer into grain yields that are viable economically. This measure is influenced by the genetic factors of the crop, the soil's ability to support plant crop growth, and the practices adopted during the fertilizer application.

$$\text{Internal Utilization Efficiency}\left(\mathbf{INE}\right) = \frac{\text{Y}}{\text{U}} = \text{kg ha}^{-1}\text{kg}^{-1}$$

Significance: It elucidates the crop's conversion efficiency of the physiological nutrient to produce grain yields that are viable economically. This estimate relies on the genetic factors of the crop and the practices adopted during the fertilizer application.

$$\text{Partial Nutrient Balance}\left(\mathbf{PNB}\right) = \frac{\text{U}_h}{\text{F}} = \text{kg ha}^{-1}\text{kg}^{-1}$$

Significance: It represents NUE compared to the level of nutrient uptake by a crop relative to the level of nutrients applied in the soil.

$$\text{Partial factor productivity}\left(\mathbf{PFP}\right) = \frac{\text{Y}}{\text{F}} = \text{kg ha}^{-1}\text{kg}^{-1}\text{ha}$$

Significance: Used to analyze crop productivity and NUE based on native availability of nutrients and applied fertilizers.

The different symbols used in the above equations are described as follows: G_f = grain yield in the fertilized plot (kg ha^{-1}); G_u = grain yield in the control (kg ha^{-1}); N_a = quantity of nutrients applied (kg ha^{-1}); N_f = nutrient uptake in the fertilized plot (kg ha^{-1}); N_u = nutrient uptake in the control (kg ha^{-1}); U_f = total nutrient uptake in above ground crop biomass with nutrient applied; U_u = Nutrient uptake in above ground crop biomass with no nutrient applied; F = amount of (fertilizer) nutrient applied (kg ha^{-1}); Y_f = biological yield of the fertilized plot (kg ha^{-1}); Y_u = biological yield of the control (kg ha^{-1}); Y = crop yield with applied nutrients (kg ha^{-1}); U = total plant nutrient uptake in aboveground biomass at maturity (kg ha^{-1}); U_h = nutrient content of harvested portion of the crop.

The consideration of the different descriptors of NUE can help assess the plant growth and development with rigor and further help in precisely monitoring the fluxes of the different nutrients through which their management can be done cautiously. Therefore, the NUE is a pertinent indicator for measuring and assessing plant growth. Moreover, there is a crucial challenge as the current scenario of climate change also limits the estimation of multiple indices or descriptors related to particular nutrients, specifically the micro-nutrients in certain staple crops, cash crops, and legumes. In this regard, it is of utmost necessity to critically develop the strategic measures through which the NUE can be harnessed to optimize plant-soil nutrition under changing climatic conditions.

4.4 Nutrient-Use Efficiency for Optimizing Plant-Soil Nutrition Under a Changing Environment

The current scenario of global climate change has led to the reemphasizing of various research and policies toward mitigation strategies for lowering its effects. One perspective is increasing and optimizing the NUE (for the major nutrients like carbon, nitrogen, phosphorus, potassium, sulfur, etc.) in the crop, forest, and agroforest lands (Ross et al. 2019; Carlson et al. 2017). The influence of ecological and environmental factors on nutrient uptake and their usage efficiency is demonstrated in Fig. 4.3. The utilization of chemical fertilizers, specifically nitrogen, phosphorus, and potassium, during crop production is crucial while handling the bimodal issues of attaining food security and safeguarding the environment. To achieve such pertinent targets and present and future demands for food, crop yields must be enhanced by optimizing the NUE via various practices in the field (Kanter et al. 2016; Gowda

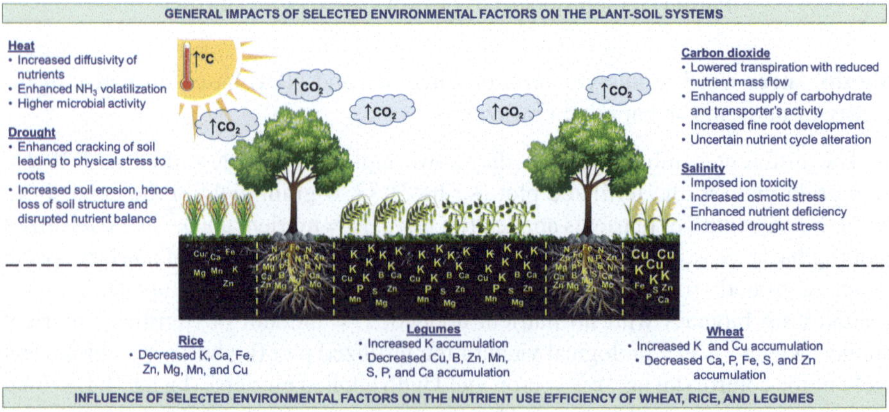

Fig. 4.3 Influence of environmental factors on nutrient accumulation and utilization efficiency. (Adapted and modified from Gowda et al. (2024), Edrisi et al. (2018))

et al. 2024). The excessive application of chemical fertilizer in the field is one of the critical sources of human-induced GHG emissions. Hence, the efficient management of fertilizer use in the cropland is crucial for addressing the pertinent role of agriculture in climate change (IPCC XE "The Intergovernmental Panel on Climate Change (IPCC)" 2019; Sapkota et al. 2021). In addition, excessive or inappropriate fertilizer use will be cost-intensive for the farmers, disturb the natural biogeochemical cycles (i.e., by creating an imbalance in the nutrient ratio of the soil), and thereby induce threat to various natural resources and ecosystem services (IPCC XE "The Intergovernmental Panel on Climate Change (IPCC)" 2019; Edrisi et al. 2018). Therefore, focusing globally on improving the NUE in the field helps resolve the tripartite issues of food insecurity, environmental deterioration, and agony in farmers' lives (Sapkota et al. 2021).

As discussed above, improving the NUE in the field will not only help in addressing those tripartite issues, but it will also underline the crucial role of plant-soil nutrition through its improvement, thereby resolving environmental concerns like sequestering GHG emissions. Therefore, it will help enable the nations to step towards achieving the ambitions of Net-Zero through such a strategy of multipurpose prospects. For instance, heat and/or drought stress is the preliminary abiotic stressors that hamper the growth and metabolism of various crops in the agriculture, forestry, and agroforestry systems (Husen and Iqbal 2023; Ansari et al. 2022; Husen 2021; Hussain et al. 2019). In addition, these stressors also limit the reproductive health of the plant species. It was observed that such stressors mutually imparted adverse effects on the maize crop while the nutrient absorption. It significantly lowered the concentration of nitrogen and potassium in the roots, phosphorus, and potassium in the stems, and even the lower levels of potassium in the leaves. It could be described by the weaker efficiency of the plants under the lower nutrient availability, transfer, and uptake during the stress conditions stimulated by the heat and drought stressors (Hussain et al. 2019). However, the excess heat stress diminished the rate of uptake of four nutrients (nitrogen, phosphorus, potassium, and iron) in the tomato plant's root rather than shoot (Giri et al. 2017). It is an established notion that under a particular range of temperatures in the soil, nitrate mineralization increases drastically along with the temperature variable. Hence, there is an urge to focus on considering temperature variables as a crucial factor while estimating the rates of nitrogen mineralization during the planting season, which is critical during the development of sustainable farming practices (Miller and Geisseler 2018).

Other stressors like salinity in the soil are one of the global components that limit crop growth and development. It is such a crucial factor that it stimulates each stage of the plant's development, right from seed germination to the plant's reproduction. The salinity induces physiological dryness in the rhizosphere regions of the plant roots, which hinders or even disables nutrient absorption mechanisms.

In addition, it also raises concerns of acute osmotic stress due to the hyper-accumulation of the toxic ions, resulting in the deficiencies of various macro- and micro-nutrients (nitrogen, phosphorus, potassium, calcium, iron, and zinc) (Corwin 2021; Gowda et al. 2024). Alterations in the temperature and rainfall have a significant influence on the salinity of soil (Ayanlade et al. 2018). It has been observed that the saline-sodic soils usually have lower or even lack iron elements, which is an important cofactor for multiple antioxidant enzymes. Therefore, iron deficiencies lead to interveinal chlorosis and deprived crop yield (Rabhi et al. 2008; Sharma et al. 2012; Kumar et al. 2010). Besides iron deficiency, salinity also has a strong impact on other macro- and micro-nutrient status of the soil, including the soil biological components, like macro-fauna and microbial diversity, thereby lowering the NUE in the field as illustrated by various scholars (Fageria et al. 2002; Kanagaraj and Desingh 2017; Zhang et al. 2019; Edrisi et al. 2020, 2023).

4.5 Status of Planting Practices to Improve Nutrient-Use Efficiency Under Net-Zero Targets

The urge to improve NUE in agricultural and other planting practices is crucial in attaining net-zero targets. There are ample pieces of evidence that highlight the significant role of integrated nutrient management (INM) practices, which encompass both inorganic and organic fertilizers for enhancing NUE. For instance, Paramesh et al. (2023) elucidated that INM not only enhances crop yield but also reduces GHG emissions, contributing to more sustainable farming systems. Precision agriculture (PA) techniques have also attracted attention due to their capacity to optimize nutrient application. Advances in remote sensing and soil nutrient mapping enable the site-specific application of fertilizer, thereby lowering nutrient losses and improving crop uptake efficiency. Hedley (2015) underpinned the effectiveness of PA in minimizing the environmental footprint of agricultural practices while sustaining high productivity levels.

Moreover, incorporating cover crops and crop rotation practices has been revealed to improve soil health and nutrient cycling. A study revealed that these practices for longer periods not only increase soil organic matter and microbial activity but also enhance the availability and uptake of essential nutrients by crops (Ashworth et al. 2020). This approach aligns with the targets of regenerative agriculture that seeks to restore and maintain ecological balance while achieving net-zero emissions. Therefore, integrating INM, PA, and regenerative approaches such as cover cropping and crop rotation represents a comprehensive approach to optimizing nutrient-use efficiency in agriculture and other planting practices. These practices are pertinent for attaining net-zero targets by lowering GHG emissions, improving soil health, and assuring sustainable crop production. Continued research, development, and innovation in these areas are vital to further optimize these practices and realize broader adoption among farmers, foresters, and the associated stakeholders. A few approaches have been discussed separately in the below subsections, as well as the connections and related elaborations in the forthcoming chapters of this book. Moreover, mitigation strategies and technologies can help attain net-zero agricultural emissions (Rosa and Gabrielli 2023), including the agroforestry systems described in Fig. 4.4.

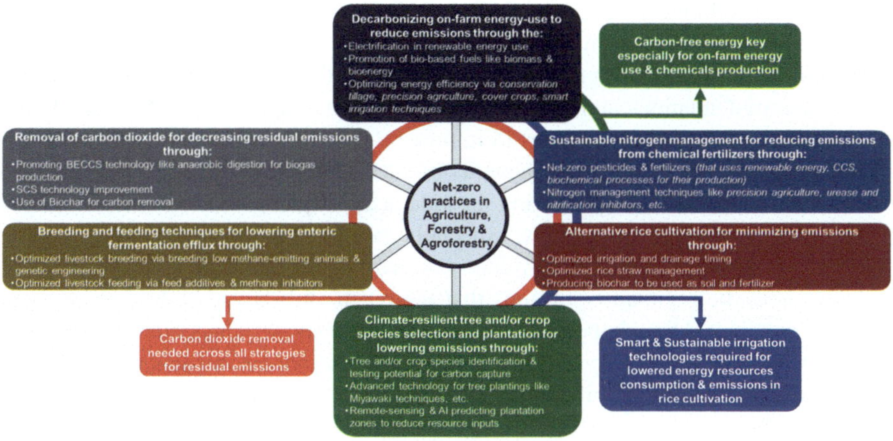

Fig. 4.4 Schematic flowchart summarizing different approaches for achieving net-zero emissions in agriculture, forestry, and agroforestry, planned to mitigate categories of emissions liberated in certain steps of the different processes involved in the dealt sectors. Connecting approaches are crucial to achieving net-zero agriculture, forestry, and agroforestry that are reported as required carbon removal (red), smart and sustainable irrigation (blue), and carbon-free energy (green connection). (Adopted and modified from Rosa and Gabrielli 2023; Roberts et al. 2022)

4.5.1 Agricultural and Agronomic Practices

Agronomic practices targeted to improve NUE are important under the framework of sustainable agriculture, especially under the scenario of burgeoning food demands and environmental concerns globally. Integrating advanced fertilization techniques, crop diversification, and soil health management are usually suggested under various scientific studies to enhance NUE. As discussed above, Jat et al. (2021) also described that precision fertilization and real-time soil and maize-wheat crop monitoring utilization significantly enhanced NUE and crop produce while minimizing nutrient leaching and environmental degradation.

Crop diversification, including the addition of legumes and other nitrogen-fixing plants under crop rotations, has also depicted substantial improvements for NUE. St. Luce et al. (2020) explored that crop diversification with rotation enhanced soil nitrogen content and soil microbial activity, resulting in improved nutrient uptake and availability by crops. This practice not only improves NUE but also lowers the dependencies on chemical fertilizers, fostering sustainable and climate-resilient agriculture. In addition, the study emphasized the inclusion of legumes during crop rotations can certainly improve soil quality through biological nitrogen fixation, thereby supporting long-term soil health (St. Luce et al. 2020).

The strategy of cover cropping has gained significant attention as it improves NUE by optimizing soil structure and nutrient retention. Sharma et al. (2018) reviewed and found that cover crops like rye and clover efficiently reduced nitrogen run-off and improved SOC content. It not only precludes nutrient loss but also makes nutrients labile for succeeding crops, hence increasing NUE as a whole. However, the efficacy of cover cropping relies on multiple parameters like crop species selection, local climate, and the adopted management strategies, highlighting the region-specific needs per the recommendations and research to optimize such approaches. A detailed discussion on the agronomic practices to improve the NUE can be found in Chaps. 10 and 12. Regardless of the proven benefits of such agricultural and agronomic practices, there are also challenges during its wide-scale implementation. There is a lack of knowledge amongst the farmers, especially those belonging to the economically weaker sections, which implies the economic barriers (Hundal et al. 2023) and the dynamicity in the regional conditions that inhibit the full-fledged adoption of such potential practices. Though the short-term studies have shown considerable outcomes, there is still a need for long-term research to understand the impacts on NUE and overall agricultural sustainability (Reich et al. 2014).

4.5.2 Agroforestry Practices

It has been suggested that to promote sustained agricultural and agronomic practices, the agroforestry system can be a supporting practice as it also has the vital potential to enhance NUE in agricultural systems along with environmental and socio-economic benefits (Mukhlis et al. 2022). This system fundamentally improves nutrient cycling, ameliorates soil health, and minimizes nutrient loss as it includes potential trees and shrubs with the particular crop species and livestock. There are studies that focus on various processes by which agroforestry can improve NUE. For example, the deep root systems of trees can hold and harness various nutrients from deep soil layers that are usually unavailable to annual crop systems and take them to make them available to the surface where the crops can harness them. Sileshi et al. (2020) described that agroforestry systems comprising deep-rooted tree species substantially lowered the nitrogen loss relative to traditional monoculture practices.

In addition, this system also contributes to enhanced SOM and microbial activities that are crucial for making nutrients available and uptake mechanisms. Trees under the agroforestry systems also provide leaf litter and enhanced root exudates that increase SOM, building a more favorable environment for below-ground microorganisms. A study demonstrated that such additions of root exudates from tree species enhanced the below-ground microbial activity and biomass, resulting in improved nitrogen mineralization and availability for the crop species (Pralhad et al. 2020). Moreover, enriching plant diversity in the agroforestry systems complements nutrient use, where different plant species harness various nutrient forms and minimize competition for the same and limited resources, thereby improving the NUE of the whole system (Gonçalves et al. 2021).

There has been a significant deployment of nitrogen-fixing tree species like *Acacia, Leucaena, Alnus,* and many others in the agroforestry systems (more discussions have been made in Chaps. 9 and 15). These tree species can fix atmospheric nitrogen and add to the soil nitrogen pool, lowering the requirement for chemical-based nitrogen fertilizers. Rosenstock et al. (2014) reviewed and reported that including nitrogen-fixing trees under the agroforestry systems enhanced nitrogen content in the soil along with increased crop yields without additional nitrogen supplements. Since tree roots are more prominent in their structure, they can hold the soil structure more significantly compared to the annual crops. Therefore. Agroforestry systems that include trees can help reduce soil erosion and nutrient loss, sustaining soil structure and quality. Research suggested that appropriate tree arrangements and planting densities in the agroforestry systems significantly lowered soil erosion and nutrient run-off as compared to conventional agricultural practices (Jafari et al. 2022).

Despite the benefits provided under this system, there are still various existing challenges with its broader implementation. The benefits of agroforestry systems also vary widely depending on the tree species selected and the management practices adopted under specific climatic conditions. There is a need to identify site-specific demands and recommend appropriate strategies accordingly for optimizing the agroforestry approaches to get benefits for longer periods under varied farming systems and regions (Sileshi et al. 2020; Fahad et al. 2022). Addressing these perspectives and hidden challenges is of utmost importance for the wide-scale adoption of agroforestry practices, which can help enhance NUE and contribute to sustainable agriculture (Schmidt et al. 2021). Therefore, policy realignment, educating the farmers, and investing in research and development (R&D) can help achieve such targets.

4.5.3 Forestry Practices

Forestry practices targeted to improve NUE are crucial regarding sustainable forest management and climate change mitigation. One of the effective practices is the harnessing of mixed-species plantations that have a proven record of enhancing nutrient cycling and reducing nutrient losses compared to monocultures. For instance, Forrester et al. (2006) demonstrated that mixed-species forests can harness nutrients more efficiently as they have established the complementary nutrient uptake mechanisms and enhanced soil nutrient availability facilitated by different transporters (Fig. 4.2). The study found that tree species diversity enhances the resilience of forest ecosystems to nutrient imbalances and promotes sustainable productivity (Forrester et al. 2006). Another pertinent practice is the controlled utilization of managed nutrient inputs through the application of organic amendments like biochar. The biochar not only improves soil structure but also enhances microbial activity and nutrient availability. Li et al. (2018) reviewed and indicated that amending forest soil with biochar led to improved nitrogen retention and reduced run-off, hence improving NUE. This strategy supports the sustainability of forest ecosystems for longer periods by maintaining soil fertility and minimizing the application of synthetic fertilizers, which can have adverse ecological effects. In addition, controlled forest thinning via selective logging can significantly affect nutrient dynamics in the forest systems. Reducing tree density can reduce competition for specific nutrients and water amongst trees, resulting in improved NUE. Bai et al. (2017) showed that controlled thinning improved soil-plant carbon content, as well as other nutrient availability and dynamics in managed forest systems. The study emphasized that these practices not only enhance tree performance but also improve overall forest health and resilience. The studies discussed above validate the significant enhancement in the NUE and thereby contribute to sustainable forest management that focuses on enhancing nutrient dynamics, lowering nutrient leaching, sustaining soil quality, and hence enabling the forest ecosystems to become healthier and more resilient.

4.6 Challenges and Policy Perspectives

There are also various impediments to improving NUE efficiently in the field under the umbrella of net-zero, thereby raising challenges that require robust policy interventions. One of the major challenges is the economic adeptness of the farmers in availing themselves of advanced and cutting-edge tools and techniques under the NUE practices. Though technologies like PA are effective, they still need significant upfront investments, which may not be affordable to small-scale farmers. It has been observed that widespread adoption is usually limited by inadequate financial support (Hundal et al. 2023). Scholars have discussed that the economic obstacles and subsidy requirements or provisions for financial incentives often help farmers adopt these technologies (Andersen and Bonnis 2021; Hundal et al. 2023).

Furthermore, various technical gaps and lacunae in the knowledge systems obstruct the deployment of effective and efficient NUE practices. Farmers often lack access to fundamental education and specialized training to harness the potential of advanced agronomic practices. Sarkar et al. (2022) highlighted the significance of extension services and training programs for peasants as a mending practice to enhance farmers' potential while implementing sustainable practices. The universal strategies will also have shortcomings under the varied climate soil types (Waqas et al. 2020). Therefore, a tailored approach, site-specific guidelines, and support systems are essential for handling such issues.

Policy frameworks also need to be realigned to focus on coupling sustainable practices under pre-existing agricultural systems. It has been suggested that current policies usually do not incentivize practices that optimize NUE, such as crop rotation and agroforestry (Srinivasarao 2021; Andersen and Bonnis 2021). Integrated approaches need to be developed to promote these practices by considering their environmental and economic impacts (Srinivasarao 2021; Brownlie et al. 2015). Furthermore, continuous monitoring and enforcement mechanisms are vital to ensure compliance and controlled progress toward NUE and net-zero targets (Srinivasarao 2021; Bumb et al. 2002). Developing an action plan to promote R&D in NUE practices is also needed. Continuous innovation is crucial for developing new approaches and tools to further enhance NUE. Rosa and Gabrielli (2023) argued that enhanced funding in R&D and public-private partnerships could push further advancements in this domain. Moreover, the dissemination of research outcomes and the execution of pilot projects can help validate the feasibility and benefits of advanced technologies.

4.7 Conclusion and Way Forward

The growing concern regarding global warming and climate change and its subsequent crises needs serious attention in every dimension of mitigation and adaptation strategies. The increasing GHG emissions, particularly from the agrarian sectors, need to be reduced significantly, as there will be the inevitable need to enhance food production in the future because of the burgeoning global population. Therefore, understanding the core mechanisms regulating plant-soil nutrition is of utmost necessity. The emphasis on improving the NUE as part of achieving net-zero emissions could be a plausible avenue to avoid the unforeseen consequences raised due to the reasons mentioned above. However, cautiously understanding the role of different transporters and exploring the new associations can lead to a significant achievement in these directions.

The practice of smart and climate-resilient agriculture supported by various agronomic, agroforestry, and forestry practices as described in the previous sections, could facilitate the realization of such ambitious targets at regional and global scales. However, strong determinations with stewardship under the enforcement mechanisms, incentivization instead of frequent subsidization, policy restructuring, and related R&Ds need serious attention. Therefore, the demand for sustainable production by reducing synthetic fertilization can be achieved if holistic approaches are adopted wisely and tailored to seek site-specific recommendations and alterations as per the varied climatic conditions. Specifically, the different planting and cropping strategies discussed above seek consistent economic support, education, and training for the associated stakeholders like farmers, region-specific guidelines, cohesive policy frameworks, and robust R&D initiatives. Addressing these issues via comprehensive policy proposals is particularly vital for promoting sustainable agricultural practices and attaining long-term environmental goals.

Acknowledgments Authors are thankful to the Director, Dean, and Head TSLAS, TIET, Patiala, India and Director, IESD, BHU, Varanasi, India for providing institutional support and encouragement. SAE is grateful to SERB for providing Start-up Research Grant (file no. SRG/2021/000281) and TIET for Seed Funding (ref. no. TU/DORSP/1596). VV is grateful to SERB for providing Junior Research Fellowship under the project. SAE is pleased to dedicate this chapter to celebrate his a decade of research in the field of restoration and sustainability science.

References

Andersen, M. S., Bonnis, G. (2021). Climate mitigation co-benefits from sustainable nutrient management in agriculture: incentives and opportunities

Ansari SA, Ansari MI, Husen A (eds) (2022) Augmenting crop productivity in stress environment. Springer, Singapore. https://doi.org/10.1007/978-981-16-6361-1

Ashworth AJ, Owens PR, Allen FL (2020) Long-term cropping systems management influences soil strength and nutrient cycling. Geoderma 361:114062

Ayanlade A, Radeny M, Morton JF, Muchaba T (2018) Rainfall variability and drought characteristics in two agro-climatic zones: an assessment of climate change challenges in Africa. Sci Total Environ 630:728–737

Bai SH, Dempsey R, Reverchon F, Blumfield TJ, Ryan S, Cernusak LA (2017) Effects of forest thinning on soil-plant carbon and nitrogen dynamics. Plant Soil 411:437–449

Baligar VC, Fageria NK (1997) Nutrient use efficiency in acid soils: nutrient management and plant use efficiency. In: Monitz AC, Furlani AMC, Fageria NK, Rosolem CA, Cantarells H (eds) Plant-soil interactions at low ph: sustainable agriculture and forestry production. Brazilian Soil Science Society Compinas, Brazil, pp 75–93

Baligar VC, Fageria NK, He ZL (2001) Nutrient use efficiency in plants. Commun Soil Sci Plant Anal 32(7–8):921–950

Blair G (1993) Nutrient efficiency—what do we really mean? In: Randall PJ, Delhaize E, Richards RA, Munns R (eds) Genetic aspects of plant mineral nutrition. Kluwer Academic Publishers, Dordrecht, pp 205–213

Brownlie WJ, Howard CM, Pasda G, Nave B, Zerulla W, Sutton MA (2015) Developing a global perspective on improving agricultural nitrogen use. Environmental Development 15:145–151

Bumb BL, Gregory D, Singh U (2002) Globalization and fertilizer use: policy and efficiency issues. Feed the soil to feed the people The role of potash in susta-inable agriculture:109

Carlson KM, Gerber JS, Mueller ND, Herrero M, MacDonald GK, Brauman KA et al (2017) Greenhouse gas emissions intensity of global croplands. Nat Clim Chang 7(1):63–68

Cassman KG, Dobermann A, Walters DT (2002) Agroecosystems, nitrogen-use efficiency, and nitrogen management. Ambio 31:132–140

Congreves KA, Otchere O, Ferland D, Farzadfar S, Williams S, Arcand MM (2021) Nitrogen use efficiency definitions of today and tomorrow. Front Plant Sci 12:637108

Corwin DL (2021) Climate change impacts on soil salinity in agricultural areas. Eur J Soil Sci 72(2):842–862

Dobermann A (2007) Nutrient use efficiency—measurement and management. In: Krauss A, Isherwood K, Heffer P (eds.) Fertilizer best management practices: General principles, strategy for their adoption and voluntary initiatives versus regulations. International Fertilizer Industry Association, Paris, France, pp 1–28.

Duncan RR (1994) Genetic manipulation. In: Wilkinson RE (ed) Plant-environment interactions. Marcel Dekker Inc., New York, pp 1–38

Edrisi SA, Abhilash PC (2021) Need of transdisciplinary research for accelerating land restoration during the UN decade on ecosystem restoration. Restor Ecol 29(8):e13531

Edrisi SA, Tripathi V, Abhilash PC (2018) Towards the sustainable restoration of marginal and degraded lands in India. Trop Ecol 59:397–416

Edrisi SA, El-Keblawy A, Abhilash PC (2020) Sustainability analysis of *Prosopis juliflora* (Sw.) DC based restoration of degraded land in North India. Land 9(2):59

Edrisi SA, Bundela AK, Verma V, Dubey PK, Abhilash PC (2023) Assessing the impact of global initiatives on current and future land restoration scenarios in India. Environ Res 216:114413

Fageria NK, Baligar VC (2003) Methodology for evaluation of lowland rice genotypes for nitrogen use efficiency. J Plant Nutr 26(6):1315–1333

Fageria NK, Baligar VC, Clark RB (2002) Micronutrients in crop production. Adv Agron 77:185–268

Fahad S, Chavan SB, Chichaghare AR, Uthappa AR, Kumar M, Kakade V et al (2022) Agroforestry systems for soil health improvement and maintenance. Sustain For 14(22):14877

Fixen P, Brentrup F, Bruulsema T, Garcia F, Norton R, Zingore S (2015) Nutrient/fertilizer use efficiency: measurement, current situation and trends. Managing water and fertilizer for sustainable agricultural intensification 270:1–30

Forrester DI, Bauhus J, Cowie AL, Vanclay JK (2006) Mixed-species plantations of eucalyptus with nitrogen-fixing trees: a review. For Ecol Manag 233(2–3):211–230

Giri A, Heckathorn S, Mishra S, Krause C (2017) Heat stress decreases levels of nutrient-uptake and-assimilation proteins in tomato roots. Plan Theory 6(1):6

Gonçalves B, Morais MC, Pereira S, Mosquera-Losada MR, Santos M (2021) Tree–crop ecological and physiological interactions within climate change contexts: a mini-review. Front Ecol Evol 9:661978

Gowda RS, Kaur M, Kaushal B, Kaur H, Kumar V, Sharma R et al (2024) Behavior, sources, uptake, interaction, and nutrient use efficiency in plant system under changing environment. In: Essential minerals in plant-soil systems. Elsevier, pp 93–127

Hawkesford MJ, Kopriva S, De Kok LJ (2016) Nutrient use efficiency in plants. Springer, Cham

Hedley C (2015) The role of precision agriculture for improved nutrient management on farms. J Sci Food Agric 95(1):12–19

Hundal GS, Laux CM, Buckmaster D, Sutton MJ, Langemeier M (2023) Exploring barriers to the adoption of internet of things-based precision agriculture practices. Agriculture 13(1):163

Husen H (2021) Plant performance under environmental stress. Springer International Publishing, Cham. https://doi.org/10.1007/978-3-030-78521-5

Husen A, Iqbal M (eds) (2023) Medicinal plants: their response to abiotic stress. Springer Nature Switzerland AG, Cham. https://doi.org/10.1007/978-981-19-5611-9

Hussain HA, Men S, Hussain S, Chen Y, Ali S, Zhang S et al (2019) Interactive effects of drought and heat stresses on morpho-physiological attributes, yield, nutrient uptake and oxidative status in maize hybrids. Sci Rep 9(1):3890

IPCC (2019) IPCC special report on climate change, desertification, land degradation, sustainable land management, food security, and greenhouse gas fluxes in terrestrial ecosystems. Summary for Policymakers. Intergovernmental Panel on Climate Change, Geneva

Jafari M, Tahmoures M, Ehteram M, Ghorbani M, Panahi F (2022) Agroforestry and its role in soil erosion biological control. In: Soil erosion control in drylands. Springer International Publishing, Cham, pp 649–700

Jat RK, Bijarniya D, Kakraliya SK, Sapkota TB, Kakraliya M, Jat ML (2021) Precision nutrient rates and placement in conservation maize-wheat system: effects on crop productivity, profitability, nutrient-use efficiency, and environmental footprints. Agronomy 11(11):2320

Kanagaraj A, Desingh R (2017) Salinity influences physiological traits of seven Sesame (Sesamum indicum L.) varieties. J Sci Agric 1:188–196

Kanter DR, Zhang X, Mauzerall DL, Malyshev S, Shevliakova E (2016) The importance of climate change and nitrogen use efficiency for future nitrous oxide emissions from agriculture. Environ Res Lett 11(9):094003

Kumar P, Tewari RK, Sharma PN (2010) Sodium nitroprusside-mediated alleviation of iron deficiency and modulation of antioxidant responses in maize plants. AoB plants, 2010, plq002

Li Y, Hu S, Chen J, Müller K, Li Y, Fu W et al (2018) Effects of biochar application in forest ecosystems on soil properties and greenhouse gas emissions: a review. J Soils Sediments 18:546–563

Lohry R (2007) Micronutrients: functions, sources and application methods, vol 15. Indiana CCA Conference Proceedings, December

Miller KS, Geisseler D (2018) Temperature sensitivity of nitrogen mineralization in agricultural soils. Biol Fertil Soils 54:853–860

Mohanavelu A, Naganna SR, Al-Ansari N (2021) Irrigation induced salinity and sodicity hazards on soil and groundwater: an overview of its causes, impacts and mitigation strategies. Agriculture 11(10):983

Mukhlis I, Rizaludin MS, Hidayah I (2022) Understanding socio-economic and environmental impacts of agroforestry on rural communities. Forests 13(4):556

Norton R, Davidson E, Roberts T (2015) Nitrogen use efficiency and nutrient performance indicators, vol 14. Global Partnership on Nutrient Management

Oyetunji O, Bolan N, Hancock G (2022) A comprehensive review on enhancing nutrient use efficiency and productivity of broadacre (arable) crops with the combined utilization of compost and fertilizers. J Environ Manag 317:115395

Paramesh V, Kumar RM, Rajanna GA, Gowda S, Nath AJ, Madival Y et al (2023) Integrated nutrient management for improving crop yields, soil properties, and reducing greenhouse gas emissions. Front Sustain Food Syst 7(10):3389

Pralhad BS, Rajendran P, Divya MP, Rajeswari R, Thangamani G, Ramaha C (2020) Assessing the effect of different agroforestry practices on soil physico-chemical properties and microbial activity. Int J Curr Microbiol App Sci 9(12):2802–2816

Roberts AJ, Crowley LM, Sadler JP, Nguyen TTT, Gardner AM, Hayward SAL, Metcalfe DB (2022) Correction: Roberts et al. effects of elevated atmospheric CO2 concentration on insect herbivory and nutrient fluxes in a mature temperate forest. Forests 13(8):998, Article 1216. https://doi.org/10.3390/f13081216

Rabhi M, Talbi O, Atia A, Abdelly C, Smaoui A (2008) Selection of a halophyte that could be used in the bioreclamation of salt-affected soils in arid and semi-arid regions. In: Biosaline agriculture and high salinity tolerance. Birkhäuser, Basel, pp 241–246

Randhawa MK, Dhaliwal SS, Sharma V, Toor AS, Sharma S, Kaur M, Verma G (2021) Nutrient use efficiency as a strong indicator of nutritional security and builders of soil nutrient status through integrated nutrient management technology in a rice-wheat system in northwestern India. Sustain For 13(8):4551

Reich M, Aghajanzadeh T, De Kok LJ (2014) Physiological basis of plant nutrient use efficiency–concepts, opportunities and challenges for its improvement. Nutrient use efficiency in plants: Concepts and approaches:1–27

Rosa L, Gabrielli P (2023) Achieving net-zero emissions in agriculture: a review. Environ Res Lett 18(6):063002

Rosenstock TS, Tully KL, Arias-Navarro C, Neufeldt H, Butterbach-Bahl K, Verchot LV (2014) Agroforestry with N2-fixing trees: sustainable development's friend or foe? Curr Opin Environ Sustain 6:15–21

Ross, K., Hite, K., Waite, R., Carter, R., Pegorsch, L., Damassa, T., Gasper, R. (2019). Enhancing NDCs: opportunities in agriculture

Sapkota TB, Jat ML, Rana DS, Khatri-Chhetri A, Jat HS, Bijarniya D et al (2021) Crop nutrient management using nutrient expert improves yield, increases farmers' income and reduces greenhouse gas emissions. Sci Rep 11(1):1564

Sarkar S, Mahra GS, Lenin V, Padaria RN, Burman RR (2022) Innovative extension approaches for diffusion of nutrient management technologies. In: Soil Management for Sustainable Agriculture. Apple Academic Press, pp 283–326

Schmidt M, Corre MD, Kim B, Morley J, Göbel L, Sharma AS et al (2021) Nutrient saturation of crop monocultures and agroforestry indicated by nutrient response efficiency. Nutr Cycl Agroecosyst 119(1):69–82

Sharma P, Jha AB, Dubey RS, Pessarakli M (2012) Reactive oxygen species, oxidative damage, and antioxidative defense mechanism in plants under stressful conditions. J Bot 2012

Sharma P, Singh A, Kahlon CS, Brar AS, Grover KK, Dia M, Steiner RL (2018) The role of cover crops towards sustainable soil health and agriculture—a review paper. Am J Plant Sci 9(9):1935–1951

Shukla AK, Behera SK, Shivay YS, Singh P, Singh AK (2012) Micronutrients and field crop production in India: a review. Ind J Agron 57(3s):123–130

Sileshi GW, Mafongoya PL, Nath AJ (2020) Agroforestry systems for improving nutrient recycling and soil fertility on degraded lands. Agrofor Degrad Landsc 1:225–253

Srinivasarao C (2021) Programmes and policies for improving fertilizer use efficiency in agriculture. Indian J Fertil 17(3):226–254

St. Luce M, Lemke R, Gan Y, McConkey B, May W, Campbell C et al (2020) Diversifying cropping systems enhances productivity, stability, and nitrogen use efficiency. Agron J 112(3):1517–1536

Tripathi V, Edrisi SA, Chen B, Gupta VK, Vilu R, Gathergood N, & Abhilash PC (2017) Biotechnological advances for restoring degraded land for sustainable development. Trends Biotechnol 35(9):847–859. https://doi.org/10.1016/j.tibtech.2017.05.001

Von Uexküll HR, Mutert E (1995) Global extent, development and economic impact of acid soils. Plant Soil 171:1–15

Waqas MA, Smith P, Wang X, Ashraf MN, Noor MA, Amou M et al (2020) The influence of nutrient management on soil organic carbon storage, crop production, and yield stability varies under different climates. J Clean Prod 268:121922

White PJ (2012) Ion uptake mechanisms of individual cells and roots: short-distance transport. In: Marschner's mineral nutrition of higher plants. Academic, pp 7–47

Xin X, Qin S, Zhang J, Zhu A, Yang W, Zhang X (2017) Yield, phosphorus use efficiency and balance response to substituting long-term chemical fertilizer use with organic manure in a wheat-maize system. Field Crop Res 208:27–33

Zhang WW, Chong WANG, Rui XUE, Wang LJ (2019) Effects of salinity on the soil microbial community and soil fertility. J Integr Agric 18(6):1360–1368

Part II
Modeling Soil Carbon and Nutrient Dynamics

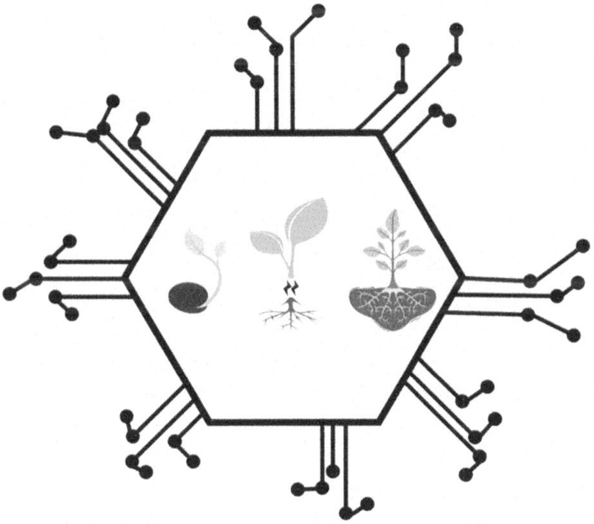

Chapter 5
Dynamism of Carbon Sequestration in Ecosystems and Its Role in Ecological Functioning

Khayali D. Vaidya and Seema B. Sharma

Abstract Organic matter is a key component of soil because it influences its structural, chemical, and biological properties, as well as its proper functioning. Although carbon is a crucial component in soil, its elevated levels in the atmosphere since pre-industrial times are warming the Earth relatively. The mechanism of harvesting CO_2 from the atmosphere is known as carbon sequestration. Enhanced SOM boosts plant productivity through improved water and nutrient retention in natural and agricultural settings. Soils hold almost three times more carbon than the atmosphere, thus becoming the largest land-based carbon pool. According to the IPCC, croplands and grasslands sequester 0.4–8.6 Gt CO_2 year^{-1}. The oceans, on the other hand, are the world's greatest sink of atmospheric carbon. However, due to the untapped potential of the different ecosystems in order to sequester atmospheric carbon, this chapter tries to understand the dynamism of carbon sequestration among different terrestrial and aquatic ecosystems and highlights some processes and amendments in the soil that can enhance carbon sequestration and discusses its impacts on the regular functioning of ecosystems and its services.

Keywords Carbon sequestration · SOC · IPCC · Ecosystems · Ecological functioning · Climate change

5.1 Introduction

Soils are closely interdependent, with soil organisms and mineral components interacting to produce a high level of variety and complexity. Soil structure and fertility are determined by the mineral makeup of the soil and the organisms that live in it. These are the largest terrestrial carbon pools and store three times more carbon than

---

K. D. Vaidya · S. B. Sharma (✉)
Department of Earth and Environmental Science, KSKV Kachchh University, Bhuj, Gujarat, India
e-mail: seemabhargavsharma@gmail.com

S. A. Edrisi et al. (eds.), *Sustainable Plant Nutrition and Soil Carbon Sequestration*, Sustainable Plant Nutrition in a Changing World,
https://doi.org/10.1007/978-3-031-53590-1_5

in the atmosphere and vegetation (Ramesh et al. 2019a). Organic and inorganic are the chief forms through which carbon is stored in soil. Soil inorganic carbon forms include carbonates of calcium and magnesium. These are found mostly in alkaline soils. Organic carbon form includes mostly every biotic living or dead component, including plant roots, soil biota, humus, litter, charcoal, and other recalcitrant material. The common element among all these is carbon; hence, a test resulting in 1% organic carbon of the soil will have approximately 1.7% (by weight) organic matter since organic matter in soil typically includes about 60% carbon (Bell and Lawrence 2009). The threshold limit of carbon in the atmosphere is about 350 parts per million. There are presently around 412 parts per million (ppm) of CO_2 in the Earth's atmosphere, and that number is steadily increasing (Stein 2022). The elevated concentration of this in the atmosphere has caused many devastating effects that lead to climate change. Earth's temperature has been increased by 1.5–5.8 °C during the twenty-first century (IPCC 2001). Prominent changes in ecosystems are the result of the rise in sea levels, which occurs due to changes in temperature. CO_2 concentrations have increased by 31% from 280 ppmv (Parts per million volume) in 1850 to 380 ppmv in 2005 and are currently increasing at 1.7 ppmv year^{-1} or 0.46 percent year^{-1} (Westerling et al. 2006). There are presently three possible ways to cut CO_2 emissions to lower the effects of global warming, and these are: (i) reducing global energy use, (ii) creating low or no-carbon fuels, and (iii) sequestering CO_2 from point sources or the atmosphere via natural and technical processes (Schrag 2007). The present chapter provides a detailed insight into various forms of carbon and its influence on terrestrial and aquatic ecosystems. In addition, it emphasizes on various studies that are done to enhance carbon storage in soil through various approaches.

5.2 Influence of Carbon on Various Ecosystem Services

An ecosystem comprises a dynamic blend of interrelated living and non-living components of nature. Costanza et al. (1997) defines ecosystem services as providers of benefits to humans through the transformation of resources into essential goods and services flow. Transformation of human well-being and all of its components are affected by ecosystem services (Edrisi et al. 2018). According to the World Business Council of Sustainable Development, the proper functioning of current business models can not be possible under the limited scope of ecosystem services. The significance of intangible services such as temperature regulation, water filtration, soil fertility, and recreational and cultural activities has become more obvious in recent years. As our understanding of human reliance on natural processes expands through time and space, so does the need to quantify and value these ecosystem services (Fig. 5.1) in economic and management terms (Department of the Environment 2009). Regular functioning of different biogeochemical cycles plays a key role in the regulation of ecosystems, and nutrient cycling is an essential part of the

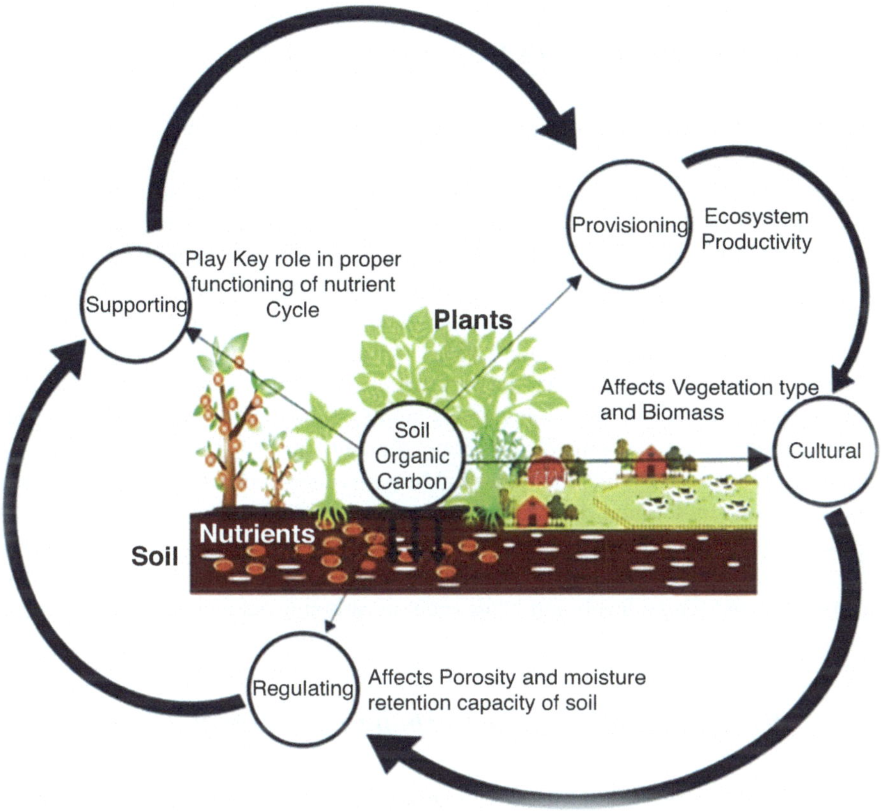

Fig. 5.1 Schematic representation of soil organic carbon influence on various ecosystem services

ecosystems. Carbon is a dynamic element and is considered a major part of an ecosystem. Its key function is to synthesize proteins, vitamins, and other essential molecules. It also provides food to plants in the form of glucose through photosynthesis. Capturing carbon from the atmosphere relies chiefly on the photosynthesis process (Paustian et al. 2019). In soil, carbon behaves as a major nutrient and forms an essential part of the humus. Carbon constitutes nearly half of the dry weight of living organisms (Green and Byrne 2004). The carbon cycle is basically comprised of carbon stocks and fluxes, which show transferring of carbon along the atmosphere, biosphere, and lithosphere. The study of the carbon cycle is gaining popularity among academics as a result of the increase in atmospheric CO_2 concentrations from 280 ppm (parts per million) before the dawn of industrialization to over 384 ppm (parts per million) in 2008. This rise reflects around half of the CO_2 emissions from anthropogenic activities; the other half has been captured by oceans and land. Figure 2.2 indicates various ways through which carbon is sequestered in soil.

5.3 Carbon Dynamics and the Sequestration in the Maintenance of Ecosystem Services

The influx and out flux of SOC are collectively regulated by (1) climatic factors like temperature and rainfall, (2) soil conditions with diverse physicochemical characteristics, and (3) biological traits such as the quantity and type of carbon inputs into the soil (Luo 2017). Therefore, a detailed study of these parameters is important to study the dynamics of SOC. Due to ever-increasing transportation and industrialization activities, the emission of carbon has increased. In soil, carbon is stored in two major forms: organic and inorganic. Dead organic matter, fallen litter, and fossil fuels are organic forms of carbon, and minerals in the soil are inorganic forms of carbon (Nelson and Sommers 2018). From total SOM, the part that is available to plants is termed as soil labile organic carbon and constitutes a very small part of soil's total organic carbon, but its decomposition has a major impact on atmospheric $CO_{2.}$ Important markers of soil labile organic carbon include microbial biomass carbon (MBC), readily oxidized organic carbon (EOC), and dissolved organic carbon (DOC). The transformation of almost 90% of SOM is chiefly regulated by microorganisms through various activities (Xu et al. 2021). Also, there are studies demonstrating a favorable correlation between soil bacterial and fungal levels and labile organic carbon (Zhang et al. 2020).

5.3.1 Distribution of Organic Carbon in Terrestrial Ecosystems

Globally, soil captures 2344 Gt (Stockmann et al. 2013) carbon in the terrestrial part. Forests cover almost one-third of the total $CO_{2.}$ Figure 5.2 shows carbon stored by different types of forests in India.

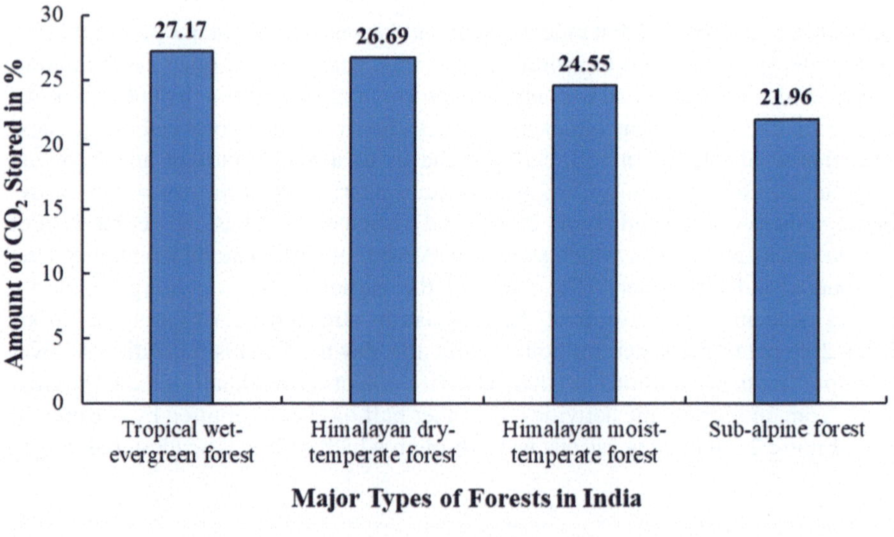

Fig. 5.2 Carbon storage efficacy of various types of forests in India (Data source: FSI 2017)

Table 5.1 Influence of different factors on SOC dynamics

	Major factors	classification	Details
Soil organic carbon	Climatic factors	Temperature	Directly impacts the microbial community and enzymatic activities
		Moisture	It directly affects the environment of plant growth and thereby decides the above- and below-ground biomass
	Edaphic factors	Soil form	SOC is influenced by nutrient availability and texture. Organic matter tends to be higher in black soils
		Soil pH	Saline soils are generally seen with low amounts of organic matter due to the presence of carbonates
		Soil moisture	Since carbon sequestration is moisture-dependent, it is a key factor for SOC decomposition
		porosity	SOC produced by microorganisms in soil pores is bound and stabilized by aggregates, which has an impact on soil carbon sequestration
		Microbial biomass	Microbial diversity helps in the decomposition of SOC

Ramesh et al. (2019a, b), Wong et al. (2010), Kadlec et al. (2012)

Due to its importance and characteristics, SOC is gaining importance day by day. Various factors, including climatic (temperature, wind, precipitation, or moisture) and edaphic (like soil type, pH, density, porosity, and microbial biomass) influence the SOC dynamics (Table 5.1). Land use practices (Ramesh et al. 2019b), deposition of plant biomass, microbial biota, infiltration, and soil erosion (Bationo et al. 2005) are key factors in studying SOC. Total organic carbon (TOC), particulate organic carbon (POC), microbial biomass carbon (MBC), dissolved organic carbon (DOC), permanganate oxidizable carbon ($KMnO_4$-C), and mineral related organic carbon are some of the pools and fractions found in the SOC (MOC). Each one has a different rate of breakdown and stability (Ramesh et al. 2019a). This is the reason why so much dynamism and variety is seen in different types of soils and also in different kinds of ecosystems (Singh et al. 2007). The transformation of plant biomass into SOC by microbial activities is also a considerable factor (Edrisi et al. 2022). Along with fertility enhancement, SOC also helps increase structural stability, increases water retention capacity, and protects against erosion due to Aeolian movements (Palmer et al. 2017).

According to research, tropical forests in India are carbon-rich, specifically in the upper layer of soil (Division 2015). Furthermore, Grasslands hold more than a third of the world's above and belowground organic carbon stores; thus, grasslands may significantly contribute to carbon sequestration if grazing management is improved (Tessema et al. 2020). On average, temperate grasslands store 331 Mg C ha^{-1}, which consists of about 12% of the world's organic carbon in soil (Schlesinger 1977). Due to the perennial nature of grassland, carbon is constantly sequestered, which results in constant carbon inputs from above-ground vegetation (Jones et al.

2016). In India, studies have reported that among carbon pools, microbial and labile carbon pools are lowered by almost 35% and 29% respectively in altered croplands then in native forestlands which shows how land use, changes the soil carbon pools (Padbhushan et al. 2020). Due to soil erosion, site disturbances brought on by fast SOC decomposition, and modifications in the quantity and quality of organic wastes given to the soil, soil carbon levels are decreased as a result of the conversion of forest land for agricultural purposes (Ramesh et al. 2019b). However, different sustainable agricultural practices like agroforestry, silvipastoral management, rotational grazing, soil erosion control, and soil nutrient management have a positive impact on the enhancement of SOC (Aryal et al. 2018). The influence of different factors on SOC dynamics (Ramesh et al. 2019a, b; Wong et al. 2010; Kadlec et al. 2012) has been compiled in Table 5.1.

5.3.2 Distribution of Organic Carbon in Coastal Ecosystems

Similarly to green carbon, which indicates carbon stored in soils and plants, blue carbon indicates carbon stored in coastal ecosystems and was originally named by Nellemann et al. (2009). Mangroves, wetlands, tidal marshes, and seagrasses sequester carbon in larger quantities than in terrestrial ecosystems (Banerjee et al. 2021). Blue Carbon initiative is a global program aiming at the conservation of coastal ecosystems (Lovelock and Duarte 2019) since these are considered highly productive ecosystems and are able to store huge amounts of carbon in them. The results of this study show that 83% of the entire carbon cycle flows through the oceans and that over half of all carbon is buried in ocean sediments. It is a natural ability of blue carbon to trap sediments and carbon from outside their natural bounds. They aid in the preservation of marine ecosystem health and stability (Domain et al. 2020). Through physical, chemical, and biological processes, the oceans have absorbed around one-third of global CO_2 emissions. Since cold waters have more dissolving capacity, cold deep waters are better sinks of carbon (McLeod et al. 2011). According to (Goldberg et al. 2020), around 62% of mangroves worldwide were destroyed between 2000 and 2016. Degradation of these coastal ecosystems becomes a reason for the emission of carbon back into the atmosphere, which is an alarming reason for global warming. Estimates suggest that damaged coastal habitats emit up to 1.02 billion tons of CO_2 each year, which is comparable to 19% of world emissions from tropical deforestation (International 2018). Similar to terrestrial ecosystems, it is difficult to create any yearly budget for blue carbon, but mangroves are the greatest at limiting it since they worldwide absorb 700 Tg C year^{-1} through gross primary production and release 525 Tg C year^{-1} (or 75% of it) back into the atmosphere as CO_2. The major supply of bio-carbon to mangrove benthos has historically been mangrove plants, and their production was designed to power the surrounding faunal populations, as well as nearby aquatic foodstuffs. In the absence of this unforeseen blue carbon, carbon emissions would have accelerated. Therefore, these ecosystems should be conserved and protected from degradation as their degradation reemits captured carbon in the atmosphere (Duarte et al. 2013).

5.4 Techniques to Sequester Carbon from the Atmosphere

Since a large amount of carbon is captured by oceans from the atmosphere, the concentration of carbon doesn't go much higher. As mentioned earlier, carbon can be sequestered using diverse natural and engineered methods, achieved either through organic matter addition or minimizing the loss of trapped organic carbon. (Laban et al. 2018). Techniques that are helpful in sequestering carbon include tillage, crop rotation, afforestation (Wang and Huang 2020), agroforestry (Lorenz and Lal 2014; Dhanwantri et al. 2014), horticulture, geological sequestration (Lal 2008), wetlands restoration. The agroforestry system was also introduced during the REDD+ program (Lorenz and Lal 2014). According to the IPCC's Second Assessment Report, between 60 and 87 Gt C might be saved or stored in forests by the year 2050, while between 23 and 44 Gt C could be stored in agricultural soils (IPCC-SAR 1995). Grasslands also contribute significantly to sequestering carbon by one-third of the total terrestrial carbon pool, which is about 412–820 million tons (Zhao et al. 2020). However, improved agriculture practices have reduced carbon emissions from agricultural lands to a certain extent only; hence, rigorous research is going to find different amendments to add more carbon to the soil.

Organic amendments are recyclable products that can enrich soil carbon and improve soil quality. Production of biodiesel hydrogen cells and other products has revolutionized the concept of carbon sequestration. The world's largest carbon-capturing plant in Iceland started in 2021, is a wonderful example of the engineered technique of carbon sequestration. It is capable of capturing 4400 t of CO_2 year^{-1} from air. Amending soils through bio amendments creates a synergistic effect in sequestering carbon. Amendments like biochar can act as recalcitrant material, thus helping sequester more carbon in soil (Smebye et al. 2016). Conant et al. (2001) studied the rates of carbon sequestration in grasslands through various improvements, which showed that carbon sequestration was highly influenced by the type of improvement, which ranged from 0.11 to 3.04 Mg C·ha^{-1} year^{-1}, with an average of 0.54 Mg C·ha^{-1} year^{-1}. Table 5.2 shows a consolidation of a few studies done to sequester carbon from atmosphere in soil through various techniques.

Studies (Wei et al. 2016) are done to assess the status of SOM after the application of organic amendments combined with fertilizers. The use of some beneficial microbes like cyanobacteria can also sequester carbon, according to studies (Zhou et al. 2017). Furthermore, there are methods like the addition of some ground silicate minerals in soils that may enhance weathering, thereby enhancing carbon storage in soil, but it is nearly difficult to strike a balance between the quantities of carbon released into the atmosphere and the amount of carbon sequestered in natural resources.

Table 5.2 Different methods or approaches to enhance carbon sequestration in soil

Sr. No.	Techniques	Dosage	Target	Amount of carbon sequestered in soil	Reference
1	Enhanced agronomy	–	–	1.6 Pg $year^{-1}$	Smith et al. (2008)
2	Croplands	–	–	0.90–1.85 Pg $year^{-1}$	Zomer et al. (2017)
3	Organic fertilizer, rice straw, and rice straw biochar	3.5 mg ha^{-1}, 9.0 mg ha^{-1}, 8.0 mg/ha^{-1}	Topsoil soil (0–20 cm) aquic Inseptisol as FAO	34, 17, and 57.4% increase in the top layer	Wu et al. (2021)
4	Pig manure	22.5 mg ha^{-1} $year^{-1.}$	Rice barley	11% increase	Wang et al. (2015)
5	Green manure	–	Agriculture soils under various climatic conditions	Average 35.4% increase	Gross and Glaser (2021)
6	Cow dung	2 t C ha^{-1} $season^{-1}$	Transplanted Aman rice	22%	Rahman et al. (2022)
7	Vermicompost		–	29%	
8	Rice straw		–	20%	
9	Poultry manure		–	26%	
10	Seaweed aquaculture (*Eucheuma denticulatum*)	–	–	16.08–68.43 ton C ha^{-1} $year^{-1}$	Erlania and Nyoman Radiarta (2015)

5.5 Conclusions

Carbon is widely regarded as the most diverse element due to its high level of dynamism in properties. It has been thoroughly researched and revealed to be extremely important in the life of any living organism. Elevation in the Earth's temperature is the result of the CO_2 emissions that have skyrocketed through anthropogenic interventions of increasing transportation and industrial expansion. Inadequate agricultural services and shifting land use patterns have also contributed significantly to growing CO_2 emissions. It is critical to lower carbon concentrations in the atmosphere by removing it through natural reservoirs and also by adding suitable amendments. Since carbon is a component that is involved in maintaining ecological processes both directly and indirectly, carbon sequestration is crucial to achieving sustainable development goals and building a brighter future for all living things.

References

Aryal DR, Ruiz DEM, Marroquín CNT, Ruiz RP, Hernández FG, Venegas JAV, Mendoza AP, López GV, Lugo FC, Larramendi LAR, de Coss AL, López AH, Jonapá FJM, Sanabria CAV, Aguilar AA, Chi IE (2018) Soil organic carbon depletion from forests to grasslands conversion in Mexico: a review. Agric 8:1–15. https://doi.org/10.3390/agriculture8110181

Banerjee K, Mitra A, Villasante S (2021) Carbon cycling in mangrove ecosystem of western bay of Bengal (India). Sustain For 13. https://doi.org/10.3390/su13126740

Bationo A, Vanlauwe B, Kihara J, Kimetu J (2005) Soil organic carbon dynamics, functions and Management in West African Agro-Ecosystems. Agric Syst 94(1):13–25

Bell M, Lawrence D (2009) Soil carbon sequestration—myths and mysteries. Trop Grasslands 43:227–231

Conant RT, Paustian K, Elliott ET (2001) Grassland management and conversion into grassland: effects on soil carbon. Ecol Appl 11(2):343–355

Costanza R, D'Arge R, de Groot R, Farber S, Grasso M, Hannon B, Limburg K, Naeem S, O'Neil (1997) The value of the world's ecosystem services and natural capital. Nature 387:253–260

Department of the Environment WH and A (2009) Ecosystem services: key concepts and applications

Dhanwantri K, Sharma P, Mehta S, Prakash P (2014) Carbon sequestration, its methods and significance. Environ Sustain 151:151–157

Division F (2015) Estimating status of soil organic carbon in tropical forests of Narmada estimating status of soil organic carbon in tropical forests of Narmada Forest Division, Gujarat, pp 10–15

Domain BC, Regulator P, Change C, Rusa U, Mitra A (2020) Blue carbon potential Regulator of climate Change "

Duarte CM, Losada IJ, Hendriks IE, Mazarrasa I, Marbà N (2013) The role of coastal plant communities for climate change mitigation and adaptation. Nat Clim Chang 3:961–968. https://doi.org/10.1038/nclimate1970

Edrisi SA, Tripathi V, Abhilash PC (2018) Towards the sustainable restoration of marginal and degraded lands in India. Trop Ecol 59:397–416

Edrisi SA, Tripathi V, Dubey PK, Abhilash PC (2022) Carbon sequestration and harnessing biomaterials from terrestrial plantations for mitigating climate change impacts. In: Biomass, biofuels, biochemicals. Elsevier, pp 299–313

Erlania NR (2015) The use of seaweeds aquaculture for carbon sequestration: a strategy for climate Change mitigation. J Geod Geomatics Eng 2. https://doi.org/10.17265/2332-8223/2015.06.006

Forest Survey of India (FSI) (2017) Carbon stock in India's Forest. Isfr:121–136

Goldberg L, Lagomasino D, Thomas N, Fatoyinbo T (2020) Global declines in human-driven mangrove loss. Glob Chang Biol 26:5844–5855

Green C, Byrne K (2004) Biomass: impact on carbon cycle and greenhouse gas emissions. Encycl Energy 1:223–236. https://doi.org/10.1016/B0-12-176480-X/00418-6

Gross A, Glaser B (2021) Meta-analysis on how manure application changes soil organic carbon storage. Sci Rep 11:1–13. https://doi.org/10.1038/s41598-021-82739-7

International C (2018) The blue carbon initiative 15

IPCC 2001: Climate change (2001) In: Houghton JT, Ding Y, Griggs DJ, Noguer M, Linden PJ, Van der Dai X, Maskell K, Johnson CA (eds) Contribution of working group I to the third assessment report of the intergovernmental panel on climate change. Cambridge Univ Press, p 94

IPCC-SAR (1995) Climate Change 1995: a report of the intergovernmental panel on climate change. Environ Sci Technol 48:4596–4603

Jones MB, Zimmermann J, Clifton-Brown J (2016) Long-term yields and soil carbon sequestration from Miscanthus: a review. Perenn biomass Crop a Resour world:43–49

Kadlec V, Holubík O, Procházková E, Urbanová J, Tippl M (2012) Soil organic carbon dynamics and its influence on the soil erodibility factor. Soil Water Res 7:97–108. https://doi.org/10.17221/3/2012-swr

Laban P, Metternicht G, Davies J (2018) Soil biodiversity and soil organic carbon: keeping drylands alive

Lal R (2008) Sequestration of atmospheric CO2 in global carbon pools. Energy Environ Sci 1:86–100. https://doi.org/10.1039/b809492f

Lorenz K, Lal R (2014) Soil organic carbon sequestration in agroforestry systems. A review Agron Sustain Dev 34:443–454. https://doi.org/10.1007/s13593-014-0212-y

Lovelock CE, Duarte CM (2019) Dimensions of blue carbon and emerging perspectives. Biol Lett 15:1–5. https://doi.org/10.1098/rsbl.2018.0781

Luo Z (2017) Soil organic carbon dynamics jointly controlled by climate, carbon inputs, soil properties and soil carbon fractions. https://doi.org/10.1111/ijlh.12426

McLeod E, Chmura GL, Bouillon S, Salm R, Björk M, Duarte CM, Lovelock CE, Schlesinger WH, Silliman BR (2011) A blueprint for blue carbon: toward an improved understanding of the role of vegetated coastal habitats in sequestering CO2. Front Ecol Environ 9:552–560. https://doi.org/10.1890/110004

Nellemann C, Corcoran E, Duarte CM, Valdés L, De Young C, Fonseca L, Grimsditch G (2009) Blue carbon: a rapid response assessment

Nelson DW, Sommers LE (2018) Total carbon, organic carbon, and organic matter. Methods Soil Anal Part 3 Chem Methods, pp 961–1010. https://doi.org/10.2136/sssabookser5.3.c34

Padbhushan R, Sharma S, Rana DS, Kumar U, Kohli A, Kumar R (2020) Delineate soil characteristics and carbon pools in grassland compared to native forestland of India: a meta-analysis. Agronomy 10. https://doi.org/10.3390/agronomy10121969

Palmer J, Thorburn PJ, Biggs JS, Dominati EJ, Probert ME, Meier EA, Huth NI, Dodd M, Snow V, Larsen JR, Parton W (2017) Nitrogen cycling from increased soil organic carbon contributes both positively and negatively to ecosystem services in wheat agro-ecosystems. Front Plant Sci 8. https://doi.org/10.3389/fpls.2017.00731

Paustian K, Larson E, Kent J, Marx E, Swan A (2019) Soil C sequestration as a biological negative emission strategy. Front Clim 1:1–11. https://doi.org/10.3389/fclim.2019.00008

Rahman MM, Kamal MZU, Ranamukhaarachchi S, Alam MS, Alam MK, Khan MAR, Islam MM, Alam MA, Jiban SI, Al Mamun MA, Abdullah HM, Biswas JC, Akhter S, Naher UA, Maniruzzaman M, Haque MM, Ahmed F (2022) Effects of organic amendments on soil aggregate stability, carbon sequestration, and energy use efficiency in wetland paddy cultivation. Sustain For 14. https://doi.org/10.3390/su14084475

Ramesh T, Bolan NS, Kirkham MB, Wijesekara H, Kanchikerimath M, Srinivasa Rao C, Sandeep S, Rinklebe J, Ok YS, Choudhury BU, Wang H, Tang C, Wang X, Song Z, Freeman OW II (2019a) In: Sparks DL (ed) Chapter one—soil organic carbon dynamics: impact of land use changes and management practices: a review. Academic Press, pp 1–107

Ramesh T, Bolan NS, Kirkham MB, Wijesekara H, Kanchikerimath M, Srinivasa Rao C, Sandeep S, Rinklebe J, Ok YS, Choudhury BU, Wang H, Tang C, Wang X, Song Z, Freeman OW (2019b) Soil organic carbon dynamics: impact of land use changes and management practices: a review. Adv Agron 156:1–107. https://doi.org/10.1016/bs.agron.2019.02.001

Schlesinger WH (1977) Carbon balance in terrestrial detritus. Annu Rev Ecol Syst 8:51–81. https://doi.org/10.1146/annurev.es.08.110177.000411

Schrag D (2007) Preparing to capture carbon. Science 315:812–813. https://doi.org/10.1126/science.1137632

Singh SK, Singh AK, Sharma BK, Tarafdar JC (2007) Carbon stock and organic carbon dynamics in soils of Rajasthan, India. J Arid Environ 68:408–421. https://doi.org/10.1016/j.jaridenv.2006.06.005

Smebye A, Alling V, Vogt RD, Gadmar TC, Mulder J, Cornelissen G, Hale SE (2016) Biochar amendment to soil changes dissolved organic matter content and composition. Chemosphere 142:100–105. https://doi.org/10.1016/j.chemosphere.2015.04.087

Smith P, Martino D, Cai Z, Gwary D, Janzen H, Kumar P, McCarl B, Ogle S, O'Mara F, Rice C, Scholes B, Sirotenko O, Howden M, McAllister T, Pan G, Romanenkov V, Schneider U, Towprayoon S, Wattenbach M, Smith J (2008) Greenhouse gas mitigation in agriculture. Philos Trans R Soc London Ser B, Biol Sci 363:789–813. https://doi.org/10.1098/rstb.2007.2184

Stein T (2022) Carbon dioxide now more than 50 higher than pre-industrial levels. Jun 3(2022):1–6

Stockmann U, Adams MA, Crawford JW, Field DJ, Henakaarchchi N, Jenkins M, Minasny B, McBratney AB, de Remy, de Courcelles V, Singh K, Wheeler I, Abbott L, Angers DA, Baldock J, Bird M, Brookes PC, Chenu C, Jastrow JD, Lal R, Lehmann J, O AG, Parton WJ, Whitehead D, Zimmermann M (2013) Author's personal copy the knowns, known unknowns and unknowns of sequestration of soil organic carbon Author's personal copy. Ecosyst Environ 164:80–99

Tessema B, Sommer R, Piikki K, Söderström M, Namirembe S, Notenbaert A, Tamene L, Nyawira S, Paul B (2020) Potential for soil organic carbon sequestration in grasslands in East African countries: a review. Grassl Sci 66:135–144. https://doi.org/10.1111/grs.12267

Wang S, Huang Y (2020) Determinants of soil organic carbon sequestration and its contribution to ecosystem carbon sinks of planted forests. Glob Chang Biol 26:3163–3173. https://doi.org/10.1111/gcb.15036

Wang Y, Hu N, Xu M, Li Z, Lou Y, Chen Y, Wu C, Wang ZL (2015) 23-year manure and fertilizer application increases soil organic carbon sequestration of a rice–barley cropping system. Biol Fertil Soils 51:583–591. https://doi.org/10.1007/s00374-015-1007-2

Wei W, Yan Y, Cao J, Christie P, Zhang F, Fan M (2016) Effects of combined application of organic amendments and fertilizers on crop yield and soil organic matter: an integrated analysis of long-term experiments. Agric Ecosyst Environ 225:86–92. https://doi.org/10.1016/j.agee.2016.04.004

Westerling AL, Hidalgo HG, Cayan DR, Swetnam TW (2006) Warming and earlier spring increase Western U.S. forest wildfire activity. Science(80-) 313:940–943. https://doi.org/10.1126/science.1128834

Wong VNL, Greene RSB, Dalal RC, Murphy BW (2010) Soil carbon dynamics in saline and sodic soils: a review. Soil Use Manag 26:2–11. https://doi.org/10.1111/j.1475-2743.2009.00251.x

Wu L, Zhang S, Ma R, Chen M, Wei W, Ding X (2021) Carbon sequestration under different organic amendments in saline-alkaline soils. Catena 196:104882. https://doi.org/10.1016/j.catena.2020.104882

Xu C, Yang, Du C, Shi JJ, Hou L, Kang WZ, Wang Q, Chao GZ (2021) The interplay of labile organic carbon, enzyme activities and microbial communities of two forest soils across seasons. Sci Rep 11:1–12. https://doi.org/10.1038/s41598-021-84217-6

Zhang L, Chen X, Xu Y, Jin M, Ye X, Gao H, Chu W, Mao J, Thompson ML (2020) Soil labile organic carbon fractions and soil enzyme activities after 10 years of continuous fertilization and wheat residue incorporation. Sci Rep 10:1–10. https://doi.org/10.1038/s41598-020-68163-3

Zhao Y, Liu Z, Wu J (2020) Grassland ecosystem services: a systematic review of research advances and future directions. Landsc Ecol 35:793–814. https://doi.org/10.1007/s10980-020-00980-3

Zhou W, Wang J, Chen P, Ji C, Kang Q, Lu B, Li K, Liu J, Ruan R (2017) Bio-mitigation of carbon dioxide using microalgal systems: advances and perspectives. Renew Sust Energ Rev 76:1163–1175. https://doi.org/10.1016/j.rser.2017.03.065

Zomer RJ, Bossio DA, Sommer R, Verchot LV (2017) Global sequestration potential of increased organic carbon in cropland soils. Sci Rep 7:1–8. https://doi.org/10.1038/s41598-017-15794-8

Chapter 6
Modeling Plant Growth, Nutrition, and Dynamics of Soil Organic Carbon Under Changing Climate and Land Use

Raghavendra Nargund, Hanamant Halli, Devideen Yadav, Amresh Chaudhary, and Sudhir K. Rajpoot

Abstract SOC and nutrients are important natural resources that play a vital role in soil fertility management and ecosystem stabilization. Improvement in SOC and soil nutrients with optimum availability to plants further enhances the growth and productivity of different crops and cropping systems under changing climate and land use systems as a mitigating option. Cropping systems and other SOC models have evolved over the last four decades in response to the demand for modeling of SOC and soil nutrient dynamics for their quantification in different crops and cropping systems under varied management practices. Crop growth simulation or cropping system models are computer-based software programs designed to mimic crop growth, nutrients, and organic carbon dynamic process in the soil to compute the relationship among soil nutrients, organic carbon, plant growth, and soil water content in the soil-plant-atmosphere system under spatial and temporal variability. Cropping system models, DSSAT (Decision Support System for Agro Technology Transfer) and APSIM (Agricultural Production System Simulator) are the important process-based models used for the quantification of long-term change in SOC, plant nutrients, and their effect on the growth and development of crop plants across the world. Therefore, the present chapter attempts to briefly elaborate upon the impact of climate change on plant nutrient dynamics, SOC, and application, and evaluation

R. Nargund (✉)
ICAR-Indian Institute of Soybean Research, Indore, Madhya Pradesh, India
e-mail: raghavendra4449@gmail.com

H. Halli
ICAR-Indian Grassland and Fodder Research Institute, Jhansi, Uttar Pradesh, India

D. Yadav
ICAR-Indian Institute of Soil and Water Conservation, Dehradun, Uttarakhand, India

A. Chaudhary
ICAR-National Institute of Abiotic Stress Management, Baramati, Maharashtra, India

S. K. Rajpoot
Institute of Agricultural Sciences, Banaras Hindu University, Varanasi, Uttar Pradesh, India

S. A. Edrisi et al. (eds.), *Sustainable Plant Nutrition and Soil Carbon Sequestration*, Sustainable Plant Nutrition in a Changing World,
https://doi.org/10.1007/978-3-031-53590-1_6

of different models available for SOC and soil nutrient dynamics under different crops and cropping systems with future prospectus of carbon and nutrients model.

Keywords Climate change · Land use change system · Nutrient dynamics · Simulation models · SOC pools

6.1 Introduction

Global climate change is resulting in major changes to the earth's geological, ecological, and biological systems and posing serious threats to the existence of human civilization and the sustenance of agricultural productivity and food security (Ramesh et al. 2019). Anthropogenic and agricultural activities, along with forestry and land use changes, have resulted in increasing GHGs emissions, such as CO_2, CH_4, and N_2O on the earth's surface and are one of the important growing concerns of this century. Therefore, many countries in the world have started developing various policies, technologies, and methods to mitigate GHGs (Ross et al. 2016). Among the various mitigation solutions developed, organic carbon storage in the soil as well as in the terrestrial biosphere through proper land use and management practices are important strategic options (Zhang and Ni 2017). Soil is an important sink for carbon stock, with global carbon storage of 1550 Pg (Batjes 1996), and accounts for 2–3 times more carbon than atmosphere and vegetation (Scharlemann et al. 2014). Hence, little change in terrestrial carbon pools has a significant impact on climate change (Zhang et al. 2016). SOM is a vital indicator of soil quality; therefore, maintaining SOM with quality and quantity is essential for sustaining long-term soil fertility (Zhao et al. 2015).

There are several pools and fractions of SOM, such as total organic carbon, particulate and mineral-associated organic carbon, dissolved organic carbon, extractable organic carbon, and microbial biomass carbon, which may be useful in the study of short-term and long-term influences of land use and management on SOC dynamics. Soil nutrients are the chemical elements vital to biological functions, i.e., for the survival and reproduction of any living organisms in soil (DeAngelis et al. 1989). Further, nutrient dynamics is broadly defined as the way nutrients are taken up, retained, transferred, and cycled over time and distance in an ecosystem (Harrison 2016). Therefore, soil nutrients and organic carbon are interlinked to each other. It is well documented that the capacity of soils to sequester carbon is strongly influenced by nitrogen and phosphorus availability because of stoichiometric links between carbon, nitrogen, and phosphorus biochemical cycles. Human disturbances (Deposition, fertilization, and nutrient mining) continuously cause large imbalances between the biogeochemical cycles and changing nutrient availabilities identified as key uncertainties in predicting future ecosystem carbon sequestration (Macdonald et al. 2018). Land use change (LUC) is a major driving feature for the balance of SOC stocks and the global carbon cycle (Poeplau et al. 2011).

Furthermore, land use and land use change are human activities with a major impact on the balance between the input and outputs of SOC and atmospheric CO_2 concentration (Houghton 2003). A better understanding of LUC on the world's terrestrial carbon balance is thus a necessary part of the global effort to mitigate climate change (The Terrestrial Carbon Group 2010). Therefore, it is necessary to understand the mechanisms that drive the interactions between carbon and nutrient cycles. This understanding will enable the long-term predictions of changes in current and future organic carbon and nutrients in the soil through crop-growth simulation models under varied management practices. Therefore, it could pave the way to enhance the productivity of different crops and cropping systems by maintaining soil quality under changing climate and land use change in the current and future scenarios.

Crop simulation models (CSM) are computer programs that mimic or simulate the growth, development, and yield of crops using various inputs such as soil water, carbon and nitrogen processes, and agronomic management practices as a function of the soil-plant-atmosphere (Radha and Murthy 2004; Sachin et al. 2019). Thus, a crop-growth simulation model not only predicts the final state of crop production or harvestable yield but also contains quantitative information about major processes involved in the growth and development of the crop. To quantify organic carbon and nutrients in the soil and their effects on productivity, the process-based models serve as a robust tool to keep track of multiple plant and soil factors and their interactions affecting soil carbon and nutrient dynamics (Ghaley et al. 2018).

Several crop-growth simulation models are employed to predict changes in SOC dynamics and soil nutrients, especially nitrogen, across the globe. For example, the DSSAT-CENTURY model, APSIM crop and pasture module, WHCNS model, ROTHC, DNDC, EPIC, DAYCENT, DAISY, LPJ, INCA, CASA, ECOSYST, etc. are the models that simulate SOC and nutrients dynamics. Among them, the CENTURY model in DSSAT has been most widely used to predict SOC dynamics in soil (Campbell and Paustian 2015). DSSAT, APSIM, QUEFTS, and WHCNS models gave better results for the estimation of soil nutrient dynamics (Pathak et al. 2003; Probert and Dimes 2004; Liang et al. 2016; Hoogenboom et al. 2019; Yan et al. 2020). Therefore, this chapter has tried to link the impact of climate change on nutrients, organic carbon, and nutrient dynamics in the soil through different models.

6.2 Potential Impact of Climate Change on Plant Nutrients and Soil Organic Carbon

SOC is very important in the global carbon cycle since it constitutes the largest terrestrial reservoir of this element (Chabra et al. 2003). Soils store at least three times more carbon (in the form of SOC) than that found in the atmosphere or living plants (Schmidt et al. 2011). This reservoir of SOC is very sensitive to climate change, and how climate change is likely to impact SOC has always been a matter of considerable debate. On the one hand, it is recognized that global warming and increasing

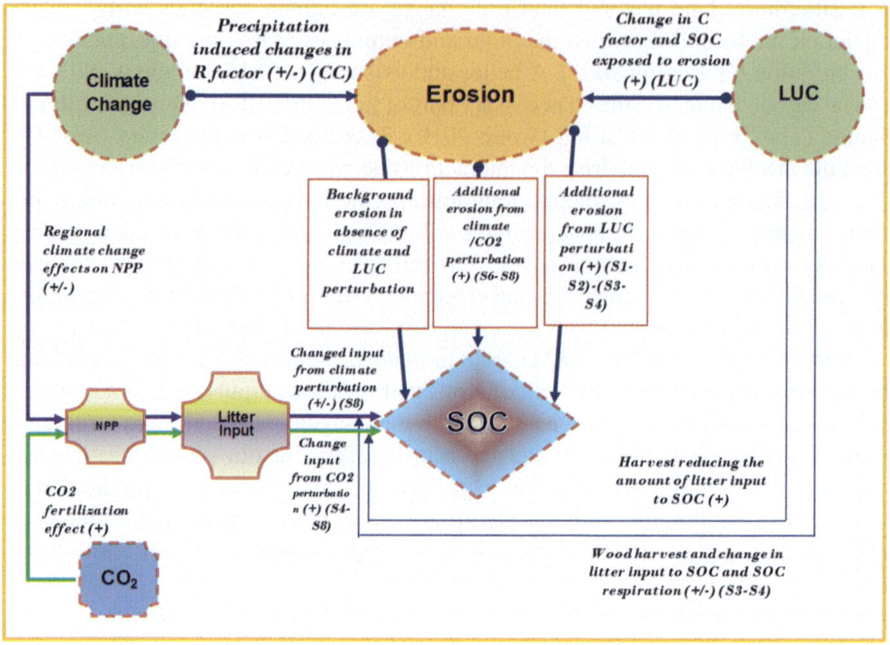

Fig. 6.1 Conceptual model on the impact of climate change and land use change on soil organic carbon. (Modified from Chabra et al. 2003)

CO_2 levels in the atmosphere can favor increased plant growth, which in turn could provide more organic matter for the soil. On the other hand, a rise in air temperature will increase the decomposition and loss of SOM. There is thus significant interest in the fate of such carbon, particularly the extent to which soils and land use can be used to regulate the sequestration of carbon from the atmosphere or the loss of SOC to the atmosphere (Fig. 6.1).

6.2.1 *Effect of Climate Change on Soil Organic Matter*

6.2.1.1 Addition of Organic Matter in the Soil

Increases in atmospheric CO_2 concentration and temperature are expected to increase net primary production (i.e., plant biomass) in areas where nutrients and water are not limiting because photosynthesis is limited by atmospheric CO_2 concentration. Climate change is expected to alter precipitation patterns, resulting in potential changes in soil moisture balance. Increases in precipitation in currently water-limited areas are expected to increase primary production. This could result in a net increase of carbon inputs to the soil, though changes in allocation (above-versus belowground) and species composition are both likely to affect the amount, location, and quality of carbon inputs.

6.2.1.2 Soil Organic Matter Decomposition

SOM is highly susceptible to land use change, soil temperature, and moisture. As a complex series of enzymatically controlled biochemical reactions, the rate of SOM decomposition is highly regulated by temperature (among other variables). Warmer temperature generally results in accelerated decomposition when other factors (i.e., soil moisture, oxygen concentration) are not limiting. The increase in decomposition with temperature is regulated by the rate at which substrate becomes available in the environment, the rate at which substrates are depolymerized by extracellular enzymes, and the rate at which substrates diffuse and are taken up by microbes. One of the mechanisms by which the temperature sensitivity of SOM decomposition is expected to change with time is through the depletion of readily mineralized components, which is frequently described as a change in overall SOM quality or stability. Kinetic theory suggests that decomposition rates of biochemically complex SOM components that decompose slowly should be more sensitive to temperature (Davidson et al. 2006).

6.2.1.3 Soil Organic Matter Stabilization

Like decomposition, protection of SOM by occlusion within aggregates, by association with mineral surfaces are chemical processes, and should thus be subject to kinetic theory. Isolating the effects of temperature on stabilization or destabilization independent of SOM decomposition has proven difficult. Data on chemical stabilization suggests a complex relationship in which warmer temperatures induce a net SOM release from mineral surfaces. However, warmer temperatures may simultaneously enhance the creation of slower reactions in which SOM becomes more tightly bound to mineral surfaces. Thus, warmer temperatures could lead to a net release of SOM (and subsequent CO_2 efflux) while also resulting in the creation of tighter mineral-SOM bonds for the remaining pool of chemically protected SOM.

6.2.2 Climate Change Effects on Organic Carbon and Nutrient Dynamics in Soil

The potential changes in the soil-forming factors directly resulting from global climate change would be in the organic matter supply, temperature regimes, hydrology, and changes in the potential evapotranspiration. Both the organic matter and carbon to nitrogen (C: N) ratio will diminish in a warmer soil temperature regime. Drier soil conditions will suppress both root growth and decomposition of organic matter and will increase vulnerability to erosion. Increased evaporation from the soil and accelerated transpiration from the plants themselves will cause soil- moisture stress. Climate change is likely to have a variety of impacts on soil quality and nutrient dynamics. The amount, intensity, timing, and kind of precipitation influence soil formation. Seasonal and daily changes in temperature affect moisture effectiveness, biological activity, rates of chemical reactions, and kinds of vegetation. A changing climate could also influence the workability of mineral soils and susceptibility to erosion, compaction, and reduced water-holding capacity. In areas where winter rainfall becomes heavier, some soils may become more susceptible to erosion. Other changes include the washing away of organic matter and leaching of nutrients. Not only does climate influence soil properties, but it also regulates climate via the uptake and release of GHGs such as CO_2, methane, and nitrous oxide. Soil can act as a source and sink for carbon, depending on land use and climatic conditions.

Many soil physical properties form the foundation of other chemical and biological processes, which may be further governed by climate, landscape position, and land use. Nutrient cycling, especially N, is intimately linked with SOC cycling, and hence drivers of climate change such as elevated temperatures, variable precipitation, and atmospheric N deposition are likely to impact N cycling and possibly the cycling of other plant available nutrients such as phosphorus and sulfur. In soil, the availability of nutrients to the crops relies on the presence of cation exchange capacity (CEC), an inherent soil property that may be strongly affected by elevated CO_2. An increase in CEC under elevated CO_2 can result from an increase in organic matter content, roots, and mycorrhizal biomass and exudation and soil aggregation status. Crops and trees growing in elevated CO_2 atmosphere might, therefore be able to (indirectly) affect the soil's capacity for nutrient provision and thus, regulate nutrient availability in the long term. Elements with biologically controlled cycles (such as C and N) might show different reactions to changing temperature than elements with cycles controlled both by biological and geological processes (such as P, S, and K) or elements with predominantly geologically controlled cycles (such as K, Ca, Mg, S or micronutrients, as shown in Fig. 6.2.

Biological transformation from organic to inorganic pools is largely influenced by soil moisture and temperature, and thus, global climate change may strongly influence soil solution concentrations of nitrogen as well as Sulphur. Some scientists have speculated that soil carbon pool will not change as increased soil respiration and decomposition caused by soil warming will be moderated by the increased carbon supply below ground (Kirschbaum 2000). Pendall et al. (2004) suggest that increased CO_2 does not have any direct effect on N mineralization but that associated warming due to climate change can cause increased N mineralization, leading to increased concentration of N in soil solutions. The adsorption/desorption reactions will also accelerate with increasing temperature, and changes in soil moisture may further modify reactions by altering the ionic strength of the soil solution.

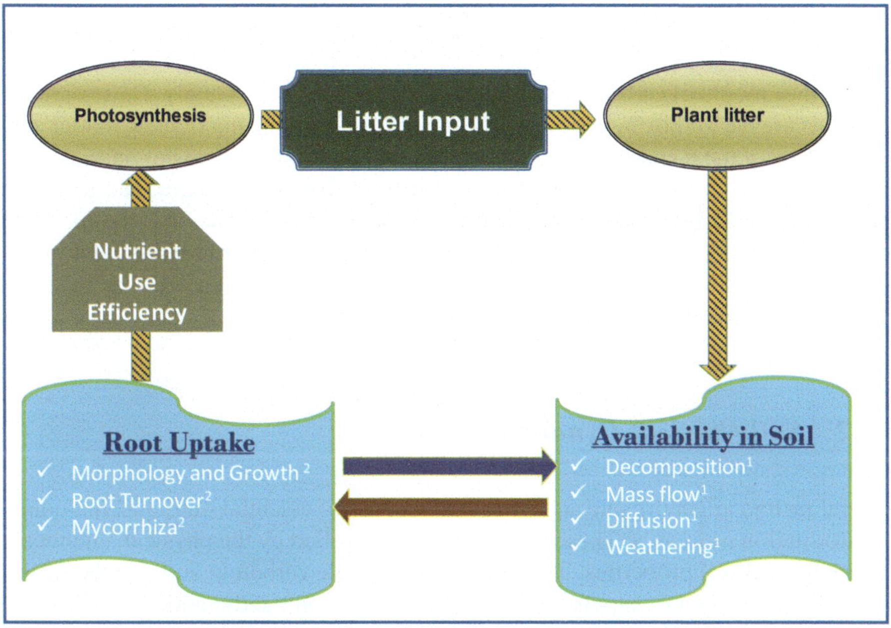

Fig. 6.2 Process of ecosystem nutrient cycle, directly[1] and indirectly[2] affected by climate change. (Adopted and modified from Wood et al. (2006), Watanabe et al. (2007), Lukac et al. (2010))

6.3 Dynamics of Plant Nutrients and Organic Carbon in Soil

SOM comprises both living and dead organic components in soil. It includes an infinite number of organic compounds that range from simple, easily mineralized organic residues to complex, unconventional products and microbial biomass. The fraction of carbon stored in this organic matter represents the total organic carbon, and SOM contains 58% carbon. SOM differs in size and solubility and is broadly grouped into three main active, slow, and passive carbon complexes (Tan et al. 2007). Active pools represent modifiable forms of carbon that are highly sensitive to change, with an average residence time of about 1–5 years. This carbon group has a pivotal role in feeding the soil food web and affects a variety of soil functions and processes, including nutrient cycling to maintaining soil quality (Majumder et al. 2008). Slow pools of SOC have an average residence time of 20 to 40 years, while the average residence time for passive pools SOC ranges from 200 to 1500 years. The stable carbon fractions have a high resistance to microbial activity and thus hardly serve as a reliable indicator of soil quality (Majumder et al. 2008) but contribute to carbon fixation.

6.3.1 *Total Organic Carbon (TOC)*

TOC is the amount of carbon associated with soil organic compounds. These organic materials are derived from internal sources such as decomposing organic matter and by-products of the metabolic activities of microbial or living organisms and external sources such as fertilizers, compost, bio-solids, fertilizers, pesticides, etc. Physical fractionation classified SOM is divided into three broad groups: coarse organic matter, containing organic fragments <250 μm; fine organic matter comprising organic materials from 53 to 250 μm; and mineral-bound organic matter <53 μm in size). The other important SOC fractions are given in Table 6.1.

6.3.2 *Soil Nutrient Dynamics*

Several factors affect organic and inorganic sources of essential nutrients, including transformation processes. These processes are controlled by the physical, chemical, and biological soil properties. The quantity of organic carbon in soil and the quantity and type of organic inputs have profound effects on the dynamics of nutrients. Most of the crop models are successful in predicting major nutrients (N, P, and K) under different agroecosystems. Therefore, major nutrient dynamics in soil are discussed in detail.

6.3.2.1 Soil Nitrogen (N) Dynamics

Soil inorganic N is derived from fertilizer N, and soil organic N mineralization is the main form of plant-available N. Four processes were found to influence the inorganic N pathways in soils, including biotic immobilization-re-mineralization, abiotic immobilization, soil organic N mineralization, and plant residue organic N mineralization (Fig. 6.3). The microbial immobilized inorganic N supports microbial proliferation and can be reused in suitable environments. This process is referred to as re-mineralization, which mainly results from microbial death.

6.3.2.1.1 Forms of N in Soil

There are three major forms of N in soil: organic N (Org-N), ammonium N (NH_4^+ -N), and nitrate N ($NO_3 - N$). Plants cannot use the N in the organic form. Plants can only use ammonium and nitrate forms of N. **Mineralization** occurs when organic N is broken down to form ammonium N, which is available for plant use. **Nitrification** occurs as ammonium is further changed to the nitrate form by microorganisms, which is also available to plants.

Table 6.1 Soil organic carbon fractions and their important features

S. No.	Fractions of SOC	Features	References
1.	**Particulate organic matter carbon (POM-C)**	POM-C is the material that passes through a 2 mm sieve but is kept on a 53 um sieve, minus the sand. Important for aggregate stability, nutrient cycling, and water infiltration. Abundant in topsoil compared to lower strata. Prone to change management practices.	Cambardella et al. (2001)
2.	**Mineral organic carbon (MOC)**	MOC includes the carbon fractions of the complex SOM that have been both physically and chemically stabilized. Negative carbon sinks with relatively longer spin times are represented.	Yadav et al. (2021)
3.	**Dissolved organic carbon (DOC)**	DOC originates from a multitude of sources, such as plant litter, root exudates, soil humus, or microbial biomass. It is organic molecules of different sizes that pass through a 0.45 μm filter. DOC has been further divided into labile, semi-labile, and non-labile parts. **Labile carbon**: Called extractable organic carbon. The primary energy source that can be readily degradable or consumed quickly by soil microorganisms. It is also identified as a short-lived carbon pool. Simple sugars (i.e., glucose, fructose) and protein degradation products (i.e., amino acids) are labile carbon compounds. **Semi-labile carbon:** The breakdown of intermediate products of cellulose or hemicellulose is an example of a **semi-labile** organic compound. This fraction can be decomposed into the labile carbon fraction with time. **Non-labile carbon:** It is one of the most highly molecular-weight humic substances. This fraction is resistant to microbial decomposition and can persist in the soil environment for several years.	Marschner and Kalbitz (2003), Marschner et al. (2008)
4.	**Microbial biomass carbon (MBC)**	It represents the living SOC fraction, which is considered an estimate of biological activity in the soil. MBC has been proposed as an index of soil stress and disturbance, and its measurement is often essential for soil ecological studies.	Hanson et al. (2000)

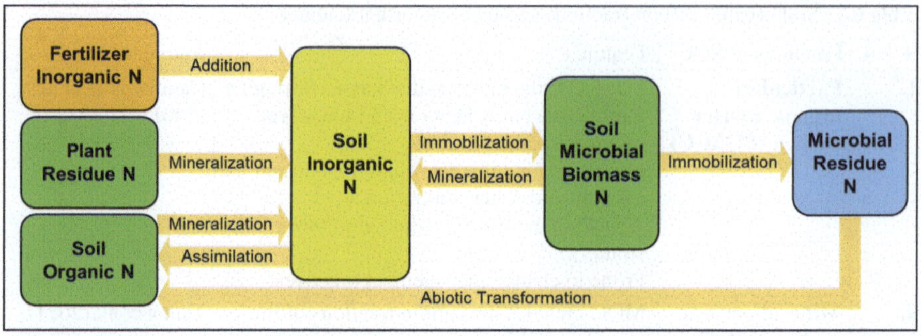

Fig. 6.3 Nitrogen (N) dynamics in soil. (Adapted and modified from Abbadie and Lata (2006), Luce et al. (2011))

(a) *Ammonium N* ($NH_4^+ - N$)

It is the first plant usable form released when microbes break down organic N. This process, called mineralization, occurs at a faster rate in the late spring, summer, and early fall than in colder months because warm, moist soils promote microbial activity.

(b) *Nitrate-N* ($NO_3 - N$)

Nitrification is the process by which a specialized group of soil microorganisms changes ammonium N to nitrate N. Like ammonium N, plants can use nitrate N. Unlike ammonium, nitrate has a negative charge, which causes it to be repelled by negatively charged soil particles.

6.3.2.2 Phosphorus (P) Dynamics in Soil

Soil P exists in various chemical forms, including inorganic P (Pi) and organic P (Po). (Turner et al. 2007). Pi usually accounts for 35% to 70% of the total P in soil (Harrison 1987). In acidic soils, P can be dominantly adsorbed by Al or Fe oxides and hydroxides and thereby become unavailable to plants. In neutral-to-calcareous soils, P retention is dominated by precipitation reactions, where phosphate can precipitate with Ca, generating dicalcium phosphate (DCP) that is available to plants. Ultimately, DCP can be transformed into more stable forms, such as octacalcium phosphate and hydroxyapatite, which are less available to plants at alkaline pH. Phosphates generally accounts for 30% to 65% of the total P in soils. Soil phosphates mainly exists in stabilized forms as inositol phosphates and phosphonates and active forms as orthophosphate diesters, labile orthophosphate monoesters, and organic polyphosphates (Turner et al. 2002). The phosphates can be released through mineralization processes mediated by soil organisms and plant roots in association with phosphatase secretion.

6.3.2.3 Potassium (K) Dynamics in Soil

Potassium (K) is the most abundant macronutrient in soils. The total K content of Indian soils varies from 0.5 to 3.0%. Out of the total K present in soils, over 98% occurs in primary and secondary minerals, and the rest occurs in either fixed or non-exchangeable form, exchangeable (adsorbed) form, or in soil solution. There are four forms of K in soil: (1) Solution K (2) Exchangeable K (3) Fixed or non-exchangeable K (4) Structural or mineral K.

6.3.2.3.1 Soil Solution, Exchangeable, Fixed or Non-exchangeable, and Structural K

This is the K present in the soil solution. Water-soluble K in soil may vary from 1 to 10 ppm. Solution K concentration is important for successful crop production, especially the maturity condition of the crop. Exchangeable K is a potassium held in the exchange complex of 2:1 layer silicates. The amount of exchangeable K in soil may vary from 40 to 600 ppm. Non-exchangeable K is distinct from mineral Kin that is not bonded covalently within the crystal structure of soil mineral particles; instead, it is held between adjacent tetrahedrals such as micas, vermiculites, and inter-graded minerals. K changes from a non-exchangeable form to an exchangeable form. Nonexchangeable K, fixed K in the soil varies from 50 to 750 ppm in soil. The structural or mineral K is a bearing mineral in soil (orthoclase, microcline, muscovite-mica, and biotite bear K in mineral form. The amount may vary generally from 5000 to 25,000 ppm in soil.

6.4 Status and Applications of Models with Respect to Soil Nutrients and Carbon

There are several methods of estimating changes in SOC. The IPCC Guidelines (IPCC 2006) provided a 3-tiered approach. Tiers 1 and 2 are regression-based. The tier 3 approach, with a dynamic model and spatially explicit data, is better if the data and model are available. One of the most promising tools for the estimation of SOC stock is modeling. There are several types of models for the estimation of SOC stock (SOMM, ITE, Verberne, RothC, CANDY, DNDC, CENTURY, DAISY, NCSOIL). The Rothamsted Carbon Model is a leading model that is widely used, as are the CENTURY (Parton 1996) and DNDC (Li et al. 1992) models. The RothC model has also been used for nation-scale estimation of carbon sequestration. RothC (RothC-26.3) and CENTURY 4.0 are the most extensively used SOC simulation models. However, simulation models are increasingly used in more 'applied' settings to measure human impacts on ecosystem function and to manage SOM for GHG mitigation, improved soil health, and sustainable use as a natural resource.

Table 6.2 Use of different models for SOM management and nutrient dynamics

Model	Applications/Features	References
SOLVEG-II	Has 0.25 h time step, immediate to long-term simulations Works up to 5.5 m, 27 varying layer depths Temperature, moisture, and soil texture are the drivers Can have exponential function across depth	Ota et al. (2013)
C-TOOL	Can get monthly, medium to long-term simulations Works up to 100 cm, 0–25 cm and 25–100 cm Temperature, clay content, soil C/N, and OM are the drivers Gives exponential function across the depth	Taghizadeh-Toosi et al. (2014)
SOMPROF	It is used for monthly, medium to long-term simulations Can use up to 0.7–2 m, variable layer depths Temperature and moisture are the drivers	Braakhekke et al. (2013)
RothPC-1	It's used for monthly, medium to long-term simulations Used for 92 cm: in 0–23, 23–46, 46–69, and 69–92 cm depth Temperature, moisture, and clay content are the drivers Consider the microbial biomass, humus, and inert OM pools	Jenkinson and Coleman (2008)
DSSAT-CENTURY	The model simulation showed that N application (N_{150}) increased wheat yield compared to no N application (N_0) DSSAT can be a useful tool for assessing and predicting wheat yield, grain N concentration, and SOC trends under wheat-maize cropping systems	Liu et al. (2017)
RothC	It's simple due to the availability of the input data It can be used for testing the efficiency to predict the development of SOC stock	Barancikova et al. (2010)
RothC version 26.3	Crop rotation without fallowing provided an initial increase in SOC stocks by 3% and 6% stocks in different scenarios Simulation demonstrates difficulties in increasing OC concentration in Chernozems under intensive farming and the potential capacity to raise OC stock through yield management.	Husniev et al. (2020)
DAISY	Building SOC enhances agronomic productivity at only 0–100 kg N ha^{-1} Management can maximize the realization of SOC benefits by building up SOC and maintaining N rates to reduce the off-farm N losses depending on the environmental zones, land use, and the production system.	Ghaley et al. (2018)
DSSAT CENTURY	The model confirmed that the use of continuous tillage is a major threat to SOC building and soil fertility restoration in the tropics. Adopting conservation agriculture is critical for future generations This model is a potential tool for predicting SOC dynamics in low-input farming systems	Musinguzi et al. (2014)

The recent developments in SOM modeling include microbial roles in SOM stabilization, modeling SOM saturation kinetics, temperature controls on decomposition, SOM dynamics in deep soil layers, and SOM representation in earth system models (Campbell and Paustian 2015). For decades, simulation models have played a crucial role in SOM research by providing an explicit mathematical framework to integrate hypotheses for soil processes and supporting hypothesis testing by predicting SOM dynamics (Table 6.2). The model's importance in SOM research is evidenced by citations on the Web of Science. For instance, three of the five highest cited publications under the search phrase 'SOM' are model analyses. The highest cited named model (the CENTURY model, used in 8 of the 34 multi-model comparison publications) showed a dominant focus on multi-year time periods and site- and regional spatial scales (Campbell and Paustian 2015). The impact of key publications on SOM model development is further supported by the citations averaged across years. Out of the publications ranked in the top ten for yearly average citations, seven (including all three CENTURY model publications) were published before 1995, and all but one of these shows consistent or increasing annual citations in recent years. Only ~5% of publications represent 35.5% of total citations, and CENTURY model publications outrank every other model in three publications and account for almost 11% of total citations.

6.5 Soil Carbon Sequestration Models

Soil Carbon Sequestration models, or more precisely, SOM models, represent the transformations of carbon in soil-plant systems with the quantity of soil organic components as a state variable. They may or may not be mechanistic in their treatment of organic carbon addition to and transformation within soils. They subsume, on the one hand, multi-compartment simulation models that aim to represent the flow of mass between compartments and, on the other, analytical mathematical models that represent the size of a few state variables over time (McGill 1996). Models of SOM change have been published since the 1940s, at least. Generally, all SOM models are process-oriented models that focus on processes controlling energy and matter transformations. Organisms are represented in process models as a form of generic biomass or active organic matter (McGill 1996). The carbon pools in these models generally follow the first order of decomposition. Some of the important SOC and nutrient models are discussed in the following subheads.

6.5.1 Rothamsted Carbon (ROTHC) Model

RothC is a model for the turnover of organic carbon in non-waterlogged top soils that allows for the effects of soil type, temperature, moisture content, and plant cover on the turnover process. It uses a monthly time step to calculate total organic carbon (t ha^{-1}), microbial biomass carbon (t ha^{-1}), and D14C (from which the equivalent radiocarbon age of the soil can be calculated) on a year-to-century timescale. (Jenkinson et al. 1987, 1992) It needs few inputs, and those it needs are easily obtainable. The RothC model was originally developed and parameterized to simulate the turnover of organic carbon in arable top soils from the Rothamsted Long Term Field Experiments – hence the name. Later, it was extended to model turnover in grassland and woodland and to operate in different soils and under different climates. The data required to run the model are: (1) Monthly rainfall (mm), (2)

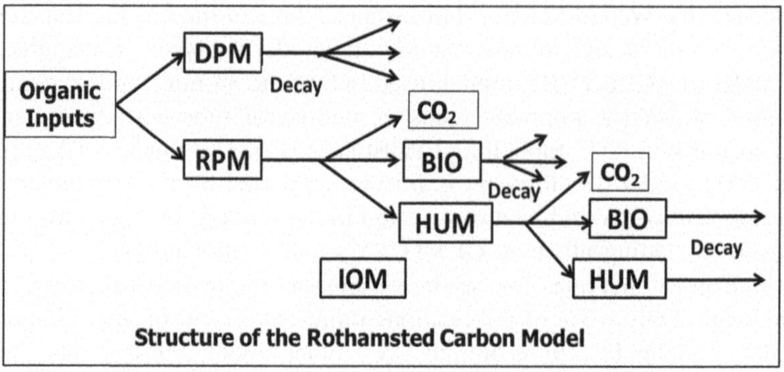

Fig. 6.4 Structure of Roth-C model; DPM decomposable plant material; RPM resistant Plant Material; HUM humified organic Material; BIO soil microbial biomass. (Adapted from Coleman and Jenkinson (1999))

Monthly evapotranspiration (mm), (3) Average monthly mean air temperature (°C), (4) Clay content of the soil (as a percentage), (5) An estimate of the decomposability of the incoming plant material—the DPM/RPM ratio, (6) Soil cover – Is the soil bare or vegetated in a particular month, (7) Monthly input of plant residues (t C ha^{-1}), (8) Monthly input of farmyard manure (FYM) (t C ha^{-1}), if any and (9) Depth of soil layer sampled (cm). Furthermore, SOC is split into four active compartments and a small amount of inert organic matter (IOM). The four active compartments are Decomposable Plant Material (DPM), Resistant Plant Material (RPM), Microbial Biomass (BIO), and HUmified organic Matter (HUM). Each compartment decomposes by a first-order process with its own characteristic rate. The IOM compartment is resistant to decomposition (Coleman and Jenkinson 1999). The structure of the model is shown in Fig. 6.4.

6.5.2 *Century Model*

The CENTURY model simulates the dynamics of carbon (C), nitrogen (N), phosphorus (P), and sulfur (S) for different plant-soil systems. The model can simulate the dynamics of grassland systems, agricultural crop systems, forest systems, and savanna systems. The grassland/crop and forest systems have different plant production sub-models, which are linked to a common SOM sub-model. The SOM sub-model simulates the flow of C, N, P, and S through plant litter and the different inorganic and organic pools in the soil. The model runs using a monthly time step, and the major input variables for the model include (I) monthly average maximum and minimum air temperature, (2) monthly precipitation, (3) lignin N, p. and S content of plant material, (4) soil texture, and (5) atmospheric and soil N inputs (Parton 1996). However, a detailed application of all SOC and nutrient models in different crops and cropping systems is described in Table 6.2.

6.6 Application of Nutrients and Soil Carbon Models Under Changing Climate and Land Use

SOC present in soil has a profound influence on many soil properties. So, a change in SOC content and its quality can have far-reaching impacts on global carbon biogeochemistry and climate change (Smith et al. 2016). Management enforced by agronomic practices such as residue management, no-tillage, cropping systems, application of organic materials (FYM, compost, vermicomposting), fertilizer application, and Integrated nutrient management can change SOC over a longer period of time. Therefore, quantification of significant changes in SOC takes decades to measure them. Besides, the feasibility of conducting long-term experiments under the full range of environments and situations may consume resources and time. Therefore, models are the only practical way of making projections at places other than experimental plots (Coleman et al. 1997). Therefore, several Direct and Indirect methods are used for detecting and scaling Soil Carbon Changes.

6.6.1 Direct Methods Include

(a) Field sampling and laboratory measurements: In this method, a soil sample is collected from the field and analyzed in the laboratory for its carbon content. The amount of carbon is analyzed by wet or dry combustion techniques like the Walkley and Black method or loss of carbon by dry ignition. The *in–situ* measurements can also be done by instruments like mid- or near-infrared spectroscopy, laser-induced breakdown spectroscopy, inelastic neutron scattering, etc.
(b) Carbon dioxide (CO_2) flux monitoring by Eddy covariance: In this method, the vertical component of air movements over a vegetated surface (eddies) is isolated and quantitatively measured as CO_2 concentration associated with each eddy. By correlating eddy size and CO2 concentration for each upward and downward-moving eddy, it is possible to calculate the net uptake or release of carbon from soil and vegetation – the net ecosystem exchange. The integration of fluxes is carried out over long enough periods to estimate changes in ecosystem C.

6.6.2 Indirect Methods Include

(a) Accounting, (b) Remote sensing, and (c) Simulation models. The direct methods provide the foundation for the indirect methods. The indirect methods are required to project the outcome of direct point measurements in space and time and, thus, obtain aggregate information on soil carbon changes at regional and national levels.

6.6.3 Applications of SOC Modeling (Bandyopadhyay 2012)

1. It helps in understanding the complex interaction of the changing climate and land use upon the organic carbon status of the soil.
2. Changes over time in the amount and distribution of SOC fractions with different turnover rates can be estimated through it.
3. It also predicts changes in SOC in response to changes in land use, management, or climate.
4. It helps farmers make management decisions and also in carbon accounting by simulating a wider variety of agricultural systems, crops, crop rotations, etc., and avail benefits from the carbon market.
5. It helps in developing management strategies for the mitigation of GHG emissions.
6. It offers the time-saving prospect of testing and extending the application of experimental results on management responses to many sites.

6.6.4 Limitations of SOC Simulation Modeling (Bandyopadhyay 2012)

1. Simulation models usually provide less reliable short-term predictions.
2. There is always some uncertainty of carbon pool homogeneity and uniqueness required for simulation modeling.
3. There is a lack of availability of reliable databases.
4. There is a requirement for a long-term database for calibration and validation, which is difficult to obtain accurately.
5. There is a need for calibration and validation under different climatic regions and management practices in order to obtain accurate predictions.

6.7 Case Studies of Crop Simulation Models

Bandaranayake et al. (2003) used a century model for the long-term simulation of SOC dynamics under a highly managed turfgrass system. The model estimated 23–32 Mg ha^{-1} SOC sequestration in 0-20 cm soil depth for about 30 years period. Further, the model obtained results with 1–45 years' historical data of 16 golf courses, and the model predicted SOC sequestration reasonably well with an r^2 value of 0.67–0.83. Therefore, the century model can be used successfully to simulate SOC sequestration under various turf management conditions. In another study, the DAISY model was used to know the SOC effect on winter wheat production under different SOC levels at 0–0.20 m soil depth over 38 years of simulations. The measured and modeled SOC during calibration and validation showed a similar trend. Further, the result found that increasing SOC content had a significant positive effect on winter wheat grain yield and above-ground biomass (Ghaley et al. 2018).

The DSSAT century model was simulated for 20–30 years to determine the effect of different management practices on SOC dynamics under maize-bean rotation. Model calibration and validation results showed a good fit between simulated and observed values of SOC dynamics. Simulating continuous tillage with no fertilizer for 30 years antecedent and 50 years extrapolated period, the model showed a high rate of SOC decline (2129 kg ha^{-1} $year^{-1}$) in newly cultivated soil compared to degraded soil. The model study concluded that the use of continuous tillage is a major threat to SOC building and soil fertility restoration in the tropics. Therefore, conservation agriculture is needed for future generations under such conditions (Musinguzi et al. 2014). Further, DSSAT-CSM and century-based soil models were used to simulate trends of crop yield, soil nitrogen storage, and SOC dynamics by using 36 years of long-term experimental data under four fertilizer practices and four representative concentration pathways (RCPs). Model simulation results concluded that fertilizer treatments were better than no fertilizer treatments for yield, nitrogen, and SOC content in the soil. Similarly, under the same climate scenario, the application of organic fertilizer increased maize yield more clearly than with a single application of fertilizer with a concomitant increase in soil nitrogen and SOC (Yan et al. 2020).

6.8 Conclusions and Future Recommendations

Models mimic different agroecosystems accurately for predicting soil and nutrient dynamics under changing climate and land use patterns if properly calibrated and validated for different situations. Models help standardize field experimental results. Further, crop models may be useful for long-term predictions of soil carbon, nutrients, and crop yield under diverse ecosystems. Therefore, it saves time and resource materials and helps policymakers formulate future strategies to mitigate climate change under different agroecosystems across the globe. The SOM modeling approaches are diverse and continually evolving. Therefore, they become an increasingly essential consideration in the range from small-scale projects to sustainability research and policy. Simulation models provide the capacity for numeric evaluation of changes, including a comparison of predicted impacts. Therefore, there would be an expanding use of SOM models in 'applied' settings, specifically to predict SOM dynamics for the policy or to make decisions (Taghizadeh-Toosi et al. 2014). Hence, considering the potential economic and policy implications of these model applications (carbon credits, for example, or payments for changing land management practices), there is an immediate need to better connect advances in SOM understanding with SOM model development in conjunction with soil nutrients.

References

Abbadie L, Lata JC (2006) Nitrogen dynamics in the soil-plant system. In: Lamto. Springer, New York, NY, pp 277–297

Bandaranayake W, Qian YL, Parton WJ, Ojima DS, Follett RF (2003) Estimation of soil organic carbon changes in turfgrass systems using the CENTURY model. Agron J 95(3):558–563

Bandyopadhyay KK (2012) Modelling carbon sequestration. Division of Agricultural Physics, IARI, New Delhi

Barancikova G, Jan H, Monika G, JarmilaMakovnikova MN, RastislavSkalsky ZT (2010) Application of RothC model to predict soil organic carbon stock on agricultural soils of Slovakia. Soil Water Res 5(1):1–9

Batjes NH (1996) Total carbon and nitrogen in the soils of the world. Eur J Soil Sci 47:151–163

Braakhekke MC, Wutzler T, Beer C, Kattge J, Schrumpf M, Ahrens B, Schoning I, Hoosbeek MR, Kruijt B (2013) Modeling the vertical soil organic matter profile using Bayesian parameter estimation. Biogeosciences 10:399–420

Cambardella CA, Gajda AM, Doran JW, Wienhold BJ, Kettler TA, Lal R (2001) Estimation of particulate and total organic matter by weight loss-on-ignition. In: Lal R, Kimble JM, Follett RF, Stewart BA (eds) Assessment methods for soil carbon. Lewis Publishers, Washington DC, pp 349–359.

Campbell EE, Paustian K (2015) Current developments in soil organic matter modelling and the expansion of model applications: a review. Environ Res Lett 10(12):123004

Chabra A, Palria S, Dadhwal VK (2003) Soil organic carbon pool in Indian forests. Forest Ecol Manag 173:187–199

Coleman K, Jenkinson DS (1999) RothC 26.3. A model for the turnover of carbon in soil. Model description and Windows users guide. pp 4–16. IACR - Rothamsted, Harpenden, Herts.

Coleman K, Jenkinson DS, Crocker GJ, Grace PR, Kffr J, Srschens MK, Poulton PR, Richter DD (1997) Simulating trends in soil organic carbon in long-term experiments using RothC-26.3. Geoderma 81:29–44

Davidson EA, Janssens IA, Luo YQ (2006) On the variability of respiration in terrestrial ecosystems: moving beyond Q (10). Glob Chang Biol 12(2):154–164

DeAngelis DL, Mulholland PJ, Palumbo AV, Steinman AD, Huston MA, Elwood JW (1989) Nutrient dynamics and food-web stability. Annu Rev Ecol Syst 20:71–95

Ghaley BB, Wosten H, Olesen JE, Schelde K, Baby S, Karki YK, Borgesen CD, Smith P, Yeluripati J, Ferrise R, Bindi M, Kuikman P, Lesschen J-P, Porter JR (2018) Simulation of soil organic carbon effects on long-term winter wheat (*Triticuma aestivum*) production under varying fertilizer inputs. Front Plant Sci 9:1158. https://doi.org/10.3389/fpls.2018.01158

Hanson PJ, Edwards NT, Garten CT, Andrews JA (2000) Separating root and soil microbial contributions to soil respiration: a review of methods and observations. Biogeochemistry 48(1):115–146

Harrison AF (1987) Soil organic phosphorus—a review of world literature. CAB International, Wallingford, p 257

Harrison MD (2016) Nutrient dynamics. In: Kennish MJ (ed) Encyclopedia of estuaries. Encyclopedia of earth sciences series. Springer, Dordrecht. https://doi.org/10.1007/978-94-017-8801-4_75

Hoogenboom G, Porter CH, Shelia V, Boote KJ, Singh UJ, White W, Hunt LA, Ogoshi RJI, Lizaso, KJ, Asseng S, Singels A, Moreno LP, Jones JW (2019) Decision Support System for Agrotechnology Transfer (DSSAT) version 4.7.5 (https://DSSAT.net). DSSAT Foundation, Gainesville

Houghton RA (2003) Revised estimates of the annual net flux of carbon to the atmosphere from changes in land use and land management 1850–2000. Tellus B 55:378–390

Husniev I, Romanenkov V, Minakova O, Krasilnikov P (2020) Modelling and prediction of organic carbon dynamics in arable soils based on a 62-year field experiment in the Voronezh region. European Russia Agronomy 10:1607

IPCC (2006) IPCC guidelines for national gas inventories. http://www.ipcc.nggip.iges.or.jp

Jenkinson D, Coleman K (2008) The turnover of organic carbon in subsoils: II. Modelling carbon turnover. Eur J Soil Sci 59:400–413

Jenkinson DS, Hart PBS, Rayner JH, Parry LC (1987) Modelling the turnover of organic matter in long-term experiments at Rothamsted. INTECOL Bull 15:1–8

Jenkinson DS, Harkness DD, Vance ED, Adams DE, Harrison AF (1992) Calculating net primary production and annual input of organic matter to soil from the amount and radiocarbon content of soil organic matter. Soil Biol Biochem 24(4):295–308

Kirschbaum MUF (2000) Will changes in soil organic carbon act as a positive or negative feedback on global warming? Biogeochemistry 48(1):21–51

Li CS, Frolking S, Frolking TA (1992) A model of nitrous-oxide evolution from soil driven by rainfall events. 2. Model applications. J Geophys Res 97(9):9777–9783

Liang H, Hu K, Batchelor WD, Qi Z, Li B (2016) An integrated soil-crop system model for water and nitrogen management in North China. Sci Rep 6(1):1–20

Liu H-l, Hong-bin Liu LEI, Qiu-liang Z L-m, Hong-yuan W, Ji-zong Z, Ye-ping Z, Sheng-ping L, Shi-juan L, Jing-suo Z, Xiao-xia L (2017) Using the DSSAT model to simulate wheat yield and soil organic carbon under a wheat-maize cropping system in the North China Plain. J Integr Agric 16(10):2300–2307

Luce MS, Whalen JK, Ziadi N, Zebarth BJ (2011) Nitrogen dynamics and indices to predict soil nitrogen supply in humid temperate soils. Adv Agron 112:55–102

Lukac M, Calfapietra C, Lagomarsino A, Loreto F (2010) Global climate change and tree nutrition: effects of elevated CO2 and temperature. Tree Physiol 30(9):1209–1220

Macdonald CA, Delgado-Baquerizo M, Reay DS, Hicks LC, Singh BK (2018) Soil nutrients and soil carbon storage: modulators and mechanisms. In: Soil carbon storage. Academic Press, pp 167–205

Majumder B, Mandal B, Bandyopadhyay PK (2008) Soil organic carbon pools and productivity in relation to nutrient management in a 20-year-old rice berseem agroecosystem. Biol Fertil Soils 44:451–461

Marschner B, Kalbitz K (2003) Controls of bioavailability and biodegradability of dissolved organic matter in soils. Geoderma 113:211–235

Marschner B, Brodowski S, Dreves A, Gleixner G, Gude A et al (2008) How relevant is recalcitrance for the stabilization of organic matter in soils? J Plant Nutr Soil Sci 171:91–110

McGill WB (1996) Review and classification of ten soil organic matter (SOM) models. NATO ASI Ser I:38

Musinguzi P, Ebanyat P, Tenywa JS, MajaliwaMwanjalolo TA, Basamba MMT, Porter C (2014) Using DSSAT-CENTURY model to simulate soil organic carbon dynamics under a low-input maize cropping system. J Agric Sci 6(5):120–131

Ota M, Nagai H, Koarashi J (2013) Root and dissolved organic carbon controls on subsurface soil carbon dynamics: a model approach. J Geophys Res Biogeosci 118:1646–1659

Parton WJ (1996) The CENTURY model. In: Powlson D.S., Smith P, Smith JU (eds) Evaluation of soil organic matter models. NATO ASI series (Series I: Global environmental change), vol 38. Springer, Berlin, Heidelberg. https://doi.org/10.1007/978-3-642-61094-3_23
Pathak H, Aggarwal PK, Roetter R, Kalra N, Bandyopadhaya SK, Prasad S, Van Keulen H (2003) Modelling the quantitative evaluation of soil nutrient supply, nutrient use efficiency, and fertilizer requirements of wheat in India. Nutr Cycl Agroecosyst 65(2):105–113
Pendall E et al (2004) Below-ground process responses to elevated CO2 and temperature: a discussion of observations, measurement methods, and models. New Phytol 162(2):311–322
Poeplau C, Don A, Vesterdal L, Leifeld J, Van Wesemael BAS, Schumacher J, Gensior A (2011) Temporal dynamics of soil organic carbon after land-use change in the temperate zone–carbon response functions as a model approach. Glob Chang Biol 17(7):2415–2427
Probert ME, Dimes JP (2004). Modelling release of nutrients from organic resources using APSIM. In ACIAR Proceedings (pp 25–31). ACIAR; 1998.
Radha V, Murthy K (2004) Crop growth modeling and its applications in agricultural meteorology. http://www.wamis.org/agm/pubs/agm8/Paper-12.pdf, 11 March 2020.
Ramesh T, Bolan NS, Kirkham MB, Wijesekara H, Kanchikerimath M, Rao CS, Sandeep S, Rinklebe J, Ok YS, Choudhury BU, Wang H (2019) Soil organic carbon dynamics: impact of land use changes and management practices: a review. Adv Agron 156:1–107
Ross CW, Grunwald S, Myers DB, Xiong X (2016) Land use, land use change and soil carbon sequestration in the St. Johns River basin, Florida, USA. Geoderma Reg 7(1):19–28
Sachin G, Mohammed Ahamed J, Nagajothi K, Rana M, Murugan BS (2019) Automation of the DSSAT crop growth simulation model. Int Arch Photogram Remote Sens Spat Inform Sci 42:251–256
Scharlemann JP, Tanner EV, Hiederer R, Kapos V (2014) Global soil carbon: understanding and managing the largest terrestrial carbon pool. Carbon Manage 5(1):81–91
Schmidt MW, Torn MS, Abiven S, Dittmar T et al (2011) Persistence of soil organic matter Asan ecosystem property. Nature 478(7367):49–56
Smith P, Wilson BR, Jagadeesh B, Daniel H, Ghosh S (2016) Comments on "Modelling soil organic carbon storage with RothC in irrigated Vertisols under cotton cropping systems in the sub-tropics" (Authors: Nimai Senapati, Nilantha R. Hulugalle, Pete Smith, Brian R. Wilson, Jagadeesh B. Yeluripati, Heiko Daniel, Subhadip Ghosh, Peter Lockwood. Soil & Tillage Research 143 (2014) 38-49). Soil Till Res 155:532–534. https://doi.org/10.1016/j.still.2015.04.012
Taghizadeh-Toosi A, Christensen BT, Hutchings NJ, Vejlin J, Kätterer T, Glendining M, Olesen JE (2014) C-TOOL: a simple model for simulating whole-profile carbon storage in temperate agricultural soils. Ecol Model 292:11–25
Tan Z, Lal R, Owens L, Izaurralde RC (2007) Distribution of light and heavy fractions of soil organic carbon as related to land use and tillage practice. Soil Tillage Res 92:53–59
The Terrestrial Carbon Group (2010) Roadmap for Terrestrial Carbon Science—Research needs for carbon management in agriculture, forestry and other land uses, pp. 1–84. The Terrestrial Carbon Group, UN-REDD agencies, the World Bank and CGIAR institutions.
Turner BL, Papházy MJ, Haygarth PM, McKelvie ID (2002) Inositol phosphates in the environment. Philos Trans R Soc Lond Ser B Biol Sci 357:449–469
Turner BL, Richardson AE, Mullaney EJ (2007) Inositol phosphates: linking agriculture and the environment. CAB International, Wallingford, p 304
Yadav D, Vishwakarma AK, Sharma NK, Biswas AK, Ojasvi PR, Kumar D, Kumawat A, Singh D (2021) Sustaining the properties of black soil of Central India through crop residue management in the conservation-agriculture based soybean-wheat system. Land Degrad Develop 32:2906–2921. https://doi.org/10.1002/ldr.3891
Yan W, Jiang W, Han X, Hua W, Yang J, Luo P (2020) Simulating and predicting crop yield and soil fertility under climate change with fertilizer management in Northeast China based on the decision support system for agrotechnology transfer model. Sustain For 12(6):2194

Zhang GS, Ni ZW (2017) Winter tillage impacts on soil organic carbon, aggregation and CO2 emission in a rainfed vegetable cropping system of the mid–Yunnan plateau. China Soil Tillage Res 165:294–301

Zhang L, Zhuang Q, He Y, Liu Y, Yu D, Zhao Q, Wang G (2016) Toward optimal soil organic carbon sequestration with effects of agricultural management practices and climate change in Tai-Lake paddy soils of China. Geoderma 275:28–39

Zhao YG, Liu XF, Wang ZL, Zhao SW (2015) Soil organic carbon fractions and sequestration across a 150-yr secondary forest chronosequence on the loess plateau, China. Catena 133:303–308

Chapter 7
Remote Sensing and Modeling Soil Organic Carbon Sequestration: A Regime in Global Climate Change

Rupak Dey and Seema B. Sharma

Abstract Soil is a complex entity of organic and inorganic materials produced from rock protolith, plays a prime role in the carbon cycle, and is surveyed to be the largest carbon reservoir apart from the oceans. SOC is a prime measure of soil fertility and is biologically, physically, and chemically associated with soil as it serves as a dominant root of nutrients for plants, enhances soil structure, improves water-holding capacity, and is also a notable part of the global carbon cycle. A decline in the amount of SOC is frequently seen as an imminent threat to the soil's longevity since it affects the processes of multiple agrochemical procedures in the soil. Conventional methods of soil monitoring and interpretation of SOC hotspots, such as peatlands, grasslands, and black soils, take a long time, require much labor, and are expensive because they are difficult to access. They also don't provide enough data to support landscape decision-making or the monitoring of carbon storage or loss over time. With the evolution of GIS and remote sensing (RS) technologies, predictive soil mapping techniques and models provide a cost-effective, reliable, and efficient approach to gathering regionally distributed data on SOC as well as monitoring that can mitigate Climate Change in the near future. RS techniques in the Visible-Near Infrared-Shortwave Infrared (VNIR-SWIR, 400–2500 nm) range potentially help in a more advanced, efficient, and timely way to simulate key indicators for soil processes. While it has been noted that accuracy errors vary with the satellite bands and sensors, improvements in machine-learning approaches may help even more in the development and elevation of calibration models. Additionally, there are still some drawbacks with vegetation cover, soil moisture, and sharpness to be resolved, as well as some problems with geometric, radiometric, and atmospheric corrections. The objective of this chapter is to outline the methodology and conclusion of each technique that was conducted to estimate SOC using RS satellites.

R. Dey · S. B. Sharma (✉)
Department of Earth and Environmental Science, KSKV Kachchh University, Bhuj, Gujarat, India
e-mail: seemabhargavsharma@gmail.com

S. A. Edrisi et al. (eds.), *Sustainable Plant Nutrition and Soil Carbon Sequestration*, Sustainable Plant Nutrition in a Changing World,
https://doi.org/10.1007/978-3-031-53590-1_7

Keywords Remote sensing · Satellite sensors · Geographic information system · Soil mapping · Carbon storage capacity

7.1 Introduction

Soil is a thin surface layer that covers the majority of land and has a mixture of both inorganic and organic components. It provides mechanical support to plants as well as nutrients and water, theredby supporting various ecosystem services (Edrisi and Tripathi 2018). Soil formation is a long-term process that takes millions of years to build a thin layer of soil. They are generated from rock protoliths through the processes of weathering and erosion. The process through which soil is formed is termed pedogenesis. The physicochemical properties of the parent rock, the extent and duration of weathering, the environment, and other factors all influence the formation of specific kinds of soil. Weakly formed, moderately formed, and highly evolved soils are produced by a combination of five key components, viz. parent rocks, climate, living organisms, relief, temperature, and water (Fig. 7.1). This combined action of various soil-forming components under varying environmental circumstances provides a roadmap to certain recognized soil-formation mechanisms

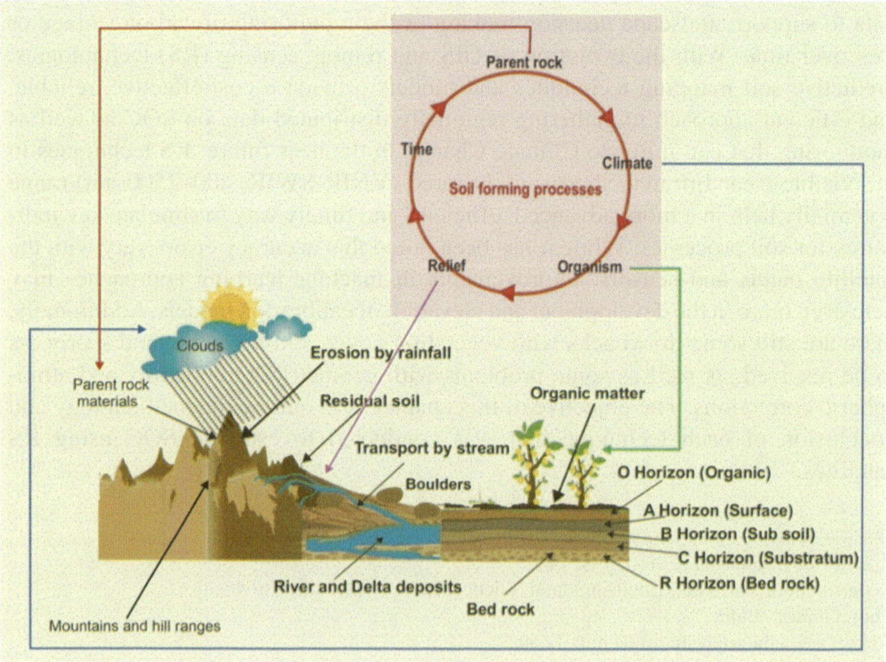

Fig. 7.1 Soil profiles, soil-forming process, and its influencing drivers. (Adopted from Balasubramanian 2017)

(Balasubramanian 2017). Soil, being regarded as the third most principal global carbon sink (> 2500 pg), is an essential feature connected to biological, physical, and chemical aspects that represent a chief contributor to the earth's carbon cycle and also contains nearly twice the carbon that is present in the air (750 pg) and land living biomass (560 pg) (Bangelesa et al. 2020).

Soil carbon reservoirs are categorized into two types: organic carbon and inorganic carbon. Inorganic carbon is a substance found in the atmosphere that is mainly composed of CO_2. The accumulation of CO_2 has continuously increased since the Industrial Revolution. SOC is a portion of SOM that accounts for 2–10% of the bulk of most soils. It contributes a significant part in the physicochemical and biological functionality of soils, including nutrient retention, soil texture, water infiltration, and accessibility, breakdown of pollutants and carbon sequestration (Blanco et al. 2013). The use of fossil fuels has been a major contributor to the rise in GHG levels. Land use changes and deforestation have also contributed to the increase in CO_2. The concept of SOC sequestration is seen as one way to reduce atmospheric CO_2. It is also believed that small increase in SOC over large areas will reduce GHG emissions (Zomer et al. 2017).

Olson (2010) proposed the best definition of SOC sequestration as "*the process of transferring CO_2 from the atmosphere into the soil through plants, plant residues, and other organic solids, which are stored or retained as part of SOM (humus).*" The retention time of sequestered carbon in the soil can range from short-term (not immediately released back to the atmosphere) to long-term (millennia) storage. The sequestered carbon process should increase the net SOC storage during and at the end of the investigation to above the previous pre-treatment baseline levels and result in a net reduction in the CO_2 levels in the atmosphere" (Olson 2010). Thus, some of the processes does not considered as the sequestration. These processes include the deposition and burial of carbon through erosion. Land use with carbon-enhanced amendments (biochar, mulch, manure, etc.) is also not considered. The burial metamorphism and diagenesis (generally at a low temperature of 300 °C, 600 MPa pressure, and a depth of about less than 20 km) of biomass in deep mines or on the seafloor brought in from outside the land units by tectonic activity and plate tectonics concepts are not considered as sequestration processes. Soil erosions accelerate the course of carbon recovery and modify the SOC flux (Ogunseitan 2005).

The carbon particles are transferred to the atmosphere and seas and dispersed throughout the terrain. Part of the eroded carbon, 10%, is transferred into the seas, 20–30% is released into the air, and 60–70 percent is distributed throughout the land (Sunquist and Broecker 1986). Climate change mitigation strategies are frequently focused on GHG emissions from livestock production systems, which provide considerable potential for carbon sequestration in the soil to counteract GHG emissions. CO_2 is a heat-storing gas that is produced both naturally and as a result of human activity. Man-made CO_2 may be generated by igniting coal, natural gas, or oil to create electricity. Biological CO_2 can be produced by decaying organic materials, bushfires, and other changes in land use patterns. Because the accumulation of CO_2 and other GHGs in the atmosphere may trap heat and contribute to climate change, it is vital to anticipate, model, and track these GHGs for our long-term and future planning purposes. A variety of methods can be used to assess the carbon stocks in the aboveground plant biomass. These include contact field measurements and remote sensing (RS) (Edrisi et al. 2022, 2023).

The basic technique involved in contact approach includes forestry inventory, table production and Eddy Covariance method. The table production technique relies on connecting self-classified regions with pre-prepared carbon stock production tables for each category based on a predefined classification. The figures in the table are obtained from prior contact measurements and literary data. The Eddy covariance method is a simple method for estimating the energy flows in a given area. It uses direct measurement of CO_2 and energy flows. Moreover, there are many dynamic (Environmental variables vary with time) and static (Environmental Variables remain constant) models like single homogenous compartment models and double compartment models. The dynamic models can be additionally divided into four types viz.; Process-oriented, Organism-oriented, Integrated, and Cohort models. Models may be assessed at a local, regional, or global level. There are now 250 models available for forecasting SOC dynamics and nutrient turnover(Sierra et al. 2012). The fundamental assumptions and the mechanism responsible for SOM breakdown vary across these models. Among the SOC models, the Rothamsted Carbon model is being extensively utilized globally and is used for Carbon measuring. It comprises five reservoirs of SOM: decomposable plant material, resistant plant material, microbial biomass, humified SOM, and inert organic matter that is resistant to biological processes (Shirato and Yokozawa 2005; Morais et al. 2019; Liu et al. 2021).

A plethora of methods are under research for sequestering atmospheric carbon. These may be handled under three broad subgroups:

(a) Ocean Sequestration: Carbon sequestration in the ocean means that plankton (floating tiny organisms) on the surface of the ocean convert CO_2 into oxygen through photosynthesis in the photic zone, just as trees and land plants do on land. The oceans absorb approximately 25% of the GHG CO_2 produced by humans each year. When the ocean releases CO_2, it creates a positive atmospheric flow(Stallinga 2018). However, if the oceans become more acidic owing to the emission of CO_2, they may potentially alter their chemistry in the future.
(b) Geological Sequestration: This type of carbon storage works by storing the CO_2 of industrial sources underground. This method involves using porous rocks to store the CO_2. Injection into deep-sea basalt formations offers distinct and substantial benefits over other possible geological storage methods. Climate scientists have long suspected that highly reactive basaltic rocks might provide a solution to the carbon storage issue. Basalts are found all over the globe and contain significant quantities of calcium and magnesium ions that combine with CO_2 to form magnesite, dolomite, and calcite. The aforementioned process makes it possible to use fossil fuels while lowering GHG emissions. Geologic sequestration can be done by the following types:

   (i) Hydrodynamic Trapping
   (ii) Solubility Trapping
   (iii) Mineral Carbonation

(c) Terrestrial Sequestration: Soils and plants, which serve as natural carbon sinks, store a considerable amount of carbon. Increased carbon fixing via photosynthesis, slowing down or decreasing decomposition of organic matter, and altering land use practices may increase carbon absorption in these natural sinks. Terrestrial sequestration can be done in the following ways:

   (i) By Forests: Forests absorb about 25% of the global GHG emissions (Brack 2019). When the trees die, their leaves and branches fall to the ground, releasing the carbon into the soil.
   (ii) In Soils: plants through photosynthesis can sequester the carbon in soil. Unfortunately, some agroecosystems can deplete the organic carbon levels in the soil. Fortunately, soil can also store carbon through carbonates.
   (iii) By Grasslands: Grasslands are more resilient than forests when it comes to storing carbon. They can also store more carbon underground due to their porous soil.

Since traditional techniques for monitoring SOC are laborious, time-consuming, and costly, scientists have been exploring the various approaches that can be used to minimize this cost. Due to the vast region where contact techniques are not possible and with the development and advancement in science and technology, RS techniques have been utilized to solve this problem. This summarized chapter will enable many better procedures and much more precise regular SOC mapping and monitoring using RS techniques, which is a part of the climate change mitigation framework.

7.2 Carbon Sinks

Carbon sinks are environmental processes or any natural or man-made reservoir that absorbs and deposits certain carbon-containing chemical molecules for a boundless amount of time, lowering the levels of CO_2 in the air. Plants, oceans, and soil are the primary natural carbon sinks. Since the 1997 Kyoto Protocol, which promotes the use of CO_2 sinks as a form of carbon offset, public understanding of their significance has expanded. Plants absorb CO_2 from the air and utilize it in photosynthesis; a chunk of it is transferred to the soil when plants die and decompose. A significant mechanism for storing CO_2 is the oceans. While some CO_2 basically dissolves in saltwater, marine organisms also absorb it for photosynthesis. The ocean, atmosphere, soil, and forests are the largest carbon sinks on Earth. Combating climate change and maintaining a stable climate depends on protecting these important ecosystems. They are, however, under growing threat.

7.2.1 Ocean as a Part of the Biosphere to Combat Climate Change

Marine life and people are both threatened by climate change. According to a study published by the IPCC, restoring blue carbon habitats may help reduce GHG emissions (Wigley et al. 1981). Ocean transportation, according to a study published by the US Department of Transportation, is the largest contributor to GHG emissions worldwide. To decrease GHG emissions, the study recommends using alternative fuels and modifying ships. Reducing emissions from maritime traffic will also reduce sulfur dioxide, which is hazardous to those who live near the shore (Shaheen and Lipman 2007). It can also help to decrease the rate of acidification in the water. While injecting CO_2 into the seabed involves certain dangers, it may be a viable option for reducing GHG emissions. The ocean is one of the biosphere's most important components. It offers us a variety of functions, including oxygen, seafood, and weather forecasting. Humans are already feeling the impacts of climate change on the ocean. It bears the brunt of climate change's impacts, including rising sea levels, pollution, and temperature fluctuations. Since the Industrial Revolution, the quantity of CO_2 in the air has risen dramatically. The Oceans filter out pollutants and assist in fighting climate change by absorbing them.

7.2.2 Vegetation and Their Role in Mitigating Climate Change and Sequestration of Carbon

Forests and other Vegetation are important carbon pools that continuously exchange CO_2 with the atmosphere. Understanding their carbon cycle is very useful for mitigating the greenhouse effect.

7.2.2.1 Suitable Plants for Increasing Oxygen

Although all trees filter impurities from the air, some trees are more effective than others at removing GHGs. Plants and the ocean absorb enough CO_2 to offset about half of the 40 billion tons of CO_2 pollution produced by people each year (Sunquist and Broecker 1986). However, this is insufficient to capture and keep up with the rate of CO2 rise in the air, which has been continuously increasing. The suitable Plants differ by location, and the best tree to trap carbon in each of the world's major countries is listed below (Table 7.1). The plants that absorb the most carbon are listed below (Nowak 2010):

(i) East Palatka Holly ≈ 7321 CO_2
(ii) Slash pine ≈ 4930 CO_2
(iii) Live oak ≈ 10,994 CO_2

Table 7.1 Suitable Plants depending on location for carbon capture based on climatic factors

Sl. No.	Country	Suitable plants and their scientific names
1	Russia	Larch (*Larix*), Fir (*Abies*), Pines (*Pinus*), Spruces (*Picea*), Cedars (*Cedrus*), etc.
2	Canada	Sugar maple (*Acer saccharum*), white pine (*Pinus strobus*), Paper birch (*Betula papyrifera*), Douglas fir (*Pseudotsuga menziesii*), Okanagan Sunflower (*Balsamorhiza sagittata*), etc.
3	China	Lucky bamboo (*Dracaena sanderiana*), Dove tree (*Davidia involucrata*), Money tree (*Pachira aquatica*), Orchids (*Orchidaceae*), Bonsai (*Prunus mume*), Peace lily (*Spathiphyllum*), etc.
4	United States of America	Aloe plant (*Aloe vera*), Coneflower (*Echinacea*), wintergreen (*Gaultheria procumbens*), Iris setosa (*Iris setosa*), etc.
5	Brazil	Tillandsia Air plant(*Tillandsia*), Philodendron (*Philodendron*), Tropical water lily (*Nymphaeaceae*), Amazon lily pads (*Victoria amazonica*), bamboo culm stripes (*Bambusa vulgaris*), etc.
6	Australia	Acacia(*Acacia*), Eucalypts(*Eucalyptus*), Wildflowers, etc.
7	India	Neem (*Azadirachta indica*), Banyan (*Ficus benghalensis*), peepal (*Ficus religiosa*), Arjuna (*Terminalia arjuna*), Bamboo (*Bambusa*), Jamun (*Syzygium cumini*), Curry tree (*Murraya koenigii),* Snake plant (*Dracaena trifasciata*) etc.
8	Argentina	Red Vein Abutilon (*Abutilon striatum*), Autumn Zephyr Lily (*Zephyranthes candida*), Darwin's Barberry (*Berberis darwinii*), Yerba Mate (*Ilex paraguariensis*) etc.
9	Kazakhstan	European Bedstraw (*Galium rubioides*), Starfruit (*Averrhoa carambola*), Great Water Parsnip (*Sium latifolium*), Lily (*Lilium*) etc.
10	Algeria	Slender Flower Thistle (*Carduus tenuiflorus*), Algerian Fir (*Abies numidica*), Saharan Cypress (*Cupressus dupreziana*), Algerian Peony (*Paeonia algeriensis*) etc.

(iv) Southern magnolia ≈ 5000 t yr.$^{-1}$
(v) Bald cypress ≈ 4872 CO_2
(vi) Jacaranda ≈ 4900 t yr.$^{-1}$
(vii) Evergreen ash ≈ 3500 t yr.$^{-1}$
(viii) Velvet ash ≈ 3500 t yr.$^{-1}$
(ix) Camphor tree ≈ 3200 t yr.$^{-1}$
(x) London Plane tree ≈ 2200 t yr.$^{-1}$
(xi) Carrotwood ≈ 2900 t yr.$^{-1}$
(xii) Olive ≈ 2200 t yr.$^{-1}$
(xiii) Sweetgum ≈ 2000 t yr.$^{-1}$
(xiv) Italian Cypress ≈ 2000 t yr.$^{-1}$

7.2.2.2 Climate Change and Surprising Products from Carbon

Using cutting-edge technology, scientists are studying novel techniques for extracting and storing carbon from the atmosphere. Some of the products that can be made from carbon are mentioned below:

(i) Graphene: CO_2 may be used as a primary ingredient to manufacture graphene, a technology-based compound that has widespread use in the modern world(Xu and Gao 2015). Graphene can be used in the production of screens for mobile phones and other electrical gadgets. Graphene production is restricted to certain sectors, but it is a prime example of how CO_2 may be used as a commodity and a remedy to reduce environmental pollution.
(ii) C2CNT: C2CNT's process is simple and cost-effective. It uses CO_2 from the flue streams of power plants and cement factories to convert carbon into pure carbon nanotubes(Licht 2017).
(iii) Nanoparticles for plastics, concrete, and coatings: Using nanoparticles produced with collected CO_2 not only prevents those emissions from entering the environment but also lowers the need to use costly, carbon-intensive conventional materials such as plastics and concrete.
(iv) Bioplastic: Scientists from California's Huntington Beach have discovered a method to collect CO_2 or methane from a farm or power plant's flue stream. They then mix the carbon with oxygen and hydrogen to create a biopolymer that's produced from PHA.
(v) Methanol: A team from India is working on an artificial photosynthesis process that will convert CO_2 into methanol, which is a key component of the fuel industry.
(vi) C4X: Wayne Song, a specialist in wood-plastic composite and plastic fiber, is the head of C4X, a company in China that focuses on the production of bioplastic composite materials by capturing and converting CO_2. Aside from wood, various natural fibers such as rice hulls, palm fiber, and flax can also be used to make bio-composite materials. These are becoming more prevalent in the construction industry.

7.2.3 *Soil as Carbon Sink*

Soils are among the world's largest reservoirs of carbon. They have the potential to reduce atmospheric CO_2 concentration by up to 10%. Soil carbon sequestration occurs when plants take up the CO_2 from the air and store it in their soil carbon pool. In semi-arid climates, it can also be achieved by converting air into soil organic compounds. SOC plays a crucial role in terrestrial ecosystems. Its storage depends on the interactions among various ecological processes. Over the last 150 years, human activities have led to the depletion of SOC (Ontl 2018). Soils are among the planet's largest sources of carbon storage. They have the potential to provide a short-term solution to reducing atmospheric CO_2 concentration.

7.3 Carbon Sequestration and the Sixth Mass Extinction

Mass extinction is a process that occurs when the Earth loses most of its species in a short amount of time. During the geological history of the earth, there have been five major mass extinctions. The Holocene extinction or the Anthropocene extinction, also known as the sixth mass extinction, is a series of events that are occurring at present. It is believed that about 100 to 1000 species are already extinct (Cowie et al. 2022). Many of these extinctions are believed to have been undocumented, and are still more to come. The current extinction issue is all our responsibility since habitat loss, pollution, the expansion of alien species, overharvest wilderness, climate change, rise in population, and other anthropogenic activities have pushed nature to the verge of extinction. Overcoming the extinction issue will need the development of novel approaches. Keeping this in mind, most climate scientists believe that we must limit GHG output to avoid disastrous climate change. However, some argue that we also need to remove and monitor CO_2 from the air by carbon capture in large quantities to keep the planet from going back to its pre-industrial conditions. Due to technical advantages, such worldwide tracking and monitoring may be accomplished by utilizing RS and GIS models.

7.4 Satellite Sensors, Tools, and Data for Modeling of Soil Organic Carbon

Modern techniques such as GPS, RS, and Geographic Information Systems have enabled multi-source data (Table 7.2) to be used to estimate the carbon storage capacity of forests, grasslands, and soils. Currently, three kinds of datasets are utilized to assess vegetation and soil carbon storage, mainly optical RS, synthetic aperture radar (SAR), and Lidar data.

7.4.1 Optical Remote Sensing Data

It employs short-wave infrared sensors, near-infrared sensors, and visible light to detect solar radiation reflected from objects on the ground and generate imagery of the Earth's surface. Different substances and materials reflect and absorb light differentially at various wavelengths (Awange et al. 2013). Thus, the spectral reflectance characteristics of the objects in remotely sensed photographs may be utilized

Table 7.2 Modern technologies are used to find the relationship and estimate the carbon storage capacity of forests, grasslands, and soils

Sl. No.	Remote Sensing broad methods and approaches	Description	References
1	Aerial photography	An aerial photograph, in general terms, is any image obtained from the air.	Dahdouh et al. (2000)
2	Infrared imagery	Infrared satellite imaging acts as a kind of thermal map.	Ghabi and Das (2018)
3	Microwave radiation	The portion of the EMR with wavelengths between 1 mm and 1 m is commonly referred to as microwave radiation.	Wang and Wang (2022)
4	Lidar (light detection and ranging)	It is a technique for creating high-resolution ground topography models and maps with a vertical precision of 10 cm.	Mitchell et al. (2017)
5	Optical	It detects solar energy reflected from objects on the ground using visible, near-infrared, and short-wave infrared sensors to create images of the Earth's surface.	Dopper et al. (2022)
6	Radar (radio detection and ranging)	A detecting technique that employs radio waves to calculate object distance, angle, and velocity. Aircraft, ships, satellites, guided missiles, motor vehicles, weather patterns, and topography may all be detected with it.	Ottinger and Kuenzer (2020)
7	Spaceborne laser scanning	Other information on a broad range of EEarth surface characteristics and processes is provided by a space-borne laser scanner.	Sun and Liu (2000)
8	Airborne laser scanning	ALS is being used to gather high-resolution data for the creation of a digital terrain model.	Lim et al. (2003)
9	Airborne mapping	Airborne mapping is a technique of accumulating imagery by flying cameras or other sensors over a study area using passenger planes, choppers, or Spacecraft.	Levi et al. (2022)
10	GLAS	The Geoscience Laser Altimeter System is the first laser-ranging equipment for continuous global monitoring of Earth.	Mitchell et al. (2017)
11	Satellite imagery	Satellite images reveal how the human impact is evolving throughout the world, including fast-expanding towns, urban expansion, and unplanned settlements.	Nowak (2010), Chen et al. (2013), Mitchell et al. (2017)
12	Earth Observations	It is the collection of data on the planet's physical, chemical, and biological processes using RS technology, often satellites equipped with imaging sensors. Earth observation is used to monitor and analyze the state of the natural and man-made environments, as well as changes in them.	Dech et al. (2021)
13	Satellite laser altimetry	A laser altimeter is a technology used onboard planet-orbiting satellites to survey a planet's geography.	Chen (2010)

(continued)

Table 7.2 (continued)

Sl. No.	Remote Sensing broad methods and approaches	Description	References
14	SRTM	SRTM (Shuttle Radar Topography Mission) is placed on a Space Shuttle and acquires Earth surface data through RS technology employing a synthetic aperture radar.	Reddy et al. (2013), Demarez et al. (2019)
15	Decision tree approach	It can be applied to large amounts of data due to its computational simplicity, used in satellite imagery analysis, and managed on thousands of pixels.	Holloway et al. (2019)
16	Regression tree model	It can be employed with Landsat data sets to calculate the present tree cover, and it produces a rule-based model for predicting a single continuous response variable from one or more explanatory variables.	Bhunia et al. (2017)
17	Laser Scanner (terrestrial, ground-based)	A ground-based, active imaging approach quickly gathers precise, dense 3D point clouds of object surfaces by laser range finding.	Mayr et al. (2019)
18	Full waveform-Neural networks, Support vector machines, Hyperspectral	This combined RS method can be made to work together to do research on many aspects.	Park et al. (2021)

to differentiate them. Optical RS systems are classified into the following categories based on how many spectral bands are used during the imaging process.

(i) Panchromatic imaging system: A panchromatic sensor is a device that uses a wide spectral range to detect the appearance of objects in space. The resulting image will resemble a black-and-white photograph if the range is within the viewing range (Shmilovich et al. 2020). IKONOS PAN and SPOT HRV-PAN are two examples of Panchromatic Imaging systems.

(ii) Multispectral Imaging System: It is a multi-channel detector that has several portions of the spectrum. The generated picture is layered and comprises the brightness and color information of the targets (Abady et al. 2020). IKONOS MS, SPOT HRV-XS, LANDSAT TM, and LANDSAT MSS are examples of multispectral systems.

(iii) Super spectral Imaging system: A super spectral image sensor contains several spectral channels (usually more than ten) than a multispectral sensor. The lower bandwidths of the bands enable the sensor to capture finer spectral features of the objects(Dhingra and Kumar 2020). Superspectral imaging technologies include MERIS and MODIS.

(iv) Hyperspectral Imaging System: An "imaging spectrometer" is another name for a hyperspectral imaging system. It captures pictures in several hundred adjoining spectral bands. The exact spectrum information included in a hyperspectral picture allows for improved target characterization and identification

(Shmilovich et al. 2020; Varade et al. 2019; Bangelesa et al. 2020). Precision agriculture (e.g., assessing crop kinds, health, moisture condition, and maturity), coastal management, and other areas may benefit from hyperspectral pictures (e.g., tracking and assessment of phytoplanktons, pollution, and bathymetry changes). One example is Hyperion on the EO1 satellite.

7.4.2 Synthetic Aperture Radar Satellite Data

Surface imagery of the Earth has been generated using radio-sensing methods known as SAR since the 1960s (Zhou et al. 2009). This technique is frequently used to map the aboveground biomass. SARs, unlike other sensors, may function at any time of day or night. They may also pass through woods and disperse microwave radiation into the sky. The wavelength of an aboveground biomass component determines its sensitivity. The X and C bands are usually sensitive to various leaf and stem sizes, while the P and L bands are sensitive to big branches and leaves. In addition, interferometric SAR (InSAR) is utilized to enhance the recovery of aboveground biomass through plant height proxies(Kumar 2017). Some of these satellites are operated by organizations in Japan, Europe, and Canada, such as TerraSAR-X. (X-band), ENVISAT/ASAR (C-band) and ALOS/PALSAR (L-band).

7.4.3 Light Detection and Ranging (Lidar Data)

A lidar uses a laser's pulse of energy to monitor vegetation continuously. Although it cannot penetrate through clouds, it is capable of taking precise measurements of vegetation's three-dimensional structure. Unlike lidar, which has only been used for forestry operations, satellite-based lidar has been extensively used to monitor vegetation(Lim et al. 2003). Although there are few instruments that specifically target vegetation characterization, none of them are designed to operate from space. Lidar missions are planned for launch in the next couple of years. One such mission would be carried out on DESDynI, which is a project of the European Space Agency (ESA).

7.4.4 Remote Sensing and GIS Techniques and Models for SOC Retrieval

For the calculation of forest Above Ground Biomass and carbon stock, several ground-based techniques are employed. Inventory techniques, plots, and field data measurements are examples of these. However, the techniques employed to estimate biomass are impractical for a large-scale evaluation. RS, on the other hand, has been shown to be both practical and cost-effective. It can give real-time data across vast regions and evaluate inaccessible locations. An overview of the most used RS and GIS models and techniques for SOC retrieval is given below:

7.4.5 Traditional Neural Network (TNN) and Deep Learning (DL)

Neural networks, which have several "hidden" layers that transform input data to outputs by comprehending properties at a higher level, are the foundation of deep learning (DL) models. Traditional neural networks (TNN) vary from deep learning (DL) in that they have shallow layers (one or two hidden layers), but a neural network with several hidden layers is referred to as "deep," thus the name "deep learning." Several studies have successfully used RS and neural network models to predict SOM and SOC (Odebiri et al. 2021). While neural network (particularly TNN) models have been extensively utilized, they are usually unpopular, owing in part to high computational needs.

7.4.6 InVEST 3: Carbon Model

It is designed to forecast current carbon storage and its future development by utilizing four carbon reservoirs (aboveground biomass, underground biomass, dead organic matter, and soil carbon) (Pechanec et al. 2017). Based on the mapping of current land cover and topography, the InVEST model identifies unique carbon pools for each investigated landscape segment. The percentage representation of natural, close to natural, and more anthropically influenced environments is used in each sector (pixel). The relevant carbon coefficient (amount of carbon per square meter) is computed and multiplied by the area of each pool before being totaled for each sector based on the percentage representation of habitats in each pool. Future carbon stock predictions are reliant on future land cover changes. The integration of this model for calculating ecosystem services with the land cover prediction model utilizing sophisticated modeling methods such as the Markov chain for calculating the trend of change and Cellular automata for modeling neighbor behavior is a unique aspect of this approach.

7.4.7 *IDRISI SELVA*

Idrisi Selva is a geospatial software created by Clark Labs that provides over 300 analytical tools for the manipulation of geographical information for the study of our constantly changing environment (Huang et al. 2020). It is mainly geared towards raster data and has the following main approaches:

(i) IDRISI IMAGE PROCESSING: It is a set of tools for RS-based image restoration, augmentation, transformation, and classification.
(ii) LAND CHANGE MODELER: LCM is a cutting-edge land modeling and judgment software tool for land change planning and prediction, with a special capability for project-level modeling of Reducing Emissions from Deforestation and Forest Degradation (REDD).
(iii) Habitat and Biodiversity Modeler: It is also known as (HBM) is a vertical program that may be used to evaluate habitat, analyze landscape patterns, and simulate biodiversity.
(iv) Ecosystem service modeler: The (ESM) is a geospatial choice tool for estimating the value of natural assets for long-term development, and it also predicts carbon emissions (Fig. 7.2).
(v) Earth Trend Modeler: ETM is a package of tools for evaluating photographic time – series coupled with remotely sensed Earth Observation photographs.
(vi) Climate Change Adaptation Modeler: CCAM is a suite of tools for forecasting future climate and assessing its implications on sea level rise, agricultural adaptation, and species distributions

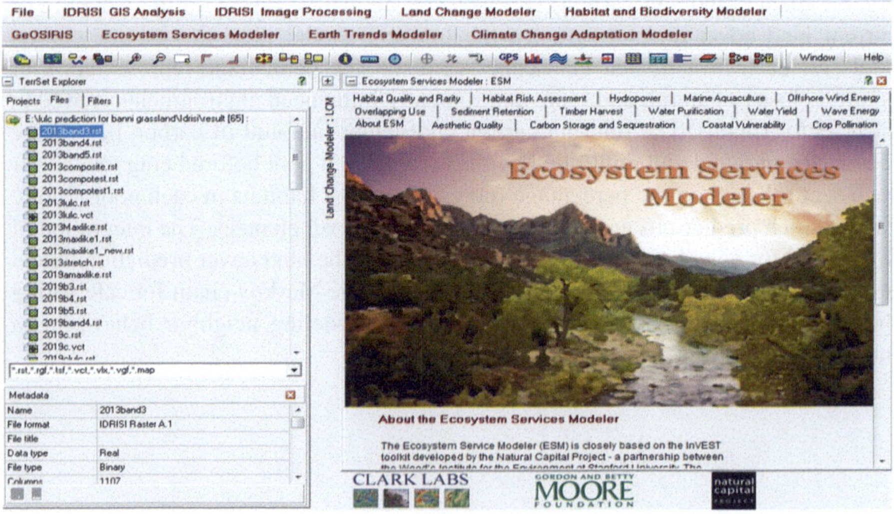

Fig. 7.2 Idrisi Selva's Ecosystem Service Modeler tool interface for assessing carbon storage, sequestration and emissions

7.4.8 Multi Regression Models

This section contains soil estimations from both the field and satellite photos. The Land Use Land Cover (LULC) result evaluates the research area's use for several purposes. The LULC map displays information on the present status of the landscape, including its current condition and how it is changing. The present purposes are quantified in this process (Donmez et al. 2015). The digital SOC map is created using the NDVI map in combination with the band math calculations. This SOC-generated map offers digital values for the same geographical locations as the field-collected sample locations, which are compared using regression analysis.

7.4.9 LUCAS Topsoil and SENTINEL 2B

The LUCAS comprises a massive collection of soil samples from 25 different nations. The soil samples' spectra and physical characteristics are included in the database. Using information from the whole LUCAS dataset and field sample analysis, the laboratory spectra of the selected LUCAS soil samples are re-sampled in accordance with the spectral bands of the MSI/Sentinel-2B satellite for the calibration dataset. As the SOC concentration of the soil increases, its albedo decreases. In the NIR and SWIR bands of the MSI/Sentinel-2, this impact is very pronounced. As a result, the spectra must be re-sampled to MSI/Sentinel-2 bands in order to provide a reference for soil brightness. Next, the correlation of all pairings of two bands within the SOC content must be calculated (Castaldi et al. 2019). In an exponential regression model, these band (B) combinations are employed as SOC indices (CI) to calculate SOC.

7.4.10 Regression Kriging (RK) and Geographically Weighted Regression Kriging (GWRK) Models

This method is typically used to quantify SOC density, which has many characteristics with SOC and is widely used to evaluate SOC storage. Many techniques, such as regression kriging (RK) along with geographically weighted regression kriging (GWRK), have been suggested to address the shortcomings of conventional methods based on the spatial variability of soil characteristics and auxiliary factors. While the ordinary kriging (OK) interpolation methodology highlights the spatial heterogeneity of simulated elements, the multiple linear regression (MLR) method mimics soil properties by employing supplementary variables as predictive factors. RK combines the advantages of the two techniques while avoiding the shortcomings of MLR, which is prone to producing larger residuals and disregards the influence of environmental factors on SOC (Liu et al. 2020). As a result, RK can make more comprehensive and precise forecasts. A global spatial regression model called MLR ignores the effects of spatial heterogeneity and environmental variable instability on SOC in particular places while accounting for the overall influence of environmental variables on SOC spatial distribution over the entire investigation region. Using the locally weighted least squares method for a pointwise parameter estimate and the idea of local smoothing, geographically weighted regression (GWR) proposed a method for incorporating spatial locations into regression parameters. As a local spatial regression model, GWR considers both the spatial dependence of the effect components and the spatial variability of soil properties when predicting the geographic distribution of soil attributes. Like the MLR model, the residual problem of GWR has an impact on forecast precision. In order to address the spatial weight consistency of GWR at various locations and eliminate the residual GWR issue, GWRK, an upgraded GWR, uses the kriging approach. Therefore, simulation outcomes may be predicted more accurately.

7.4.11 Specifications of the Optical Remote Sensing (RS) Sensors Most Frequently Used for Estimating Above-Ground Biomass (AGB)

Natural resource management and biomass assessment have recently benefited from the use of geospatial technology. At a reasonable cost and with sufficient accuracy, RS can collect biomass information across vast regions with repeated coverages. Numerous research have utilized and tested various methods and sensors. RS, both active and passive, is among the most time- and cost-effective methods for estimating aboveground biomass on a regional and national scale. Following are the most popular sensors used to estimate above-ground biomass, along with their specifications and details:

7.4.12 AVHRR

The Advanced Very-High-Resolution Radiometer (AVHRR) is a spatially based device that detects the reflectance of the Earth in five wide spectral bands by twenty-first-century norms. The device scans many channels, two of which are focused on the red (0.6 micrometers) and near-infrared (0.9 micrometers) regions, a third on the 3.5-micrometer range, and two more on the 11 and 12-micrometer ranges, which are used to detect the planet's heat radiation (Balew and Korme 2020). It is used in operational meteorology to detect and measure cloud properties (phase, temperature, and height); in inland analysis to calculate vegetation indices, soil moisture, snow cover, and land surface temperature (skin); and in ocean analysis to calculate sea surface temperature and sea ice concentration, as well as aerosols and fire detection. AVHRR bands 1, 2, and 3A are used for monitoring urban vegetation, snow/ice, daylight clouds, land/water boundaries, snow/ice, floods, and dust (Dech et al. 2021).

7.4.13 *MODIS*

Both the Terra and Aqua satellites use the Moderate Resolution Imaging Spectroradiometer (MODIS) sensor. On December 18, 1999, the Terra satellite was launched, and on May 4, 2002, the Aqua satellite was launched. It has a viewing swath of 2330 kilometers and takes one to two days to observe the whole surface of the Earth (Blackett 2017). It detects 36 spectral bands and has detectors that can record data at three different spatial resolutions: 250 m, 500 m, and 1000 m.

7.4.14 *SPOT VEG*

SPOT 4 and 5, which were released in 1998 and 2002, respectively, are equipped with the SPOT Vegetation sensor. The sensor has a 1.15 km ifov and is intended to track and monitor the Earth on a regional and global scale. Every day, it is capable of picturing the whole Earth. It's especially useful for researching cropland, deforestation, and various large-scale vegetation changes. Vegetation 1 (found in SPOT 4 Satellite) and Vegetation 2 (placed in SPOT 5 Satellite) are the two current Vegetation sensors aboard Spot satellites. SPOT Vegetation products are collected, stored, and made accessible online in a methodical manner. Some of the items are free, while others will cost money. SPOT Vegetation gathers information in four spectral bands. The normalized differential vegetation index is calculated using bands 2 and 3. (NDVI). Band 0 (blue) is used only for atmospheric adjustments(Onyia and Balzter 2018). The sensor has the same bands as the SPOT HRVIR, allowing for connected investigations at various scales. The radiometric resolution of SPOT Vegetation pictures is 8 bits (0–255).

7.4.15 *TM*

This is a multispectral (also known as Landsat 4–5) Earth resources sensor with better picture resolution, finer spectral separation, enhanced geometric fidelity, and higher radiometric accuracy with better resolution than the MSS sensor. TM data are collected in seven spectral bands at the same time. Thermal (heat) infrared radiation is detected by Band 6. Only night sceneries in band six may be acquired by Landsat (Xiao and Weng 2007).

7.4.16 ETM+

Also referred to as Landsat 7, this satellite circles the Earth for approximately 99 minutes on a sun-synchronous, near-polar trajectory at a height of 705 kilometers and an inclination angle of 98.2 degrees. The satellite operates on a 16-day cycle with an equatorial passage time of 10:00 a.m. plus/minus 15 minutes. Landsat 7 data are collected using the Worldwide Reference System-2 (WRS-2) path/row system, with swath overlap (or sidelap) ranging from 7% at the Equator to roughly 85% at high latitudes (Jat et al. 2008).

7.4.17 SPOT

The SPOT program is a constellation of optical RS satellites with the primary objective of gathering Earth pictures for uses in mapping, geographic information systems (GIS), agriculture, forestry, geology, and land use. It has established an international network of command posts, receiving stations, processing facilities, and data distributors devoted to global commercial RS. The SPOT satellites are under the management of CNES, the French Space Agency. The SPOT system provides global coverage between 87 degrees north and 87 degrees south latitude (Mas et al. 2017).

7.4.18 LANSAT 8 OLI

Two sensors on board the LANDSAT 8 TIRS/OLI are the Thermal Infrared Sensor and the Operational Land Imager (OLI). In order to observe an area of the Earth that is 185 kilometers wide swath and has a resolution of 15 to 30 meters, OLI gathers images with nine spectral bands at various visible, near-infrared, and shortwave light wavelengths. This coverage will allow for the identification of features like urban areas, farms, forests, and other land uses (Koparan 2019).

7.4.19 LISS-III (IRS)

The ISRO LISS III (Linear Imaging Self-Scanning Sensor 3) provides multispectral data in four bands. The spatial resolution for visible light (two bands) and near-infrared (one band) is 23.5 meters, with a ground swath of 141 km. The fourth band (short wave infrared band) gives a spatial resolution of 70.5 meters with a ground swath of 148 km. LISS III has a repeatability of 24 days (Jat et al. 2008).

7.4.20 *SENTINEL-2*

This satellite has an optoelectronic multispectral sensor with a resolution of 10 to 60 m in the visible, near-infrared (VNIR), and short-wave infrared (SWIR) spectral zones, with 13 spectral channels, which ensures the snap of differences in vegetation dynamics, including temporal changes (Dvorakova et al. 2021). The GMES (Global Monitoring for Environment and Security) program, which is run in conjunction with the European Commission and the ESA, uses Sentinel-2 data to provide services like land management, agricultural output, and forestry, as well as calamity monitoring and humanitarian help (Castaldi et al. 2019; Demarez et al. 2019; Varade et al. 2019).

7.4.21 *IKONOS*

IKONOS is a high-resolution satellite sensor that is operated by DigitalGlobe. Its Near-Infrared (NIR) resolution at nadir is 3.2 m multispectral and 0.82 m panchromatic. Among the applications are the mapping of natural resources and natural disasters in both urban and rural regions, tax mapping, crop cultivation and ecological research, mining, engineering, construction, and change detection. It might impart knowledge that is helpful for almost all environmental study fields (Bhagat 2013; Kumar 2017).

7.4.22 *WORLD VIEW 2*

It provides a high-resolution panchromatic band, eight multispectral bands, four standard colors (red, green, and blue), and four new bands (coastal, yellow, red edge, and near-infrared 2). These full-color images can be used for spectral assessment, geography, surveillance, land-use planning, disaster assistance, power, protection, and smartness (Jawak and Luis 2011). WorldView-2's improved agility and high altitude allow it to return to every location on Earth in 1.1 days on average, with revisit times falling below one day and never exceeding two days, giving it the most same-day passes of any commercial high-resolution satellite.

7.4.23 *QUICKBIRD*

High spatial resolution panchromatic imaging at 61 cm is possible with the 2001-launched Quickbird spaceborne sensor. The sensor has four multispectral bands as well, each having a nadir resolution of 2.44 m. The picture distinguishes man-made objects on the terrain well enough to provide an effective foundation map for GIS applications. It has mainly been utilized in rangeland management to identify plant types on a fine scale (Petrokofsky et al. 2012). Image capture may be costly with Quickbird images as compared to other high spatial resolution commercial sensors. It is a "tasking satellite," which means it only collects data on demand and does not continuously photograph and archive the globe.

7.4.24 HYMAP

It was created by Sydney, Australia-based Integrated Spectronics Inc. It has four spectrometers with wavelengths ranging from 0.45 to 2.45 micrometers, excluding the two major windows for atmospheric water absorption. The bandwidths, which range from 15 to 18 nanometers, are not constant. A light source calibration device for evaluating signal stability is also built into the scanner. The signal/noise ratio measured outside the aircraft with a sun angle of 30° and a 50% reflectance standard is greater than 500/1, with the exception of the principal atmospheric water absorption bands. The scanner is mounted on a hydraulically driven stabilized Zeiss-Jena SM 2000 platform. The platform can adjust pitch and roll by +/− 5 degrees. With +/− 8 degrees of stability, the yaw may be offset by +/− 20 degrees. The platform has a residual nadir aiming inaccuracy of less than one degree and minimizes airplane motion impacts by a ratio ranging from 10:1 to 30:1 (Kruse et al. 2000).

7.4.25 AVIRIS

A tried-and-true tool in the study of geological RS is AVIRIS. With wavelengths spanning from 400 to 2500 nanometers, it is a unique optical sensor that offers calibrated images of upwelling spectrum radiance in 224 continuous spectral bands (Weng 2009). AVIRIS made trips across Argentina, South America, Europe, and North America. The main objective of the AVIRIS project is to use molecular absorption and particle scattering properties to identify, measure, and track components of the Earth's surface and atmosphere. Understanding processes related to the global environment and climate change is the main aim of the AVIRIS data study (Patil et al. 2012; Bangelesa et al. 2020; Shmilovich et al. 2020).

7.5 The Future of Carbon Modeling Through Remote Sensing Sensors

The number of Earth-observing satellites is increasing, and the amount of remote-sensing data is growing quickly. These satellites will produce a sizable amount of RS data, which will produce data that is intricate and varied in high-dimensional properties. Over the past several decades, the globe has seen extraordinary and fast advancement in the fields of RS, geospatial information collection, and mapping (Goetz et al. 2009). The technique has gained traction in its usage and application in a variety of areas, like monitoring carbon stocks (Yadav et al. 2015; Angelopoulou et al. 2019; Dvorakova et al. 2021).

As the need for geospatial data has grown, different modern techniques like space-borne, airborne, and ground-based RS have been developed and deployed. As a result, many research institutions, government organizations, and commercial sectors are heavily relying on this technology to get timely data as they have the ability to find solutions to more complicated problems. The future use of RS applications in environmental psychology would be contingent on the availability of open satellite data and open-access data analysis tools. RS advances may aid in solving new environmental research problems by enabling studies to combine information ranging from ground-truthing-based field samples to satellite-collected RS data (Chen et al. 2020; Petrokofsky et al. 2012; Li et al. 2013; Bhagat 2013; Bera and Ahmad 2016; Andualem et al. 2018; Haylemariyam 2018; Balew and Korme 2020; Onilude and Vaz 2020).

7.6 Conclusions

The purpose of this chapter was to summarise the many approaches that can be used to monitor and model carbon using RS techniques. Several attempts have been made in various regions of the world to achieve this goal. The RS and GIS method is the most robust, cost-effective, and promising way to predict the pace and future distribution of carbon across the world. It is a program that contains spatial databases capable of collecting and executing geographically referenced data, initiates spatial analysis, integrates different kinds and forms of spatial data, generates spatial models for future scenario prediction, and allows for appropriate carbon management for sustainability. The application of these technologies may be a helpful tool, particularly for developing nations, for measuring, mapping, monitoring, modeling, and managing their carbon stored in biomass and soil, resulting in increased soil and plant productivity, increased food security, and control of land degradation. In response, by decreasing carbon emissions, these nations may play a significant role in minimizing the harmful effects of climate change.

References

Abady L, Barni M, Garzelli A, Tondi B (2020) SeptemberGAN generation of synthetic multispectral satellite images. In: Image and signal processing for remote sensing XXVI (vol 11533, pp. 122–133). SPIE

Andualem TG, Belay G, Guadie A (2018) Land use change detection using remote sensing technology. J Earth Sci Climat Change 9(10):6

Angelopoulou T, Tziolas N, Balafoutis A, Zalidis G, Bochtis D (2019) Remote sensing techniques for soil organic carbon estimation: a review. Remote Sens 11(6):676

Awange JL, KyaloKiema JB, Awange JL, KyaloKiema JB (2013) Optical remote sensing. Environmental Geoinformatics: Monitoring and Management, pp119–132

Balasubramanian A (2017) Soil forming processes soil forming factors (February), pp 1–8. https://doi.org/10.13140/RG.2.2.34636.00644

Balew A, Korme T (2020) Monitoring land surface temperature in Bahir Dar city and its surrounding using Landsat images. Egypt J Remote Sens Space Sci. National Authority for Remote Sensing and Space Sciences 23(3):371–386. https://doi.org/10.1016/j.ejrs.2020.02.001

Bangelesa F, Adam E, Knight J, Dhau I, Ramudzuli M, Mokotjomela TM (2020) Predicting soil organic carbon content using hyperspectral remote sensing in a Degraded Mountain Landscape in Lesotho. Appl Environ Soil Sci Hindawi 2020:1–11. https://doi.org/10.1155/2020/2158573

Bera S, Ahmad M (2016) Site suitability analysis using remote sensing & Gis for rain water harvesting. Int J Geol Earth Environ Sci 6(2):101–110. Available at: http://www.cibtech.org/jgee.htm

Bhagat SV (2013) Use of remote sensing techniques for robust digital change detection of land: a review. Recent Patents Space Technol 2(2):123–144. https://doi.org/10.2174/1877611611202020123

Bhunia GS, Shit PK, Pourghasemi HR (2017) Soil organic carbon mapping using remote sensing techniques and Soil organic carbon mapping using remote sensing techniques and multivariate regression model, Geocarto Int. Taylor & Francis, (September), pp 0–1. https://doi.org/10.1080/10106049.2017.1381179

Blackett M (2017) An overview of infrared remote sensing of volcanic activity. J Imaging 3(2). https://doi.org/10.3390/jimaging3020013

Blanco CH, Shapiro CA, Wortmann CS, Drijber RA, Mamo M, Shaver TM, Ferguson RB (2013) Soil organic carbon: the value to soil properties. J Soil Water Conserv 68(5):129A–134A

Brack D (2019) Forests and climate change Duncan Brack in the fourteenth session of the United Nations Forum on Forests, United Nations Forum on Forests, (c), p 56

Castaldi F, Chabrillat S, Don A, van Wesemael B (2019) Soil organic carbon mapping using LUCAS topsoil database and Sentinel-2 data: an approach to reduce soil moisture and crop residue effects. Remote Sens 11(18):2121

Chen Q (2010) Assessment of terrain elevation derived from satellite laser altimetry over mountainous forest areas using airborne lidar data. ISPRS J Photogram Remote Sens. Elsevier B.V. 65(1):111–122. https://doi.org/10.1016/j.isprsjprs.2009.09.004

Chen L, Li M, Huang F (2013) Relationships of LST to NDBI and NDVI in Wuhan City based on Landsat ETM + image (Cisp), pp 840–845

Chen M, Zhou Y, Hu M, Zhou Y (2020) Influence of urban scale and urban expansion on the urban heat Island effect in metropolitan areas: case study of Beijing–Tianjin–Hebei urban agglomeration. Remote Sens 12(21):3491

Cowie RH, Bouchet P, Fontaine B (2022) The sixth mass extinction: fact, fiction or speculation? Biol Rev 97:640–663. https://doi.org/10.1111/brv.12816

Dahdouh F, Verheyden A, De Genst W, Hettiarachchi S, Koedam N (2000) Four decade vegetation dynamics in Sri Lankan mangroves as detected from sequential aerial photography: a case study in Galle. Bull Mar Sci 67(2):741–759

Dech S, Holzwarth S, Asam S, Andresen T, Bachmann M, Boettcher M, Dietz A, Eisfelder C, Frey C, Gesell G, Gessner U (2021) Potential and challenges of harmonizing 40 years of AVHRR data: the TIMELINE experience. Remote Sens 13(18):3618

Demarez V, Helen F, Marais Sicre C, Baup F (2019) In-season mapping of irrigated crops using Landsat 8 and Sentinel-1 time series. Remote Sens 11(2):118

Dhingra S, Kumar D (2020) Natural computing based optimization for hyper spectral satellite images, In: Proceedings of the 2nd International Conference on Inventive Research in Computing Applications, ICIRCA 2020, pp 513–519. https://doi.org/10.1109/ICIRCA48905.2020.9182812

Donmez C, Berberoglu S, Erdogan MA, Tanriover AA, Cilek A (2015) Response of the regression tree model to high resolution remote sensing data for predicting percent tree cover in a Mediterranean ecosystem. Environ Monit Assess 187:1–12

Dopper V, Jagdhuber T, Holtgrave AK, Heistermann M, Francke T, Kleinschmit B, Forster M (2022) Following the cosmic-ray-neutron-sensing-based soil moisture under grassland and forest: exploring the potential of optical and SAR remote sensing. Sci Remote Sens 5:100056

Dvorakova K, Heiden U, Van Wesemael B (2021) Sentinel-2 exposed soil composite for soil organic carbon prediction (pp 1–21)

Edrisi S, Tripathi V (2018) Managing soil resources for sustainable development. J Clean Prod 174:199–200

Edrisi SA, Dubey PK, Chaturvedi RK, Abhilash PC (2022) Bioenergy crop production potential and carbon mitigation from marginal and degraded lands of India. Renew Energy 192:300–312

Edrisi SA, Singh A, Dubey PK, Abhilash PC (2023) *Tectona grandis* lf. Mediated restoration of marginal lands in eastern Uttar Pradesh, North India. Land Degrad Dev 34(4):929–942

Ghabi M, Das KK (2018) Assessment of biomass in Magra and Dhanaulti reserve Forest (Missouri Forest) using remote sensing and GIS. https://doi.org/10.13140/RG.2.2.17085.97766

Goetz SJ, Baccini A, Laporte NT, Johns T, Walker W, Kellndorfer J, Houghton RA, Sun M (2009) Mapping and monitoring carbon stocks with satellite observations: a comparison of methods. Carbon Balance Manag 4(1):1–7

Haylemariyam MB (2018) Detection of land surface temperature in relation to land use land cover change: Dire Dawa City, Ethiopia. J Remote SensGis 7(3):245

Holloway J, Helmstedt KJ, Mengersen K, Schmidt M (2019) A decision tree approach for spatially interpolating missing land cover data and classifying satellite images. Remote Sens 11(15):1796

Huang Y, Yang B, Wang M, Liu B, Yang X (2020) Analysis of the future land cover change in Beijing using CA–Markov chain model. Environ Earth Sci 79(2):60

Jat MK, Garg P, Khare D (2008) Monitoring and modelling of urban sprawl using remote sensing and GIS techniques. Int J Appl Earth Obs Geoinf 10(1):26–43. https://doi.org/10.1016/j.jag.2007.04.002

Jawak SD, Luis AJ (2011) Applications of WorldView-2 satellite data for extraction of polar spatial information and DEM of Larsemann Hills, East Antarctica (February). https://doi.org/10.13140/RG.2.1.3291.5043

Koparan MH (2019) Estimating soil organic carbon in cultivated soils using soil test data, remote sensing imagery from satellites (Landsat 8 and PlantScope), and Web Soil Survey Data

Kruse FA, Boardman JW, Lefkoff AB, Young JM, Kierein Young KS, Cocks TD, Jensen R, Cocks PA (2000) August. HyMap: an Australian hyperspectral sensor solving global problems-results from USA HyMap data acquisitions. In: Proceeding of the 10th Australasian Remote Sensing and Photogrammetry Conference (pp 18–23)

Kumar L (2017) Remote sensing of above-ground biomass (pp 1–8). https://doi.org/10.3390/rs9090935

Levi N, Karnieli A, PazKagan T (2022) Airborne imaging spectroscopy for assessing land-use effect on soil quality in drylands. ISPRS J Photogram Remote Sens Elsevier BV 186(December 2021):34–54. https://doi.org/10.1016/j.isprsjprs.2022.01.018

Licht S (2017) Co-production of cement and carbon nanotubes with a carbon negative footprint. J CO2 Util 18(August 2016):378–389. https://doi.org/10.1016/j.jcou.2017.02.011

Lim K, Treitz P, Wulder M, St Onge B, Flood M (2003) LiDAR remote sensing of forest structure. Prog Phys Geogr 27(1):88–106

Liu T, Zhang H, Shi T (2020) Modeling and predictive mapping of soil organic carbon density in a small-scale area using geographically weighted regression kriging approach. Sustain For 12(22):9330

Liu W, Zhu M, Li Y, Zhang J, Yang L, Zhang C (2021) Assessing soil organic carbon stock dynamics under future climate change scenarios in the Middle Qilian Mountains. Forests 12(12):1698

Li ZL, Wu H, Wang N, Qiu S, Sobrino JA, Wan Z, Tang BH, Yan G (2013) Land surface emissivity retrieval from satellite data. Int J Remote Sens 34(9–10):3084–3127

Mas JF, Lemoine Rodríguez R, Gonzalez Lopez R, Lopez Sanchez J, PinaGarduno A, Herrera Flores E (2017) Land use/land cover change detection combining automatic processing and visual interpretation. Eur J Remote Sens 50(1):626–635

Mayr A, Bremer M, Rutzinger M, Geitner C (2019) Unmanned aerial vehicle laser scanning for erosion monitoring in alpine grassland. ISPRS Annals of Photogrammetry, Remote Sensing & Spatial Information Sciences, p 4

Mitchell AL, Rosenqvist A, Mora B (2017) Current remote sensing approaches to monitoring forest degradation in support of countries measurement, reporting and verification (MRV) systems for REDD+. Carbon Bal Manage Springer International Publishing 12(1). https://doi.org/10.1186/s13021-017-0078-9

Morais TG, Teixeira RFM, Domingos T (2019) Detailed global modelling of soil organic carbon in cropland, grassland and forest soils. PLoS One 14(9):1–27. https://doi.org/10.1371/journal.pone.0222604

Nowak DJ (2010) Assessing urban forest effects and values: Los Angeles' urban forest. Forest Service, Northern Research Station. US Department of Agriculture

Odebiri O, Odindi J, Mutanga O (2021) International journal of applied earth observations and geoinformation basic and deep learning models in remote sensing of soil organic carbon estimation: a brief review. Int J Appl Earth Observ Geoinform Elsevier BV 102(June):102389. https://doi.org/10.1016/j.jag.2021.102389

Ogunseitan O (2005) Cycling of carbon and nitrogen

Olson KR (2010) Impacts of tillage, slope, and erosion on soil organic carbon retention. Soil Sci 175(11):562–567. https://doi.org/10.1097/SS.0b013e3181fa2837

Onilude OO, Vaz E (2020) Urban sprawl and growth prediction for Lagos using GlobeLand30 data and cellular automata model. Science 2(4):80. https://doi.org/10.3390/sci2040080

Ontl T (2018) Soil carbon storage. Soil Carbon Storage (June). https://doi.org/10.1016/c2016-0-03949-9

Onyia NN, Balzter H (2018) Normalized difference vegetation vigour index: A new remote sensing approach to biodiversity monitoring in oil polluted regions. https://doi.org/10.3390/rs10060897

Ottinger M, Kuenzer C (2020) Spaceborne L-band synthetic aperture radar data for geoscientific analyses in coastal land applications: a review. Remote Sens 12(14). https://doi.org/10.3390/rs12142228

Park SH, Jung HS, Lee S, Kim ES (2021) Mapping Forest vertical structure in Sogwang-ri Forest from full-waveform Lidar point clouds using deep neural network. Remote Sens 13(18):3736

Patil MB, Desai CG, Umrikar BN (2012) Image classification tool for land use/land cover analysis: a comparative study of maximum likelihood and minimum distance method. Int J Geol, Earth Environ Sci 2(3):189–196

Pechanec V, Purkyt J, Benc A, Nwaogu C, Lenka S (2017) Ecological informatics modelling of the carbon sequestration and its prediction under climate change

Petrokofsky G, Kanamaru H, Achard F, Goetz SJ, Joosten H, Holmgren P, Lehtonen A, Menton MC, Pullin AS, Wattenbach M (2012) Comparison of methods for measuring and assessing carbon stocks and carbon stock changes in terrestrial carbon pools. How do the accuracy and precision of current methods compare? A systematic review protocol. Environ Evid 1:1–21

Reddy GO, Nagaraju MSS, Ramteke IK, Sarkar D (2013) Terrain characterization for soil resource mapping using IRS-P6 data and GIS-A case study from basaltic terrain of Central India. J Indian Soc Remote Sens 41:331–343
Shaheen SA, Lipman TE (2007) Reducing greenhouse emissions and fuel consumption. IATSS Res 31(1):6–20. https://doi.org/10.1016/s0386-1112(14)60179-5
Shirato Y, Yokozawa M (2005) Applying the Rothamsted carbon model for long-term experiments on Japanese Paddy soils and modifying it by simple tuning of the decomposition rate. Soil Sci Plant Nutr 51(3):405–415. https://doi.org/10.1111/j.1747-0765.2005.tb00046.x
Shmilovich S, Oiknine Y, AbuLeil M, Abdulhalim I, Blumberg DG, Stern A (2020) Dual-camera design for hyperspectral and panchromatic imaging, using a wedge shaped liquid crystal as a spectral multiplexer. Sci Rep 10(1):3455
Sierra CA, Müller M, Trumbore SE (2012) Models of soil organic matter decomposition: the SoilR package, version 1.0. Geosci Model Dev 5(4):1045–1060
Stallinga P (2018) Carbon dioxide and ocean acidification. Eur Sci J, ESJ 14(18):476. https://doi.org/10.19044/esj.2018.v14n18p476
Sun W, Liu X (2020) Review on carbon storage estimation of forest ecosystem and applications in China. Forest Ecosyst 2000
Sunquist ET, Broecker WS (1986) The carbon cycle and atmospheric CO 2. EOS Trans Am Geophys Union 67(15):191. https://doi.org/10.1029/eo067i015p00191
Varade D, Sure A, Dikshit O (2019) Potential of Landsat-8 and Sentinel-2A composite for land use land cover analysis. Geocarto Int 34(14):1552–1567. https://doi.org/10.1080/10106049.2018.1497096
Wang X, Wang Z (2022) Microwave emissivity of typical vegetated land types based on AMSR2. Remote Sens 14(17). https://doi.org/10.3390/rs14174276
Weng Q (2009) Remote sensing and GIS integration. McGraw-Hill Professional Publishing, New York
Wigley TML, Jones PD, Kelly PM (1981) Global warming. Nature 291(5813):285. https://doi.org/10.1038/291285a0
Xiao H, Weng Q (2007) The impact of land use and land cover changes on land surface temperature in a karst area of China. J Environ Manag 85(1):245–257. https://doi.org/10.1016/j.jenvman.2006.07.016
Xu Z, Gao C (2015) Graphene fiber: a new trend in carbon fibers. Mater Today Elsevier Ltd 18(9):480–492. https://doi.org/10.1016/j.mattod.2015.06.009
Yadav RS, Pandya IY, Jangid MS (2015) Estimating status of soil organic carbon in tropical forests of Narmada Forest Division, Gujarat, India. Int Res J Environ Sci 4(1):19–23
Zhou X, Chang NB, Li S (2009) Applications of SAR interferometry in earth and environmental science research. Sensors 9(3):1876–1912. https://doi.org/10.3390/s90301876
Zomer RJ, Bossio DA, Sommer R, Verchot LV (2017) Global sequestration potential of increased organic carbon in cropland soils. Sci Rep 7(1):1–8

Chapter 8
Carbon Dynamics and Sequestration Under Past and Future Climate Scenarios over India in Earth System Model Simulations

Archisman Barat, Sunny Kumar, P. Parth Sarthi, and Pradeep Kumar Dubey

Abstract Carbon sequestration refers to the fixing of carbon in the soil, vegetation, oceans, etc., out of which soil is considered an important sink for carbon. CO_2 flux between the soil and atmosphere has a great impact on atmospheric CO_2 concentrations. Earth System models (ESM) are extensively used for the assessment of climate change on regional and global scales. The present study aims to assess the change in carbon storage in various sinks of carbon from ESM of Coupled Models Inter-comparison Project phase 5 (CMIP5) under the past and future time periods using the CanESM2 model data. The present study shows the spatio-temporal trend of carbon sequestration in soil, vegetation, and litter under different representative concentration pathways in the near (2031–2060) and far future (2061–2100). Changes in net ecosystem productivity (NEP) are also analyzed for past and future time periods. The ESM outputs will help the scientific community to understand the carbon flux and fate of carbon sequestration under changing climate.

Keywords Carbon sequestration · CMIP5 · Earth system models · Homogenous monsoon rainfall regions of India

A. Barat · S. Kumar · P. Parth Sarthi (✉)
Department of Environmental Science, Central University of South Bihar, Gaya, Bihar, India
e-mail: drpps@hotmail.com

P. K. Dubey
Institute of Environment and Sustainable Development, Banaras Hindu University, Varanasi, Uttar Pradesh, India

WRI India, AADI, New Delhi, India

S. A. Edrisi et al. (eds.), *Sustainable Plant Nutrition and Soil Carbon Sequestration*, Sustainable Plant Nutrition in a Changing World,
https://doi.org/10.1007/978-3-031-53590-1_8

8.1 Introduction

The increasing atmospheric carbon is a well-known climate concern in recent times. It has been estimated that atmospheric carbon has increased by at least 30% in the last sesquicentenary (ESA 2012). Being one of the major GHGs, carbon dioxide (CO_2) is significantly affecting the climate and resulting in global warming. In estimation, the carbon-di-oxide in the atmosphere is reported to increase from 277 ppm (in the year 1750) to 409.85 ppm (in the year 2019) (Joos and Spahni 2008; Dlugokencky and Tans 2020). To combat this rapid increase in atmospheric CO_2, the removal of CO_2 from the atmosphere is very much required. Carbon sequestration is one of the most important mechanisms by which anthropogenic perturbation could be mitigated. Apart from terrestrial sequestration, the oceanic sequestration by the phytoplankton also plays a great role (Shoenfelt et al. 2018), leading to the birth of theories like Iron fertilization theory in the mitigation of climate change. The last decade (2010–2019) witnessed a massive flux of 9.4 PgC (petagrams of Carbon) per year from fossil fuel burning; however, 2.5–4.3 Pg C yr.$^{-1}$ was sequestered by the vegetation (Friedlingstein et al. 2020). Carbon Sequestration is the process comprising both natural and artificial processes by which CO_2 is removed from the atmosphere (or directly captured from emission sources) and is stored in the ocean, terrestrial environments (vegetation, soils, and sediments), and geologic formations (USGS 2008). In other words, carbon sequestration refers to the phenomenon of transferring carbon from a "dilute source (atmosphere) to a concentrated source (biomass) (Grace 2001). The atmospheric pool is very small if compared with Geologic and Oceanic pools; however, due to anthropogenic emissions, the perturbation in this pool is affecting the earth's climate largely, and the need to study mitigation measures has become indispensable of late. In the vegetation pools, the wetlands and forests play a great role. In the forest, a carbon pool of around 44% is found in the living biomass, and approximately 45% is found as Soil Organic Matter (SOM), along with a fraction in litter, etc. In a recent report, the total carbon stock in forests has been found to decrease to 662 PgC in 2020 from 668 PgC in 1990 (FAO 2020). Mitsch et al. (2013) estimated that the world's wetlands are net carbon sinks of about 830 Tg/year, even after they emit methane, although they contribute to merely 5–8% of the terrestrial landscape, with an approximately 118 g C m^{-2} year^{-1} of net carbon retention. Most of this carbon retention is found to be in wetlands of subtropical and tropical regions. Grace (2001) reported that terrestrial photosynthesis can respond even in elevated CO_2 levels up to 700 ppm; hence, for future climate studies, carbon sequestration estimation is very important. Higher atmospheric CO_2 levels increase carbon uptake in plants; this increased productivity is then transferred to soils in due time (Norby et al. 2005) from the plants through litter fall (Liu et al. 2005) and biomass in roots (Pregitzer et al. 2008). However, under future warmer conditions, it is very uncertain how much of the elevated CO_2 concentrations will be sequestered in the soil carbon pool (Norby and Zak 2011) since the warming will also cause the elevation of carbon efflux in the atmosphere. Due to this, the decomposition rates will increase, causing increased

SOC losses (Hopkins et al. 2012; Hopkins et al. 2014) and respiratory losses of carbon from plant tissues and soils (Drake et al. 2011). Not only this, but warming also enhances heterotrophic respiration (Davidson and Janssens 2006). The sequestration rates vary greatly under different stages of succession, the effect of climate change and global warming, and their counter-responses vary significantly amongst the ecosystems (Beier et al. 2008; Phillips et al. 2011; Peng et al. 2013; Xia et al. 2014; Su et al. 2015). However, carbon sequestration is found to be a global solution to climate change across all major ecosystems. The policies like 4 per mille (4p 1000) by the French government opines that the carbon sequestration measures need to be strongly implemented, and if we become able to increase the carbon content of soils by 0.4% per year (or 4‰ as the name suggests), we can halt the CO_2 increase in atmospheres (www.4p1000.org), thus also can mitigate global warming. Minasny et al. (2017) estimated that the 4 per mille is based on the calculation of the global 2 m profile carbon stock, which totals to a sequestration rate of 9.6 Pg C $yr.^{-1}$; the potential to increase SOC is even more on agricultural lands, which can mitigate around 20–35% of global GHG emissions alone through carbon sequestration. It is a well-known fact that carbon sequestration can be achieved by agricultural practices (Schahczenski and Hill 2009); as India is agriculturally based, the carbon sequestration potential in India needs to be analyzed in detail, especially for the long-term future. The agro-ecosystems are evolving and important to understand with the response to climate change (Dubey et al. 2016, 2020a, b; Sarkar et al. 2020); here, the climate models may provide an insight into the future carbon dynamics. Many of the earth system models have been evaluated across the globe, like BCC_CSM1.1(m), CanESM2, IPSL-CM5A-LR, IPSL-CM5B-LR, MIROC-ESM-CHEM, MPI-ESM-LR, MPI-ESM-MR, CSM4, CESM1(CAM5), CESM1(WACCM), BNU-ESM, CESM1-BGC, GFDL-ESM2G, HadGEM2-ES, NorESM1−ME, IPSL−CM5A−MR, INM−CM4, MPI−ESM−MR by several researchers (Todd-Brown et al. 2014; Negrón-Juárez et al. 2015; Wu et al. 2018) for analyzing the carbon cycle and dynamics. Therefore, the present chapter aims to study the carbon dynamics within the pools (viz soil, litter, etc.) along with the ecosystem productivity over different zones of India under varying emission scenarios in an earth system model (CanESM2) environment.

8.2 Carbon Sequestration Through Earth System Models

It has been envisaged by many researchers that warming generally enhances both the fluxes to and from the soil (Day et al. 2008; Mahecha et al. 2010), but the net global balance between these fluxes are greatly uncertain as most of the measurements are limited to single-site experiments (Crowther et al. 2016). Atmospheric–ocean carbon measurements and field studies hint that, at present, the terrestrial biosphere is found to be behaving like a sink (Houghton 2007; Le Quéré et al. 2009; Todd-Brown et al. 2013) for carbon, although uncertainty prevails regarding the persistence of this sink in the future (Todd-Brown et al. 2013). Site-specific experiments are the most realistic estimates; however, they can neither handle spatial coverage nor give us an idea about the future period. Earth system models provide an entirely new insight by helping the scientific community to look into the future.

Earth system models (ESMs) are the basic tools for assessing the impacts of climatic effects on carbon sequestration and fluxes in carbon pools (Todd-Brown et al. 2014). The carbon cycle in Earth system models (ESMs) incorporates the effects of land use land cover changes (LULCC) and CO2 fluxes primarily based on net primary productivity (NPP) and heterotrophic respiration (Wieder et al. 2015). Coupled Model Intercomparison Project Phase 5, or the CMIP5, is a coordinated project incorporating more than 20 modeling groups across the globe (Taylor et al. 2012). To look into the possible future changes, the CMIP5 ESMs use the representative concentration pathways (hereafter, RCP) (Moss et al. 2010); each RCP suffix (viz. 2.6, 4.5, 6.0, and 8.5) depicts the change in radiative forcing of atmosphere to the pre-industrial values in watt per square meters. RCPs are not predictive tools for imagining the future directly, but they are tools for projecting future changes while incorporating associated uncertainties (Kumar and Dimri 2018).

The RCP 4.5 and 8.5 are most widely used in climate studies; the RCP 4.5 takes an assumption of CO_2 concentration equivalent of 650 ppm after 2100, while the concentration of 850 ppm for RCP 8.5 (Moss et al. 2010). Some studies using projections of modern ESMs show that the land carbon sink may fall in response to climate change by 2100 (Friedlingstein et al. 2006; Koven et al. 2011). On the other hand, a study using 11 ESMs showed a net increase of up to 59% in NPP in the future (Todd-Brown et al. 2014). Multi-model studies around the world have shown that the land carbon uptake can vary between −0.3 to1.5 PgC per year (Friedlingstein et al. 2014), while some other studies found Net primary productivity can increase in the range of 46–93 PgC per year in the future compared to the historical data (Todd-Brown et al. 2014).

8.2.1 Terrestrial Carbon Pools and Key Model Variables for Carbon Sequestration Studies

The IPCC guidance document (Penman et al. 2003) described terrestrial pools under the following

Living Biomass: This contains aboveground biomass like stems, bark, branches, seeds, bark, foliage stumps, etc., and belowground biomass like roots above 2 mm in diameter.

Dead organic matter: This pool contains dead wood and litter (including the humic layers); sometimes, a live fine root less than 2 mm in diameter is also included.

Soils: The SOM and organic carbon in minerals and organic soils, including peat. Live fine roots may also be incorporated.

The earth system model variables contain the following:

Net Ecosystem Productivity: The Net Ecosystem Productivity (NEP) is an estimate of the net carbon flux to the ecosystem by taking into account the losses by plant and heterotrophic respiration and microbial losses to the atmosphere.

$$NEP = \left(GPP - R_a\right) - R_h$$

Where GPP is gross primary productivity or the overall photosynthetic biomass gain, R_a is autotrophic respiration, and R_h is heterotrophic respiration. Hence, it is the more stable method to assess the actual carbon sequestered in the biomass from the atmosphere. In models, the NEP is expressed in kg m^{-2} s^{-1} (in this research, the variable in the model is represented by the standard name "nep" and the parameter is denoted by "NEP" hereafter). Some of the other important parameters and their standard variable names (in parenthesis) are as follows.

$C_{(Leaf)}$(cLeaf): This is the carbon mass in leaves and is expressed in kg m^{-2}
$C_{(Litter)}$(cLitter): This is the carbon mass in litter and is expressed in kg m^{-2}
$C_{(Root)}$(cRoot): This is the carbon mass in roots and is expressed in kg m^{-2}
$C_{(Soil)}$(cSoil): This is the carbon mass in the soil pool and is expressed in kg m^{-2}
$C_{(Veg)}$(cVeg): This is the carbon mass in vegetation and is expressed in kg m^{-2}
$C_{(Wood)}$(cWood): This is the carbon mass in wood and is expressed in kg m^{-2}

Along with these, variables like leaf area index (LAI), gross primary productivity (GPP), transpiration, autotrophic and heterotrophic respiration, etc., are simulated by the earth system models.

8.3 Carbon Sequestration Dynamics Through the CanESM2 Model in India

India is the third-largest emitter (Andrew 2020) of CO_2; hence, it is very important to assess the possible future changes in carbon flux dynamics and sequestration for every region of India. Carbon sequestration combating GHG emissions can be achieved by agricultural practices (Schahczenski and Hill 2009), as India is an agricultural-dependent country; hence, the carbon sequestration potential in India needs to be analyzed in detail. As we know, carbon sequestration is very much dependent on the precipitation regime (Dai et al. 2015); hence, we will look into the carbon sequestration pattern for each of the Homogenous Monsoon Rainfall Regions (Fig. 8.1) of India (Parathasarathy et al. 1994).

Carbon sequestration studies all over the world are mainly based on the particular vegetation type or biome dependent. Some are site-specific; hence, there is a much need to focus on region-specific carbon sequestration studies. The models can provide a good opportunity here. The CanESM2 model is found to be very useful in climate modeling for the Indian subcontinent. In a recent study by Kumar et al.

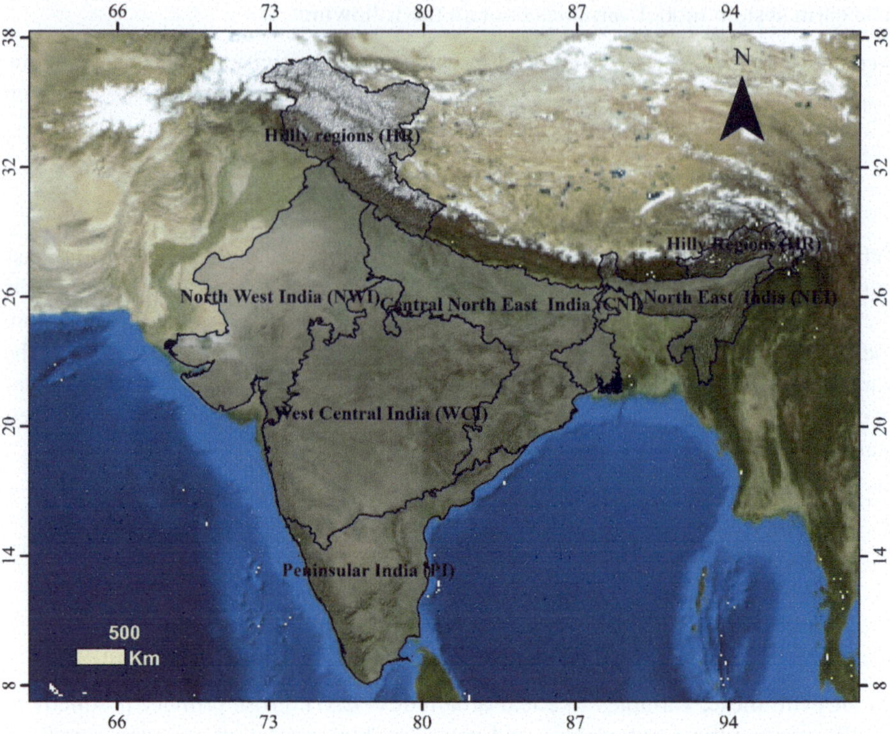

Fig. 8.1 Different categories of the homogenous monsoon regions of India. (Insight from Parathasarathy et al. (1994))

(2020b), *the CanESM2 model was found to perform* best among the other models chosen for most of the zones in India. India is a very fast-developing nation, resulting in massive land use and land cover changes (LULCC) in recent decades. LULCC is often linked with global climatic changes as they alter albedo, heat, and moisture fluxes (Kharol et al. 2013) and evapotranspiration rates at regional levels, which in turn also affect the global climate. LULCC is also infamous for modifying hydrology (Aggarwal et al. 2012; Garg et al. 2012) and has an important effect on ecosystems (Mishra et al. 2014; Karwaria and Tripathi 2012). LULCC changes are so notorious that they can severely affect biogeochemical cycles and influence climate change from regional to global scales (Tian et al. 2014). Houghton (2012) reported a 35% contribution of LULC changes in anthropogenic CO_2 emissions across the globe. Hence, region-wise studies on spatially resolved data are also very important to understand the dynamics of carbon sequestration in the Indian context. The vast spatial studies on carbon flux are almost impossible without the use of ESMs, as carbon flux calculations in the field are very expensive and gigantic tasks. Some of the researchers in India in the past used the simpler biogeochemical models like BIOME-BGC (Kumar and Raghubanshi 2011) or DNDC (Singh and Benbi 2020); however, rigorous evaluation on a country scale is yet to be done for ESMs in India due to coarse resolution and computational limitations associated with ESMs. However, some very recent studies used multi-model analysis to evaluate carbon fluxes in India (Rao et al. 2019). It is a known fact that the future climate will show a warming effect over India (Sarthi 2019), so it will be very interesting to study the future dimension of carbon fluxes in the changing climate. In the present study, we will try to visualize the carbon dynamics in the ESM environment using a state-of-the-art CanESM2 model.

8.3.1 Carbon Dynamics in Different Homogenous Monsoon Rainfall Regions of India in the Recent Past and Future

In the present research, carbon dynamics in the study area are investigated using ESM data. In this modeling approach, the key factor was to select the variables of interest, which can give an inkling of possible C-dynamics in the future. For this purpose, variables representing carbon in soil, litter, and vegetation (total biomass) pool have been considered. Furthermore, NEP was also considered, as it is a more accurate measure of actual carbon sequestered than GPP or NPP.

8.3.1.1 Datasets Used and Processing of the Datasets

In the present research, the CMIP5 dataset for the CanESM2 model is used by the Canadian Centre for Climate Modeling and Analysis (CCCma), Government of Canada (https://climate-modelling.canada.ca) for historical and RCP 4.5and RCP 8.5 is used. The details are given in Table 8.1.

The data are extracted and quantified on the 5D environment of the Grid analysis and display system (GrADS) using built-in functions over each of the HMRRI zones (hereafter, zones). The plotting is done on R software, and the maps are created using ArcGIS 10.1 and GrADS.

8.3.1.2 Model Variables and Multicollinearity Analysis

The $C_{(Soil)}$, $C_{(litter)}$, $C_{(Veg)}$, and NEP are selected and analyzed to envisage multicollinearity in the data series extracted over each of the zones. The extracted data for each model variable (cSoil, cLitter, cVeg & nep) over each zone is correlated with each other for intra-annual (monthly climatology) cycle Fig. 8.2a and inter-annual time series Fig. 8.2b for the historical experiment. It was observed that the intra-annual variations follow a specific pattern within and between the variables, or in simple terms, the seasonal dynamics or variation within a year for carbon pools is well represented by the model.

Table 8.1 Specifications of the CanESM2 model

Model name	Spatial resolution	Land carbon component[a]	Ocean carbon component[a]	Land use emissions[a]	Experiment	Year considered (Present study)	Variables considered
Second Generation Canadian Earth System Model (CanESM2)	2.79×2.81[b]	Canadian Terrestrial Ecosystem Model (CTEM)	Canadian Model of Ocean Carbon (CMOC)	Computed	Historical	1961–2005	cSoil, cLitter, cVeg, NEP
					RCP 4.5	2031–2100	
					RCP 8.5	2031–2100	

[a]Arora et al. (2011)
[b]Todd-brown et al. (2013)

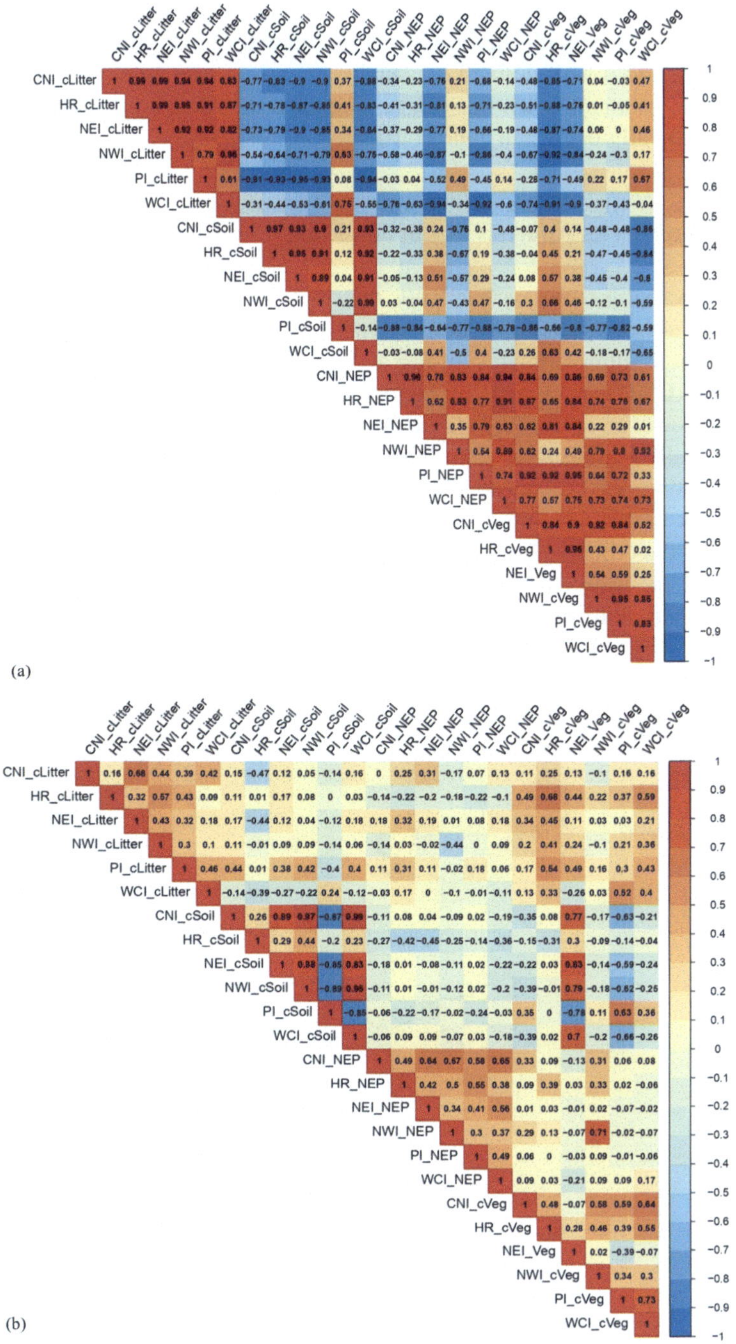

Fig. 8.2 Correlation matrix showing multicollinearity analysis for (**a**) Intra-annual (**b**) Inter-annual series of the selected variables for historical simulation of CanESM2 model

Inter-zonal agreement in intra-annual series is also seen for most of the variables, which may be linked with the rainfall and temperature annual pattern over India, which also shows a well-defined seasonality. The inter-annual series, on the other hand, shows lesser similarity, thus suggesting the trend of carbon dynamics must be studied specifically over each zone considered.

8.3.1.3 Time Series Analysis of Carbon Flux

Out of the variables, the carbon pool in soil (cSoil), carbon pool in vegetation (cVeg), and net ecosystem productivity (nep) are major variables to study carbon sequestration in future climates. In Fig. 8.3a–f, we can see that the cSoil in Central North India (CNI) is decreasing in historical simulation. The trend will continue in the future; however, under RCP 4.5, a minor gain will be noticed in the far future.

For the Hilly regions (HR), a higher emission scenario is showing a great loss in the soil carbon pool, whereas the RCP 4.5 scenario projects a more or less similar carbon stock in the near future. However, things will deteriorate in the far future. The NEI, on the other hand, is showing great sequestration potential in the future if the RCP 4.5 pathway can be maintained. However, this potent situation is somewhere inhibited in the business as usual (RCP 8.5) scenario, although an increase is noticed in both scenarios. While North West India (NWI) shows a net loss in both

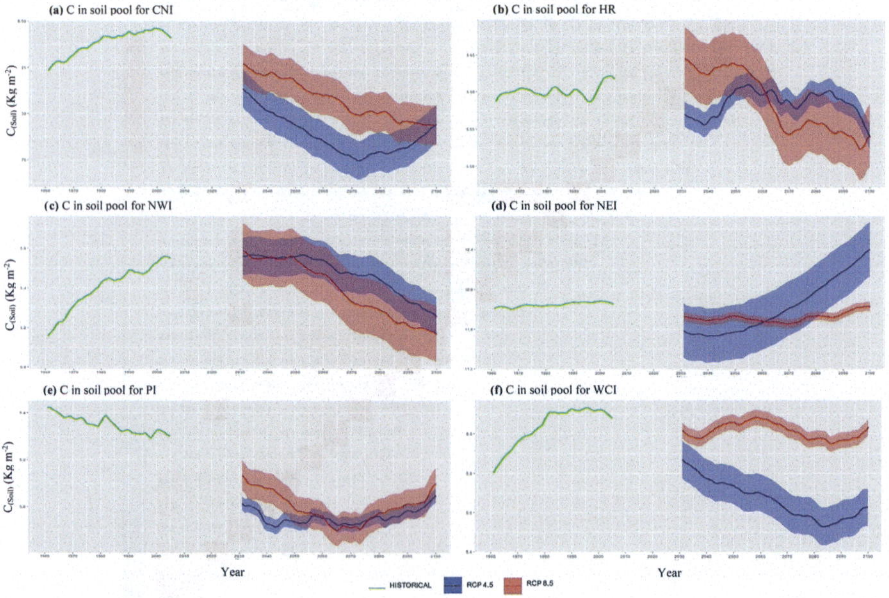

Fig. 8.3 Time Series analysis of carbon in soil pool (the shaded regions indicate ±1σ). (**a**) C in soil pool for CNI. (**b**) C in soil pool for HR. (**c**) C in soil pool for NWI. (**d**) C in soil pool for NEI. (**e**) C in soil pool for PI. (**f**) C in soil pool for WCI

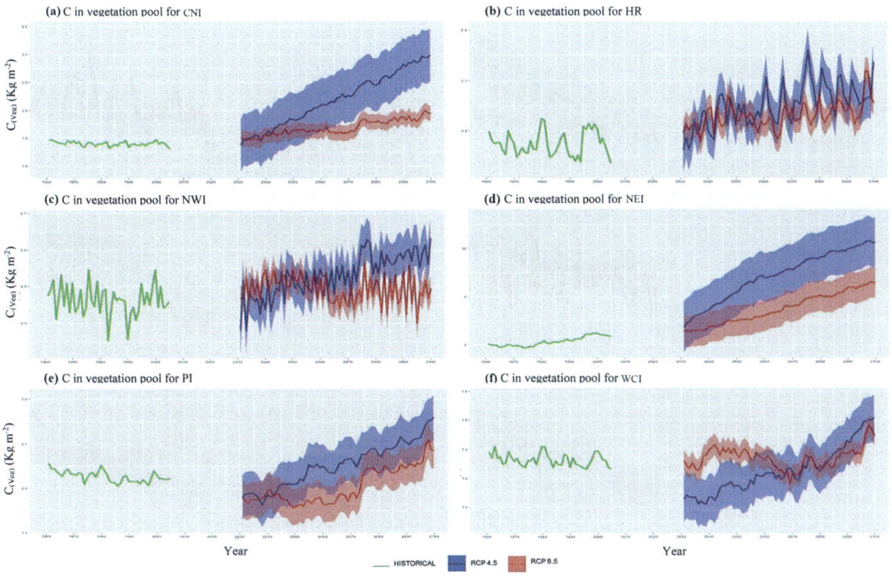

Fig. 8.4 Time Series analysis of carbon in vegetation pool (the shaded regions indicate ±1σ). (**a**) C in vegetation pool for CNI. (**b**) C in vegetation pool for HR. (**c**) C in vegetation pool for NWI. (**d**) C in vegetation pool for NEI. (**e**) C in vegetation pool for PI. (**f**) C in vegetation pool for WCI

scenarios, Peninsular India (PI) shows a gain in the far future. From Fig. 8.4a–f, We can infer that in RCP 4.5, all of the regions are showing a good increase in carbon in vegetation (cveg), which can further strengthen the fact of higher carbon sequestration and increased plant biomass under a warmer climate. However, it is also seen that if the RCP 8.5 scenario is concerned, the net carbon sequestration efficiency in vegetation will be reduced, especially in NEI and CNI.

Figure 8.5a–f shows the NEP time series. It is a known fact that the NEP is the most crucial parameter in carbon sequestration studies, as NEP considers the respiration losses also; hence, a positive NEP will give us an inkling of effective carbon sequestration even if some loss is noticed from a carbon pool. The NEP trends in different zones give a very positive sign of carbon sequestration potential under a warmer future climate. Even though the NEP is relatively lesser in higher emission scenarios, it is at a greater level than that of today, leading to the fact that carbon sequestration can play a major role in climate resilience.

8.3.2 *Trends of Carbon Pools in Past and Future Climate*

The soil carbon pool has been found to be slightly increasing over Uttar Pradesh, Madhya Pradesh, and Gujrat, but more or less, other parts of the country are exhibiting a decrease in the period of 1961–2005 (Fig 8.6a). In the future period,

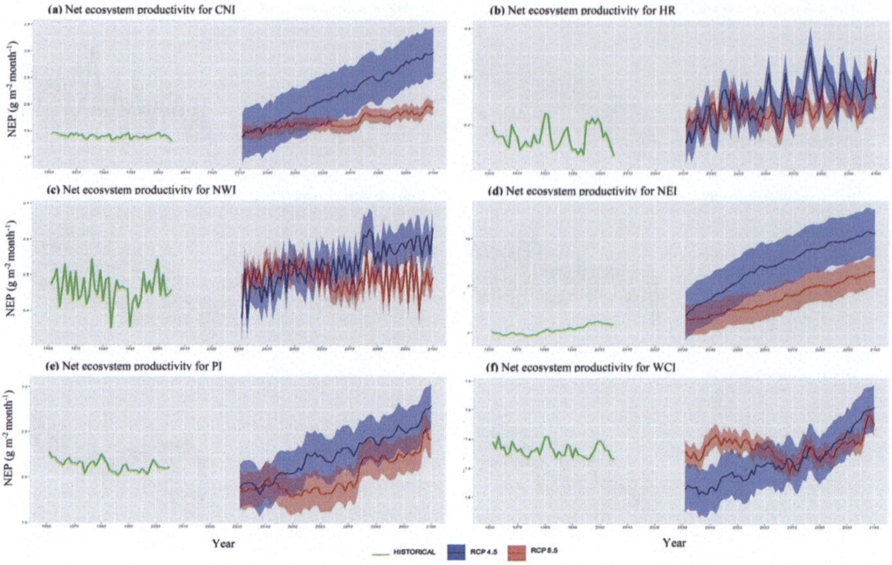

Fig. 8.5 Time Series analysis of net ecosystem productivity (the shaded regions indicate ±1σ). (**a**) Net ecosystem productivity for CNI. (**b**) Net ecosystem productivity for HR. (**c**) Net ecosystem productivity for NWI. (**d**) Net ecosystem productivity for NEI. (**e**) Net ecosystem productivity for PI. (**f**) Net ecosystem productivity for WCI

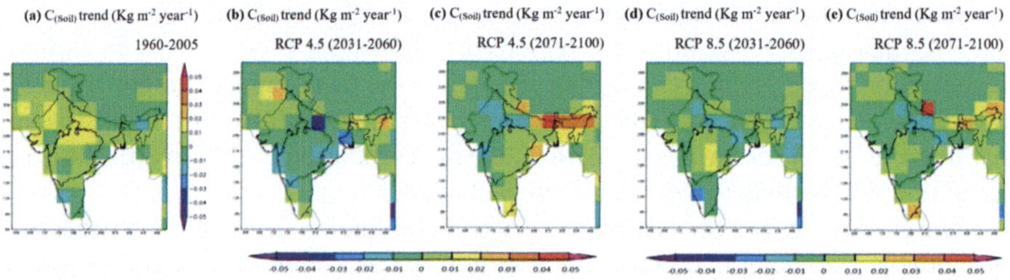

Fig. 8.6 The trend of carbon in soil pool for historical (**a**); near future (**b**, **d**); and far future (**c**, **e**) period

Fig. 8.6b–e, a little decreasing trend is noticed in the soil carbon pool. However, an increase in cSoil is witnessed by NEI for both RCPs. It is also worth mentioning that the trend in the far future is getting better than in the near future across the country except for NWI. The HR is showing a constant downward trend in all scenarios, leading to a net loss of the pool in CanESM2 model simulations. The cVeg trends are starkly different in all the scenarios Fig. 8.7a–e. However, it is found that in the near future, in a higher emission scenario, a net increase in the vegetation carbon pool will be noticed across the country. Tables 8.2, 8.3, and 8.4 further showcase a statistically significant increasing trend in vegetation carbon pool that can be noticed in most parts of the country in almost all seasons in the future in RCP 4.5, except for NWI, which is showing a decreasing trend in soil carbon pool. In a higher emission

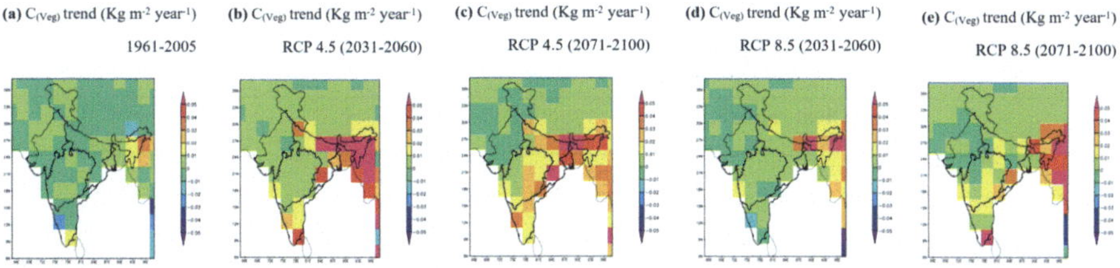

Fig. 8.7 The trend of carbon in vegetation pool for historical (**a**); near future (**b**, **d**); and far future (**c**, **e**) period

scenario, the carbon in the vegetation pool is found to be increasing in all seasons in a majority of the zones. Therefore, we can conclude that a significant increasing trend is expected in the vegetation carbon pool that, in turn, emphasizes the effectiveness of carbon sequestration in a future climate.

8.3.3 Model Limitations

All the models come with an amount of uncertainty. The present study is focused on the possible dynamics of carbon across pools in the CanESM2 model environment and not on the model evaluation; hence, the assessment of the model's limitations is beyond the scope of current research. However, it is worth mentioning that some of the previous studies found some general drawbacks in CMIP5 models; Negrón-Juárez et al. (2015) in their study found a poor allocation of biomass and productivity if specific carbon pools in leaf, root, and wood are concerned in the tropics. Wu et al. (2018) concluded from the outcomes of their global study that an underestimation of soil carbon and biomass turnover time exists in CMIP5 models; this, in turn, may be translated to an overestimation of carbon sequestration potential. Todd-Brown et al. (2014) *opined that global soils have good sequestration potential using CMIP5 model data.* This type of contrasting finding is often found across modeling studies. Hence, the uncertainty in the models must be taken into account while considering the research outputs. The previous CMIP5 model studies (Kumar et al. 2020a, b) show that the models may perform differently depending on the regional influence. Thus, the model outputs must be considered region-specific and as a tool to sense the possible changes in the future rather than considering it as an absolute value based on studies from other regions of the world or from results validated for coarser or finer systems.

8.4 Concluding Remarks

The Earth System models are a great tool for looking into future changes. However, there is a high degree of variation among the models in many global studies. Also, due to the low spatial resolution, the ESM outputs cannot be used as point

Table 8.2 Mann- Kendal trends of Carbon Flux within the CanESM2 model for the historical period (1961–2005)

Zone and Season		NEP	$C_{(Veg)}$	$C_{(Soil)}$	$C_{(Litter)}$
Hilly Regions (HR)	Winter	⊕	⊕	↑	⊕
	Pre-Monsoon	⊕	⊕	↑	⊕
	Monsoon	⊕	⊕	↑	⊕
	Post-Monsoon	⊕	⊕	↑	⊕
Central North India (CNI)	Winter	⊕	↓	↑	⊕
	Pre-Monsoon	⊕	↓	↑	⊕
	Monsoon	⊕	⊕	↑	⊕
	Post-Monsoon	⊕	⊕	↑	⊕
West Central India (WCI)	Winter	⊕	↓	↑	⊕
	Pre-Monsoon	⊕	↓	↑	⊕
	Monsoon	⊕	⊕	↑	⊕
	Post-Monsoon	⊕	⊕	↑	⊕
North East India (NEI)	Winter	⊕	↑	↑	⊕
	Pre-Monsoon	⊕	↑	↑	⊕
	Monsoon	⊕	↑	↑	⊕
	Post-Monsoon	⊕	↑	↑	⊕
North West India (NWI)	Winter	⊕	⊕	↑	⊕
	Pre-Monsoon	⊕	⊕	↑	⊕
	Monsoon	⊕	⊕	↑	⊕
	Post-Monsoon	⊕	⊕	↑	⊕
Peninsular India (PI)	Winter	⊕	↓	↓	↑
	Pre-Monsoon	⊕	↓	↓	↑
	Monsoon	⊕	↓	↓	⊕
	Post-Monsoon	⊕	↓	↓	↑

Symbols ↑, ↓, ⊕ represents increasing, decreasing and no trend respectively

references and can only be used for larger spatial areas. Despite having the aforementioned drawbacks, the ESMs are the only tool to predict the possible future changes at regional and country level scales. The CanESM2 model, which was efficient in past climate studies for HMRRI, is deployed in the present study. India is a very diverse landmass with a very heterogeneous topography and land cover; hence,

Table 8.3 Mann- Kendal trends of Carbon Flux within the CanESM2 model for the future period in RCP4.5

Zone and Season		NEP		$C_{(Veg)}$		$C_{(Soil)}$		$C_{(Litter)}$	
		NF	FF	NF	FF	NF	FF	NF	FF
Hilly Regions (HR)	Winter	⊕	⊕	↑	⊕	↑	↓	⊕	↓
	Pre-Monsoon	⊕	⊕	↑	⊕	↑	↓	⊕	↓
	Monsoon	⊕	⊕	↑	⊕	↑	↓	⊕	↓
	Post -Monsoon	⊕	⊕	↑	⊕	↑	↓	⊕	↓
Central North India (CNI)	Winter	⊕	⊕	↑	↑	↓	↓	↓	⊕
	Pre-Monsoon	⊕	⊕	↑	↑	↓	↑	↓	⊕
	Monsoon	⊕	⊕	↑	↑	↓	↑	↓	⊕
	Post -Monsoon	⊕	⊕	↑	↑	↓	↑	⊕	⊕
West Central India (WCI)	Winter	⊕	⊕	↑	↑	↓	⊕	⊕	⊕
	Pre-Monsoon	⊕	⊕	↑	↑	↓	⊕	⊕	⊕
	Monsoon	⊕	⊕	↑	↑	↓	⊕	⊕	⊕
	Post -Monsoon	⊕	↑	↑	↑	↓	↓	⊕	⊕
North East India (NEI)	Winter	⊕	↓	↑	↑	↑	↑	↓	↑
	Pre-Monsoon	⊕	↓	↑	↑	↑	↑	↓	↑
	Monsoon	⊕	⊕	↑	↑	↑	↑	↓	↑
	Post -Monsoon	⊕	⊕	↑	↑	↑	↑	↓	↑
North West India (NWI)	Winter	⊕	⊕	↑	⊕	↓	↓	⊕	↓
	Pre-Monsoon	⊕	⊕	↑	↑	↓	↓	⊕	↓
	Monsoon	⊕	⊕	↑	↑	↓	↓	⊕	↓
	Post -Monsoon	⊕	⊕	↑	↑	↓	↓	⊕	↓
Peninsular India (PI)	Winter	⊕	⊕	↑	↑	↓	↑	⊕	↑
	Pre-Monsoon	⊕	⊕	↑	↑	↓	↑	⊕	↑
	Monsoon	⊕	↑	↑	↑	⊕	↓	⊕	⊕
	Post -Monsoon	⊕	↑	↑	↑	⊕	↑	⊕	⊕

Abbreviations NF, FF represents near future (2031–2060), far future (2061–2100), while symbols ↑, ↓, ⊕ represents increasing, decreasing and no trend respectively

it is very much required to address each of the spatially distinct units separately. Carbon sequestration is very dependent on rainfall, and because India is a monsoon-dependent country, studying the carbon fluxes with respect to the rainfall homogenous regions will help the scientific community and policymakers in decision-making. In the present study, we found that the carbon sequestration behavior as fluxes and storage trend of carbon is significantly dissimilar amongst the different zones;

Table 8.4 Mann- Kendal trends of Carbon Flux within the CanESM2 model for the future period in RCP8.5

Zone and Season		NEP		$C_{(Veg)}$		$C_{(Soil)}$		$C_{(Litter)}$	
		NF	FF	NF	FF	NF	FF	NF	FF
Hilly Regions (HR)	Winter	⊕	⊕	↑	↑	↓	↓	⊕	⊕
	Pre-Monsoon	⊕	⊕	⊕	↑	↓	↓	↓	⊕
	Monsoon	⊕	⊕	⊕	↑	↓	↓	⊕	⊕
	Post -Monsoon	⊕	⊕	⊕	↑	↓	↓	⊕	⊕
Central North India (CNI)	Winter	⊕	⊕	↑	↑	↓	↓	⊕	↓
	Pre-Monsoon	⊕	⊕	↑	↑	↓	↓	⊕	⊕
	Monsoon	⊕	⊕	↑	↑	↓	↓	⊕	⊕
	Post -Monsoon	⊕	⊕	↑	↑	↓	↓	⊕	⊕
West Central India (WCI)	Winter	⊕	⊕	↑	↑	↑	⊕	↑	↑
	Pre-Monsoon	⊕	⊕	⊕	↑	↑	⊕	↑	↑
	Monsoon	⊕	⊕	⊕	↑	↑	⊕	↑	↑
	Post -Monsoon	⊕	⊕	⊕	↑	↑	⊕	↑	↑
North East India (NEI)	Winter	⊕	⊕	↑	↑	⊕	↑	⊕	↑
	Pre-Monsoon	⊕	⊕	↑	↑	⊕	↑	⊕	↑
	Monsoon	⊕	⊕	↑	↑	⊕	↑	⊕	↑
	Post -Monsoon	⊕	⊕	↑	↑	⊕	↑	⊕	↑
North West India (NWI)	Winter	⊕	⊕	⊕	⊕	↓	↓	⊕	⊕
	Pre-Monsoon	⊕	⊕	⊕	⊕	↓	↓	⊕	⊕
	Monsoon	⊕	⊕	⊕	⊕	↓	↓	⊕	⊕
	Post -Monsoon	⊕	⊕	⊕	⊕	↓	↓	⊕	⊕
Peninsular India (PI)	Winter	⊕	⊕	⊕	↑	↓	↑	⊕	↑
	Pre-Monsoon	⊕	⊕	↓	↑	↓	↑	⊕	⊕
	Monsoon	⊕	⊕	↓	↑	↓	↑	⊕	⊕
	Post -Monsoon	⊕	⊕	⊕	↑	↓	↑	↓	↑

Abbreviations NF, FF represents near future (2031–2060), far future (2061–2100), while symbols ↑, ↓, ⊕ represents increasing, decreasing and no trend respectively

hence, the studies at the country level will not be sufficient to plan or track the actual carbon storage and dynamics. It was also noticed that the net ecosystem productivity in the future would increase in all regions, giving us an inkling of the immense potential of carbon sequestration in the country. India is an agricultural powerhouse; hence, agricultural management along with soil management practices will certainly help to mitigate carbon emissions through carbon sequestration. It was found that the future carbon stocks will be in vegetation, and a depletion of the soil carbon pool is speculated, which may be linked to the effects of a warming climate. The seasonal studies for the near and far future under low and high emission scenarios also further hinted that a strong region-specific policy is needed for agricultural practices to reap the benefits of carbon sequestration potential in India. Thus, this is the time for more ESM-based studies in India, incorporating a more holistic approach on regional scales to mitigate the ill effects of climate change in the future by the tool of carbon sequestration.

Acknowledgments The authors want to thank CCCma, the CanESM2 modeling group, and the Government of Canada (https://climate-modelling.canada.ca) for providing the CMIP5-CanESM2 data for analysis. We are also thankful to the reviewers for their valuable comments, which helped improve the chapter.

References

Aggarwal SP, Garg V, Gupta PK, Nikam BR, Thakur PK (2012) Climate and LULC change scenarios to study its impact on hydrological regime. Int Arch Photogramm Remote Sens Spat Inf Sci 39:147–152

Andrew RM (2020) Timely estimates of India's annual and monthly fossil CO2 emissions. Earth Syst Sci Data 12(4):2411–2421

Arora VK, Scinocca JF, Boer GJ, Christian JR, Denman KL, Flato GM et al (2011) Carbon emission limits required to satisfy future representative concentration pathways of greenhouse gases. Geophys Res Lett 38(5)

Beier C, Emmett BA, Peñuelas J, Schmidt IK, Tietema A, Estiarte M et al (2008) Carbon and nitrogen cycles in European ecosystems respond differently to global warming. Sci Total Environ 407(1):692–697

Crowther TW, Todd-Brown KE, Rowe CW, Wieder WR, Carey JC, Machmuller MB et al (2016) Quantifying global soil carbon losses in response to warming. Nature 540(7631):104–108

Dai Z, Johnson KD, Birdsey RA, Hernandez-Stefanoni JL, Dupuy JM (2015) Assessing the effect of climate change on carbon sequestration in a Mexican dry forest in the Yucatan Peninsula. Ecol Complex 24:46–56

Davidson EA, Janssens IA (2006) Temperature sensitivity of soil carbon decomposition and feedbacks to climate change. Nature 440:165–173. https://doi.org/10.1038/nature04514

Day TA, Ruhland CT, Xiong FS (2008) Warming increases aboveground plant biomass and C stocks in vascular-plant-dominated Antarctic tundra. Glob Chang Biol 14(8):1827–1843

Dlugokencky E, Tans P (2020) Trends in atmospheric carbon dioxide. National Oceanic and Atmospheric Administration, Earth System Research Laboratory (NOAA/ESRL). Available at: http://www.esrl.noaa.gov/gmd/ccgg/trends/global.html

Drake JE, Gallet-Budynek A, Hofmockel KS, Bernhardt ES, Billings SA, Jackson RB et al (2011) Increases in the flux of carbon belowground stimulate nitrogen uptake and sustain the long-term enhancement of forest productivity under elevated CO2. Ecol Lett 14(4):349–357

Dubey PK, Singh GS, Abhilash PC (2016) Agriculture in a changing climate. J Clean Prod 113:1046–1047

Dubey PK, Singh A, Raghubanshi A, Abhilash PC (2020a) Steering the restoration of degraded agroecosystems during the United Nations Decade of Ecosystem Restoration. J Environ Manag 280. https://doi.org/10.1016/j.jenvman.2020.111798

Dubey PK, Singh GS, Abhilash PC (2020b) Adaptive agricultural practices: building resilience under changing climate. Springer Nature Switzerland, Springer Brief in Environmental Science, ISBN 978-3-030-15518-6. https://doi.org/10.1007/978-3-030-15519-3

ESA (2012). https://www.esa.org/esa/wp-content/uploads/2012/12/carbonsequestrationinsoils.pdf

FAO (2020) Global Forest Resources Assessment 2020—Key findings. https://doi.org/10.4060/ca8753en. ISBN 978-92-5-132581-0

Friedlingstein P, Cox P, Betts R, Bopp L, von Bloh W, Brovkin V, Cadule P, Doney S, Eby M, Fung I, Bala G, John J, Jones C, Joos F, Kato T, Kawamiya M, Knorr W, Lindsay K, Matthews HD, Raddatz T, Rayner P, Reick C, Roeckner E, Schnitzler K-G, Schnur R, Strassmann K, Weaver AJ, Yoshikawa C, Zeng N (2006) Climate–carbon cycle feedback analysis: results from the C4MIP model Intercomparison. J Clim 19:3337–3353. https://doi.org/10.1175/JCLI3800.1

Friedlingstein P, Meinshausen M, Arora VK, Jones CD, Anav A, Liddicoat SK, Knutti R (2014) Uncertainties in CMIP5 climate projections due to carbon cycle feedbacks. J Clim 27(2):511–526

Friedlingstein P, O'Sullivan M, Jones MW, Andrew RM, Hauck J, Olsen A et al (2020) Global carbon budget 2020. Earth Syst Sci Data 12(4):3269–3340

Garg V, Khwanchanok A, Gupta PK, Aggarwal SP, Kiriwongwattana K, Thakur PK, Nikam BR (2012) Urbanisation effect on hydrological response: a case study of Asan River watershed, India

Grace J (2001) Carbon cycle. In: Encyclopedia of biodiversity (vol 1, pp 609–629. Academic Press

Hopkins FM, Torn MS, Trumbore SE (2012) Warming accelerates decomposition of decades-old carbon in forest soils. Proc Natl Acad Sci 109(26):E1753–E1761

Hopkins FM, Filley TR, Gleixner G, Lange M, Top SM, Trumbore SE (2014) Increased belowground carbon inputs and warming promote loss of soil organic carbon through complementary microbial responses. Soil Biol Biochem 76:57–69

Houghton RA (2007) Balancing the global carbon budget. Annu Rev Earth Planet Sci 35:313–347. https://doi.org/10.1146/annurev.earth.35.031306.140057

Houghton RA (2012) Carbon emissions from land use land cover changes. Biogeosciences 9(12):5125–5142

Joos F, Spahni R (2008) Rates of change in natural and anthropogenic radiative forcing over the past 20,000 years. Proc Natl Acad Sci 105(5):1425–1430

Karwaria S, Tripathi S (2012) Land use / Land cover mapping of Achanakmar Amarkantak Biosphere Reserve, India using unsupervised classification technique. Int J Comput Eng Res 2(5):1302–1307

Kharol SK, Kaskaoutis DG, Badarinath KVS, Sharma AR, Singh RP (2013) Influence of land use/land cover (LULC) changes on atmospheric dynamics over the arid region of Rajasthan state, India. J Arid Environ, 88:90–101

Koven CD, Ringeval B, Friedlingstein P, Ciais P, Cadule P, Khvorostyanov D et al (2011) Permafrost carbon-climate feedbacks accelerate global warming. Proc Natl Acad Sci 108(36):14769–14774

Kumar D, Dimri AP (2018) Regional climate projections for Northeast India: an appraisal from CORDEX South Asia experiment. Theor Appl Climatol 134(3–4):1065–1081

Kumar M, Raghubanshi AS (2011) Sensitivity analysis of BIOME-BGC model for dry tropical forests of Vindhyan highlands, India. Int Arch Photogram Remote Sens Spatial Inf Sci 38(8/W20)

Kumar P, Sarthi PP, Kumar S, Barat A, Sinha AK (2020a) Evaluation of CORDEX-RCMS and their driving GCMs of CMIP5 in simulation of Indian summer monsoon rainfall and its future projections. Arab J Geosci 13(5):1–14

Kumar P, Kumar S, Barat A, Sarthi PP, Sinha AK (2020b) Evaluation of NASA's NEX-GDDP-simulated summer monsoon rainfall over homogeneous monsoon regions of India. Theor Appl Climatol 141(1):525–536

Le Quéré C, Raupach MR, Canadell JG, Marland G, Bopp L, Ciais P, Conway TJ, Doney SC, Feely RA, Foster P, Friedlingstein P (2009). Trends in the sources and sinks of carbon dioxide. Nat Geosci 2(12):831–836

Liu L, King JS, Giardina CP (2005) Effects of elevated concentrations of atmospheric CO2 and tropospheric O3 on leaf litter production and chemistry in trembling aspen and paper birch communities. Tree Physiol 25(12):1511–1522

Mahecha MD, Reichstein M, Carvalhais N, Lasslop G, Lange H, Seneviratne SI et al (2010) Global convergence in the temperature sensitivity of respiration at ecosystem level. Science 329(5993):838–840

Minasny B, Malone BP, McBratney AB, Angers DA, Arrouays D, Chambers A et al (2017) Soil carbon 4 per mille. Geoderma 292:59–86

Mishra VK, Rai PK, Mohan K (2014) Prediction of land use changes based on Land Change Modeler (LCM) using remote sensing : a case study of Muzaffarpur (Bihar) India. J Geogr Inst Cvijic 64(1):111–127

Mitsch WJ, Bernal B, Nahlik AM, Mander Ü, Zhang L, Anderson CJ et al (2013) Wetlands, carbon, and climate change. Landsc Ecol 28(4):583–597

Moss RH, Edmonds JA, Hibbard KA, Manning MR, Rose SK, Van Vuuren DP et al (2010) The next generation of scenarios for climate change research and assessment. Nature 463(7282):747–756

Negrón-Juárez RI, Koven CD, Riley WJ, Knox RG, Chambers JQ (2015) Observed allocations of productivity and biomass, and turnover times in tropical forests are not accurately represented in CMIP5 earth system models. Environ Res Lett 10(6):064017

Norby RJ, Zak DR (2011) Ecological lessons from free-air CO2 enrichment (FACE) experiments. Annu Rev Ecol Evol Syst 42:181–203

Norby RJ, DeLucia EH, Gielen B, Calfapietra C, Giardina CP, King JS et al (2005) Forest response to elevated CO2 is conserved across a broad range of productivity. Proc Natl Acad Sci 102(50):18052–18056

Parathasarathy B, Munot AA, Kothawale DR (1994) Droughts over homogeneous regions of India: 1871–1990. Drought Network News (1994-2001) 67. https://digitalcommons.unl.edu/droughtnetnews/67

Peng S, Piao S, Ciais P, Myneni RB, Chen A, Chevallier F et al (2013) Asymmetric effects of daytime and night-time warming on Northern Hemisphere vegetation. Nature 501(7465):88–92

Penman J, Gytarsky M, Krug T, Kruger D, Pipatti R, Buendia L et al (2003) Good Practice Guidance for Land Use, Landuse Change and Forestry (GPG-LULUCF). IPCC-IGES, Kanagawa

Phillips RP, Finzi AC, Bernhardt ES (2011) Enhanced root exudation induces microbial feedbacks to N cycling in a pine forest under long-term CO2 fumigation. Ecol Lett 14(2):187–194

Pregitzer KS, Burton AJ, King JS, Zak DR (2008) Soil respiration, root biomass, and root turnover following long-term exposure of northern forests to elevated atmospheric CO2 and tropospheric O3. New Phytol 180(1):153–161

Rao AS, Bala G, Ravindranath NH, Nemani R (2019) Multi-model assessment of trends, variability and drivers of terrestrial carbon uptake in India. J Earth Syst Sci 128(4):99

Sarkar D, Kar SK, Chattopadhyay A, Shikha RA, Tripathi VK, Dubey PK, Abhilash PC (2020) Low input sustainable agriculture: a viable climate-smart option for boosting food production in a warming world. Ecol Indic 115:106412

Sarthi PP (2019) Future changes in rainfall and temperature under emission scenarios over India for Rice and wheat production. In: Climate change and agriculture in India: impact and adaptation. Springer, Cham, pp 1–10

Schahczenski J, Hill H (2009) Agriculture, climate change and carbon sequestration. ATTRA, Melbourne, pp 14–18

Shoenfelt EM, Winckler G, Lamy F, Anderson RF, Bostick BC (2018) Highly bioavailable dust-borne iron delivered to the Southern Ocean during glacial periods. Proc Natl Acad Sci 115(44):11180–11185

Singh P, Benbi DK (2020) Modeling soil organic carbon with DNDC and RothC models in different wheat-based cropping systems in North-Western India. Commun Soil Sci Plant Anal 51(9):1184–1203

Su H, Feng J, Axmacher JC, Sang W (2015) Asymmetric warming significantly affects net primary production, but not ecosystem carbon balances of forest and grassland ecosystems in northern China. Sci Rep 5(1):1–8

Taylor KE, Stouffer RJ, Meehl GA (2012) An overview of CMIP5 and the experiment design. Bull Am Meteor Soc 93:485–498

Tian H, Banger K, Bo T, Dadhwal VK (2014) History of land use in India during 1880–2010 : Large scale land transformations from satellite data and historical archives. Glob Planet Chang 121:78–88

Todd-Brown KEO, Randerson JT, Post WM, Hoffman FM, Tarnocai C, Schuur EAG, Allison SD (2013) Causes of variation in soil carbon simulations from CMIP5 Earth system models and comparison with observations. Biogeosciences 10(3):1717–1736

Todd-Brown KEO, Randerson JT, Hopkins F, Arora V, Hajima T, Jones C et al (2014) Changes in soil organic carbon storage predicted by earth system models during the 21st century. Biogeosciences 11(8):2341–2356

Wieder WR, Cleveland CC, Smith WK, Todd-Brown K (2015) Future productivity and carbon storage limited by terrestrial nutrient availability. Nat Geosci 8(6):441–444

Wu D et al (2018) Evaluation of CMIP5 earth system models for the spatial patterns of biomass and soil carbon turnover times and their linkage with climate. J Clim 31(15):5947–5960

Xia J, Chen J, Piao S, Ciais P, Luo Y, Wan S (2014) Terrestrial carbon cycle affected by non-uniform climate warming. Nat Geosci 7(3):173–180

Part III
Strategies for Enhancing Plant Nutrition and Carbon Sequestration

Chapter 9
Agroforestry as a Potential Measure to Enhance Plant Nutrition and Carbon Sequestration

Preeti Singh, Jyotika Dhankhar, and Asha Sharma

Abstract Agroforestry systems (AFSs) are land management strategies to tackle carbon sink and soil fertility depletion along with other ecosystem services around the globe. Soils are losing their quality due to factors like oxidation of organic matter, washing away of nutrient-rich topsoil with runoff water, poor recycling back of crop residues, repeated and intensive cultivation of crops, etc. AFSs are particularly important for deep capture of nutrients, nutrient cycling, reducing nutrient loss from erosion and leaching, high SOM production and decomposition, and thus, soil fertility improvement and degraded land restoration. AFSs are capable of maintaining a good and diverse consortium of microorganisms, which is again crucial to soil health and its fertility. Soils rich in microflora have significantly higher amounts of organic carbon, nitrogen, and phosphorus than soil with poor microflora. Besides this, agroforestry has gained great recognition for enhancing carbon sinks and reducing CO_2 emissions. It is a potential land-use strategy for increasing and conserving aboveground and soil carbon sequestration. AFSs entrap more carbon from the atmosphere than agricultural crops or pastureland. The tree components in AFSs are potential carbon sinks and can boost the flexibility of the systems to combat the extreme impacts of climate change. AFSs can deposit 0.29–15.21 Mg ha^{-1} $year^{-1}$ of carbon yearly. Therefore, AFSs should be popularized and implemented throughout the world to alleviate climate change.

Keywords Agroforestry · Carbon sequestration · Carbon sink · Climate change · Mycorrhizal association · Soil fertility · Soil nutrition

P. Singh · A. Sharma (✉)
Department of Botany, Maharshi Dayanand University, Rohtak, Haryana, India
e-mail: drasha.botany@mdurohtak.ac.in

J. Dhankhar
Department of Food and Technology, Maharshi Dayanand University, Rohtak, Haryana, India

S. A. Edrisi et al. (eds.), *Sustainable Plant Nutrition and Soil Carbon Sequestration*, Sustainable Plant Nutrition in a Changing World,
https://doi.org/10.1007/978-3-031-53590-1_9

9.1 Introduction

Higher quantities of GHGs (GHGs) in the atmosphere have caused an imbalance, increasing the greenhouse effect and, consequently, global warming, which has imposed various threats to the entire ecosystem. Around 10–12% of the total global anthropogenic emission is contributed by agriculture alone, with an estimated 5120–6116 $MtCO_2eq$/year non-CO_2 GHG emission in 2005 (IPCC, 2007). Hundreds of millions of people suffer due to climate change, and the global food chain and rural livelihoods are being jeopardized. The primary factor responsible for such climatic alterations is increased atmospheric CO_2 concentration, which is rising at a faster rate over the years. Moreover, land degradation is taking place at a very high rate around the world, and land resources, e.g., fuel, food, fodder, human habitation, and other developmental actions, are continuously shrinking (Edrisi et al. 2018). Regular removal of nutrients like nitrogen (N), phosphorus (P), potassium (K), zinc (Zn), sulfur (S), etc., is mainly due to intense agricultural practices causing severe nutrient deficiency in the soil, which ultimately leads to depleted soil fertility and hence decreased crop yield (Misra 2011).

The constant deficiency of required nutrients in depleted soils and their losses due to soil erosion, etc., not only exacerbates soil degradation but also puts agricultural sustainability at risk (Ayoub 1999; Sheldrick et al. 2002). Furthermore, the ever-exponentially growing world population is predicted to reach around 8 billion by the end of the twenty-first century, while food consumption is expected to increase by 100% in developing countries and 60% globally by around 2050. According to (the Food and Agricultural Organization of the United Nations (FAO), food production needs to be approximately doubled over the next couple of decades to achieve global food security goals (Godfray et al. 2010). As a result, there is an undeniable need to protect our natural resources, such as water, soil, vegetation, etc., to meet future necessities and sustain the ever-growing world population. Climate change adaptation and mitigation are inevitable if basic survival and food security are to be ensured.

So precisely, climatic change, depleting soil fertility, low crop yield/agricultural productivity, food insufficiency, etc., are a few of the major challenges worldwide which need to be dealt with seriously. Agroforestry, a farming system where crops are grown in combination with perennial trees or livestock on the same piece of land simultaneously, is a well-known remedy against such challenges, which immensely contributes to addressing the same (Kumar et al. 2014). It is being practiced across the globe and presently covers approximately 1023 mha of land worldwide. A large portion of unproductive land consisting of agriculture, grass, and forest areas, including degraded fields, might be converted to agroforestry (Kumar et al. 2014). Agroforestry is a planting system that has more sustainability and capacity for degraded ecosystem restoration and agrobiodiversity preservation (Gonçalves et al. 2021). It has significant prospects for achieving synergy between adaptation and mitigation efforts.

In addition, it is one of the best measures of sustainable land management and serves the dual purpose of climate change mitigation through carbon sequestration and soil fertility maintenance through litter/organic matter addition, decomposition and nutrient release, nutrient flow and uptake, control soil erosion, productive usage of rainfall in the farmlands, etc. It offers great resilience to various ecological systems for dealing with adverse climate change and has become a crucially important mechanism for the conservation of various natural resources. Particularly, soil, as it protects soil from water and wind-induced soil erosions, ameliorates the severe weather effects such as extreme temperatures, precipitation, and winds on soil fertility, soil flora, and fauna, etc. Consequently, keeping the beneficial soil microbiome intact greatly aids in enhancing nutrient uptake from deeper soil layers through extensive tree root systems, including nitrogen-fixing ones, and hence improves soil productiveness. Trees play a crucial role in efficient nutrient cycling. For instance, the evaluation of *Faidherbia albida*—maize AFS in Zambia revealed that more than 18 kg N ha $year^{-1}$ were supplied by the litter inputs from *Faidherbia albida,* and the microbial diversity was improved (Yengwe et al. 2018).

In another example, in Nicaragua, Hoosbeek et al. (2018) investigated the reason behind alterations in carbon, nitrogen, and phosphorus stocks and microbial respiration with remoteness from silvopastoral systems trees consisting of grazed open plains with isolated trees. In comparison to the open grassland, they found increased C and N stocks beneath the tree canopies and in areas receiving litter. Hence, AFSs aid in the recycling of soil nutrients through litter decomposition, root residues, pruning, etc., and reduce nutrient loss by leaching through deeper tree roots and mycorrhizal systems. Various agroforestry land-use systems like Agri-silvopastoral, agro-pastoral, agro-horticulture, etc., are proficient ways through which SOM is restored/recycled (Pandey 2007). Jama et al. (2006) *reported that higher yields are* obtained with improved crop rotation practices than with continuous cropping. Agroforestry gives an assurance to food and economic security because the resistance possessed by trees against various climatic extremities like flood, cold, heat, drought, etc., ensures a sustainable availability of food, fodder, and fuel sources when crops are fully or partially destroyed. Soil and agroforestry both work synergistically to achieve soil fertility and land restoration goals (Fig. 9.1).

Various studies indicate that tree deposits in AFSs have sufficient quantities of different nutrients, like nitrogen, potassium, sulfur, calcium, etc., to meet the essential requirements of associated crops. Increased nitrogen inputs through various approaches like biological nitrogen fixation, deep nitrate capture, efficient nitrogen recycling from plant residues, etc., supply a sufficient amount of nitrogen to sustain a modest crop yield. Contrarily, AFSs are unlikely to generate enough phosphorus for agricultural yields to be sustained. Subsurface phosphorus uptake is often low, and it is unlike nitrogen, which is not fixed from the atmosphere. Due to the relatively low phosphorus content of organic materials, their recycling from organic materials is typically insufficient to fulfill the phosphorus requirements of crops. On phosphorus-deficient soils, persistent crop production with agroforestry will usually necessitate phosphorus fertilizer inputs (Buresh et al. 1995).

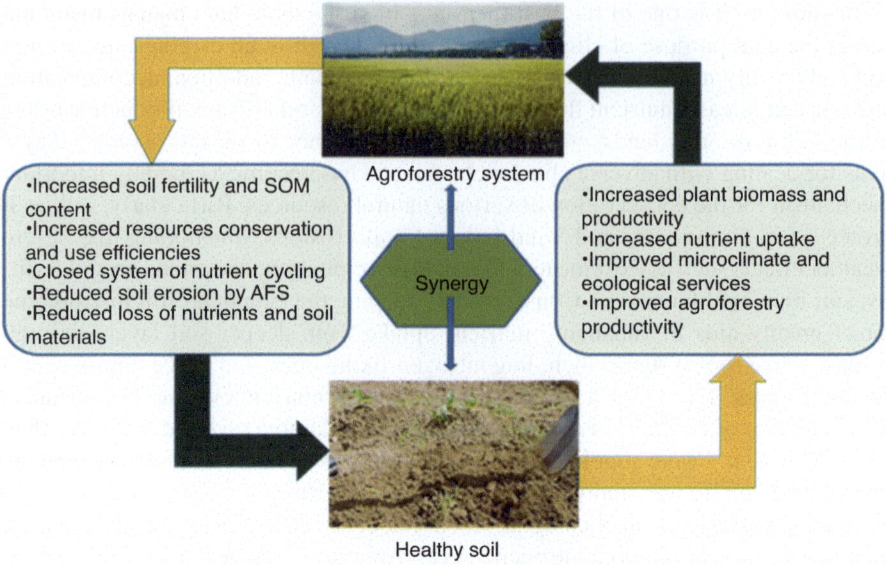

Fig. 9.1 Synergy between AFSs and soils. Agroforestry practices influence the health parameters of soil that influence soil productivity, therefore affecting crop production sustainability. (Source: Harvey et al. (2005))

Besides this, the role of AFSs in carbon sequestration has been well-recognized since time immemorial. Various studies have been conducted and reported that shifting from conventional agricultural systems to agroforestry has enhanced carbon sequestration to many folds. Planting trees and bushes together promotes carbon sequestration both above and below ground, helping to reduce GHG emissions (Verchot et al. 2007). Agroforests can accumulate a huge amount of carbon, and this potential comes from the ability of AFSs to proficiently detain and exploit growth resources, such as nutrients, water, light, etc. Tree components in AFSs act as substantial carbon sinks because of their rapid growth and high productivity. As a result, practicing and encouraging agroforestry might be one of the solutions for dealing with land use issues and carbon-induced global warming. The amount of carbon sequestered is primarily governed by the AFS itself, including its structural and functional aspects, all of which are influenced by environmental and socioeconomic variables. Carbon storage in AFSs can also be influenced by system management practices and tree species diversity (Albrecht and Kandji 2003). The estimated carbon content stored in the AFSs ranges from 0.29–15.21 Mg C ha^{-1} $year^{-1}$ aboveground and 30–300 Mg C ha^{-1} $year^{-1}$ belowground up to 1-meter depth in the soil. Even in small-scaled AFSs, the sequestered carbon amount changes from 1.5 to 3.5 Mg C ha^{-1} $year^{-1}$, thus it is undoubtedly a viable carbon sequestration strategy (Montagini and Nair 2004; Roshetko et al. 2007). The topsoil carbon storage found increased up to around 1.6 Mg C ha^{-1} $year^{-1}$ by AFSs in sub-humid tropical regions containing degraded soil (Mutuo et al. 2005). Thus, it would not be inappropriate to say that an accurate and proper design and management of AFSs can prove to be a very effective carbon sink strategy for the future. Here, we have attempted to highlight several benefits accrued from AFSs for improving soil nutrient status and sequestering carbon content.

9.2 Agroforestry and Plant Nutrition

Almost all the AFSs, such as agrisilviculture, agrosilvopastoral, improved fallows, home gardens, alley cropping, etc., are expected to enhance the biological/economic productivity of any agricultural land that is accomplished in various ways. For example, a) by increasing the nutrient inputs into the soil through a variety of mechanisms like litter production, decomposition, nutrient release, tree pruning, root-soil interactions, microbial interactions, etc., b) by reduction of nutrient loss by soil erosion and leaching, c) by proper soil moisture conservation and water quality management, etc. (Misra 2011). Figure 9.2 depicts a few of the strategies of plant nutrition improvement through agroforestry.

9.2.1 Litter Production, Decomposition, and Nutrient Release

Aerial parts of crops and trees, such as stem bark, twigs, branches, flowers, buds, leaves, and other reproductive and vague parts, form the key components of the litter (Sarvade et al. 2014). The nutrient status of the soil is improved when litter is

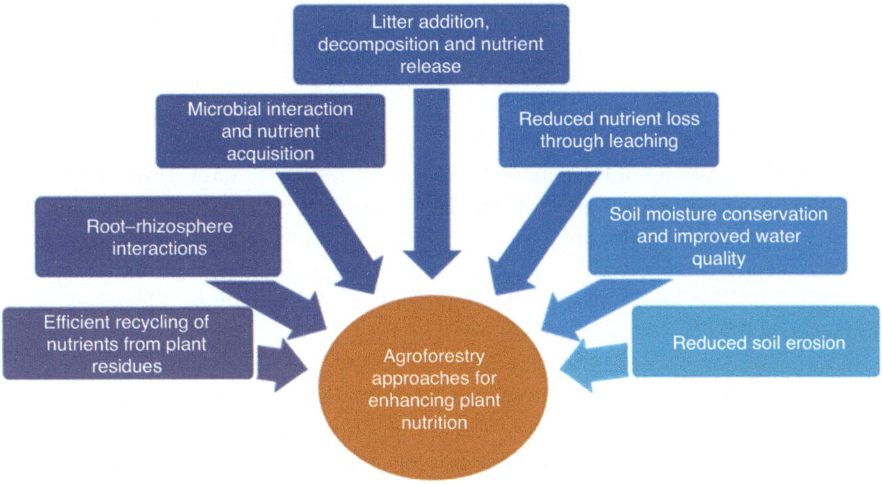

Fig. 9.2 Different ways through which plant nutrition can be enhanced. (Adopted from Misra (2011), Edrisi et al. (2018))

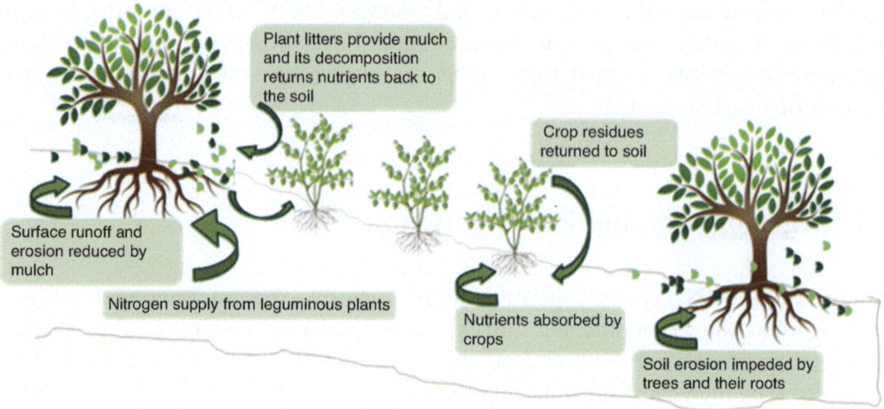

Fig. 9.3 Soil nutrient cycling through litterfall and its decomposition in AFS. Tree leaves, twigs, and branches go through the process of senescence and abscission, followed by falling to the ground as litter over a tree's lifetime. Litter decomposes and contributes to the formation of organic matter in the soil. Organic matter-rich soil has a greater capacity to absorb and hold water, making it more resistant to erosion. (Source: Kang et al. (1986))

produced and decomposed into the soil, and organic matter flows back to the soil from aboveground/aerial parts of the trees (Fig. 9.3), illustrating litterfall and its functions).

Various factors such as topographic, edaphic, climatic, plant species characteristics, and other biotic factors affecting plant growth and phenological parameters affect the production of litter in AFSs (Rawat et al. 2009). Perennial woody trees contribute the most to litter production as compared to crops. Litterfall is comparatively higher in the spring and summer seasons than in the fall or autumn seasons because the intensity of wind velocity is higher in summers and springs than in autumn, which promotes the abscission process of older leaves (Munishamappa et al. 2012; Kumar 2011; Pragasan and Parthasarathy 2005; Singh et al. 2011). Soil is protected from direct insulation and evaporation by shade trees. These approaches also preserve SOM, maintain soil productivity and moisture content, and later help enhance the microbiota of the soil. Hence, they are responsible for a high decomposition rate, resulting in a high crop yield. Alternatively, they shelter the soil from exposure to direct sunlight, aid in organic matter and maintaining soil productivity, minimize evaporation, etc. Soil moisture is beneficial to the biota and the decomposition process. The yearly litter production from *Gliricidia sepium,* a legume tree in Indonesia, is 3.9 t ha^{-1} $year^{-1}$ (Moser et al. 2010).

In terms of components and chemical framework, the term "litter" is broad, and leaf is the fraction that appears to be studied most. Because of the variations in the chemical composition of leaf litter of various species, dissimilar decomposition patterns between the species/genera are predicted. Furthermore, different decomposition patterns are followed by the leaf, branch, bark, and root litter of the same plant because of their different chemical composition. The litter forms a layer of organic matter on the surface of the soil, creating a distinct edaphic microclimate and favorable circumstances for a wider range of species. The breakdown of litter into simpler compounds in the form of SOM is known as decomposition. It regulates various processes such as the nitrogen cycle, primary productivity, and soil fertility and productivity. Nutrients and carbon are transferred from the vegetation to the soil through a major process of SOM production. SOM constitutes one of the most important elements of nutrient cycling in AFSs and contains a significant amount of nutrients essential for the growth of soil flora (Zheng 2006). Soil nutrient pool is greatly dependent upon the production, decomposition, and decomposition rate of the litter. Proper fragmentation and decomposition of litter are very crucial for efficient nutrient release. A variety of soil animals and microorganisms (soil fauna) mediate the decomposition and mineralization of litter. Interaction between numerous factors, such as decomposers, litter quality, and the environmental surroundings, regulate the rate of litter decomposition (Zheng 2006). The diversity of AFS flora generates appropriate circumstances for the soil microorganisms, which participate chiefly in nutrient release and litter decomposition (Kumar 2011). The rate of decomposition of SOM is primarily determined by the activities of soil microorganisms that, in turn, are governed by litter quality (intrinsic structural and chemical characteristics of litter). The diverse variety of decomposer organisms is encouraged by the growth of shade trees. Decomposers also play a considerable role in the nitrogen and phosphorus cycles in ecosystems. In a study conducted on soil nitrogen dynamics in cocoa AFS, Indonesia, higher rates of nitrogen mineralization, ammonia absorption, and quick return to the *Zea mays* monoculture ammonium reservoir have been reported, indicating higher N availability in AFSs. This indicates that, unlike maize monocultures, the decomposer community in cocoa AFS retains the majority of its activities through nutrient recycling. The amount of nutrient content released by litterfall and dry matter production in different cropping systems is given in Tables 9.1 and 9.2, respectively.

Soil enrichment through a tree-based agriculture system has been supported by various research findings. An increase in organic matter (nitrogen and phosphorus) content was observed in the soil under the *Prosopis* tree species as compared to nude sites. A significant amount of organic carbon, nitrogen, and phosphorus was found to be increased in AFS, which introduced the *Eucalyptus hybrid* and *Populus deltoid* canopies (Anonymous 1987). Higher amounts of organic carbon, nitrogen, and phosphorus have been reported in silvopastoral systems in comparison to open fields.

Table 9.1 Nutrient content present in aboveground vegetation under different cropping systems

		Nutrient content (kg ha^{-1} yr^{-1})		
Sn No.	Cropping systems	N	P	K
1.	*Eucalyptus tereticornis*	246	21	276
2.	*Dalbergia sissoo*	1063	68	434
3.	Agrisilvi-horticultural system	532	40	461

Mishra (2011)

Table 9.2 Nutrient release and dry matter production of some plant species under AFSs

S No.	Plant species	Dry matter (t/ha/yr)	Nutrient release (kg ha^{-1} yr^{-1})		
			N	P	K
1.	*Populus deltoids*	4.5	25	16	26
2.	*Dalbergia sissoo*	4.2	67	3	16
3.	*Eucalyptus hybrid*	4.6	22	9	7
4.	*Eucalyptus globules*	8.5	58	5	48

Mishra (2011)

9.2.2 Nutrition Enhancement Through Root-Soil Interactions in AFSs

Soil nutrient status can also be improved by the addition of nutrients in the form of root exudates from the root biomass of trees. These additions serve as the principal substrate for the huge variety of microorganisms found in soil interactions and their various biological activities, such as nitrogen fixation and absorption of different nutrients such as potassium, phosphorus, etc. A big portion of gross primary production (GPP) is allocated belowground to sustain the production, growth, and maintenance of plant roots and mycorrhizae (Nair et al. 2009). In such situations, SOC increases and perks up soil characteristics. AFSs put a great emphasis on the plant interactions with soil microbiota, such as nitrogen-fixing bacteria, mycorrhizal fungi, etc., that play a vital role in improving plant nutrition. In today's scenario where sustainable agriculture is being practiced globally, the rhizosphere plant-microbe interactions become crucial as they transform, mobilize, and solubilize the nutrients from a narrow nutrient pool, and thus help in the uptake of essential nutrients by plants.

9.2.2.1 Root–Rhizosphere Interactions

Active roots promote nutrient availability in the soil by opting for spatial proximity. These can actively modify the rhizosphere chemistry to improve nutrient inputs to the soil, for instance, by employing organic acids exudation, H+/OH−, or certain enzymes, e.g., phosphatases, for enhancing the availability of phosphorus in the soil (George et al. 2002; Hinsinger et al. 2011). These root exudates are the chief sources of nutrients and carbon in soils and possess high variability within a rooting system (Phillips et al. 2008). They make a significant contribution to the root-soil boundary (Danjon et al. 2013), and such contributions are repetitive derivatives of the tree elements in AFSs. For example, in AFSs with alluvial soil, considerably higher NO_3 content was reported in the rhizospheric zone than their monocultures (Munroe et al. 2015). Organic anion exudations and phosphatase activity of the tree roots mobilized the rhizospheric phosphorus present in the AFS consisting of maize (Zea mays)-tree on Ferralsols (Radersma and Grierson 2004). Gum acacia–wheat AFS exhibited rhizospheric soil with low pH and high inorganic phosphorus due to tree root presence (Isaac et al. 2012) etc. Moreover, pieces of evidence are reported suggesting multi-trophic interactions in the rhizosphere, which can further mediate root-root interactions in AFSs (Mommer et al. 2016), so there is a great significance of such multi-trophic interactions in rhizosphere processes. Complex interactions among the soil microbial communities in space and time are contributed by root-based organic input delivery and resource uptake throughout the root systems (Upson and Burgess 2013; Mommer et al. 2016).

9.2.2.2 Mycorrhizal Interactions

Role of Arbuscular Mycorrhizal Fungi in Plant Nutrition

Mycorrhizal associations are the mutualistic associations between plant roots and fungi found present in the soil. Arbuscular mycorrhizal fungi (AMF) and ectomycorrhizal fungi (EMF) associations with tree roots signify the most efficient nutrient acquisition strategy and direct nutrient transfer through roots among various species in the AFSs. Tree and tree-based cropping systems act as a diverse and abundant mycelial pool for the crops (Ingleby et al. 2007; Bainard et al. 2011). More than 80% of the land plants, including the majority of the tropical trees and crops like cocoa and coffee, etc. (Kähkölä et al. 2012), are known to be found in the symbiotic associations with AMF (Sanders 2003). AMF and EMF have a great contribution to the direct transfer of nitrogen and various other indirect processes, such as soil volume enlargement for more access to nutrients, modifications in phosphorus absorption and solubilization, soil microsite exploration, etc. (Carvalho et al. 2010). AMF belongs to the phylum Glomeromycota and forms the most significant group in tropical AFSs. These fungi can constitute the larger parts of the soil microbial biomass (Olsson et al. 1999) and have the capacity to colonize almost all the plants belonging to gymnosperms and angiosperms and some among bryophytes and pteridophytes (Smith and Read 2008). In a study conducted on around a hundred species of trees belonging to various families in southeastern Brazil, more than ninety species were found to be colonized by AMF alone (Carneiro et al. 1998).

The increase in the volume of the soil, which the host plant exploits, is the most apparent function of AMF in AFSs. As a result, the uptake of nutrients and water increases, which may further boost the acquisition of various other nutrients as well (Bolan 1991; Garg et al. 2006; Smith and Read 2008). Root system protection against pathogens and phytotoxic elements, e.g., aluminum, heavy metals, salinity, etc., is another concern of AMF in AFSs (Rufyikiri et al. 2000; Andrade et al. 2003; Tavares 2007; Elsen et al. 2008). Soil structure maintenance and formation, carbon inputs to the soil, etc., also show AMF involvement and help in reducing soil erosion (Rillig et al. 2001; Zhu and Miller 2003; Rillig and Mummey 2006). Plant biodiversity maintenance is also associated with AMF (Van der Heijden et al. 1998). Thus, in AFSs, mycorrhizal association (Newsham et al. 1995) performs multifunctional roles. They potentially improve the physical, chemical, and biological properties of the soil.

In Ethiopia, the survival potential of *F. alba* seedlings against water stress and AMF inoculation was investigated by they discovered that independent of the origin and water status of the inoculum, AMF inoculation boosted seedling growth. A flourishing and rich diversity of microbes is maintained by agroforestry, which is crucial to soil health and productivity. Nine AFSs consisting of *Millettia, Cordia*, or *Erythrina* linked with *Ensete* or coffee were compared based on their AMF diversity and spore density. Both spore density and AMF diversity were minor affected, with greater AMF diversity and density found to be possessed by tree-*Ensete* than tree-coffee systems.

Leguminous Trees and Biological Nitrogen Fixation

Significant quantities of nitrogen through biological nitrogen fixation (BNF) by trees or shrubs can increase soil quality and productivity in agroforestry and agricultural systems (Herridge et al. 2008). Annual N_2 fixation by pasture and fodder legumes is 12–25 Tg, with overall BNF inputs in agricultural systems being 50–70 Tg. Through litterfall, root and nodule decomposition, root exudation, and direct transfer through mycorrhizal networks, fixed nitrogen in tree biomass may be recycled back into the soil. According to 14% of the nitrogen in the fodder grass *Dichantium aristatum* came from the legume tree *Gliricidia sepium*. Same way, an average of 14% nitrogen was transmitted from *Acacia senegal* to wheat (Isaac et al. 2012). Soil, climate, tree-microbe interactions, and management circumstances all influence the number of nitrogen inputs delivered by BNF.

For decades, soil-beneficial bacteria have been used to enhance crop productivity. The key functions served by these bacteria are (a) nutrients supply to the crops; (b) improvement of plant growth, e.g., by producing plant growth regulators/hormones; (c) inhibition of plant pathogen activity; (d) soil structure improvement; and (e) bioaccumulation/bioleaching of inorganic compounds. Lately, polluted soils are being bio-remediated, or bacteria (Middledrop et al. 1990; Zhuang et al. 2007; Zaidi et al. 2008) are mineralizing organic pollutants. In the present scenario where sustainable agriculture is being practiced, the rhizosphere plant-microbe interactions become crucial as they transform, mobilize, and solubilize the nutrients from the narrow nutrient pool, and thus help in the uptake of essential nutrients by the plants. Nitrogen-fixing trees, particularly leguminous trees, supplement soil with nutrients and organic matter by fixing nitrogen into its usable forms. The amount of nitrogen fixed by tree species like *Acacia, Leucaena,* and *Alnus* has been reported to be 270 kg ha^{-1} $year^{-1}$, 400–500 kg ha^{-1} $year^{-1}$, and 100–300 kg ha^{-1} $year^{-1}$, respectively. The amount of fixed nitrogen is symbiotically beneficial to crops that are found closely associated and aids in improving soil fertility. The amount of fixed nitrogen stays long in the soil in stored organic matter form as nitrogen content added by the tree residues, prunings, or initial crops take up legumes N_2 fixation in a very low quantity, so long-term benefits are provided to the immediate crops. Different components of trees, such as the leaf, twigs, reproductive parts, wooden portions, etc., have different rates of decomposition, which helps in the distribution of released nutrients over time. A few of the vital plant species having nitrogen-fixing potential are given in Table 9.3.

Table 9.3 Some important nitrogen (N_2) fixing plant species

S. No.	Name of plant	Family	Nitrogen fixed in kg ha^{-1} yr^{-1}
1.	*Acacia mearnsii* (black wattle)	Mimosoideae	270
2.	*Faidherbia albida* (apple ring acacia)		20
3.	*Inga jinicuil* (Inga)		35–40
4.	*Leucaena leucocephala* (Subabool)		400–500
5.	*Casuarina equisetifolia* (saru, beaf wood)	Casuarinaceae	60–100
6.	*Alnus nepalensis* (Indian elder)	Betulaceae	100–300
7.	*Erythrina poeappigiana* (Erythrina)	Papilionoideae	60
8.	*Glycin max* (soybean)	Fabaceae	60–165
9.	*Sesbania sesban* (Dhaincha)		7–18
10.	*Trifolium species* (clover)		128–207
11.	*Vicia faba* (horse bean)		45–552
12.	*Crotalaria juncea* (Sunnhemp)		199–223
13.	*Cicer arietinum* (chickpea)		103
14.	*Sesbania rostrata* (new dhaincha		70–458
15.	*Desmanthus virgatus* (wild tantum)		196–226
16.	*Pisum sativum* (pea)		55–77
17.	*Lens esculenta* (lentil)		88–114
18.	*Vigna mungo* (mung bean)		63–342
19.	*Cajanus cajan* (pigeon pea)		68–88
20.	*Indigofera tinctoria* (true indigo)		79
21.	*Gliricidia sepium* (Gliricida)		13
22.	*Vigna sinensis* (cowpea)		73–354
23.	*Medicago sativa* (alfalfa)		229–290
24.	*Phaseolus vulgaris* (bean)		40–70
25.	*Arachis hypogeal* (groundnut)		72–124
26.	*Stylosantes spp* (Stylo)		20–263

Source: Nair (1993)

There are two ways of biological nitrogen fixation: symbiotic and non-symbiotic. In symbiotic fixation plants, roots are found directly associated with nitrogen-fixing soil microorganisms. Several leguminous trees form a close association with *Rhizobium* bacteria, while the non-leguminous symbiotic species belong to the *Frankia* genus of the Actinomycetes family. Free-living soil microorganisms affect non-symbiotic fixation and can play a very significant role in ecosystems having comparatively modest nitrogen requirements from the outside systems (Nair 1993).

9.2.3 Reduced Leaching

Excessive fertilizer usage in farmland monocultures, along with declined fertilizer use efficiency, has resulted in severe nutrient loss by leaching, particularly for nitrate, in recent decades. As a result, groundwater pollution is becoming widespread, and the European Union has begun acknowledging it. Rows of unfertilized trees alternating with crop rows (for example, alley-cropping AFSs) are thought to operate as a safety net by absorbing excess nutrients below the crop-rooting zone. Evaluation of nitrogen, phosphorus, and potassium leaching losses in AFSs and neighboring monocultures at three sites in Germany was conducted, covering a wide variety of soil properties across two growing seasons. When compared to crop monocultures, leaching losses of N, P, and K were reported to be lower under agroforestry tree rows at all sites. In comparison to monocultures, agroforestry reduced nitrate-leaching losses by up to 82%, although P and K losses were equivalent (Schmidt et al. 2020).

Sesbania fallows had considerably lower subsurface nitrate than maize, implying that rotating deep-rooting and fast-growing trees with maize can accumulate and cycle subsoil nitrogen, especially on deep soils having anion sorption capability. However, the typically identical subsoil nitrate under *Sesbania* and natural fallows lift issues about whether short-duration, planted-tree fallows are more successful in capturing and cycling subsurface nutrients than natural fallows (Hartemink et al. 1996). Even though subsoil nitrate levels in *Sesbania* and natural fallow were frequently comparable, the lower subsoil water content and the much deeper rooting network in *Sesbania* fallow indicate that the potential for nutrient leaching is significantly lower in *Sesbania* than in natural fallow. *Sesbania's* low soil water content and nitrate were attributed to the fast-growing *Sesbania*'s high water and nitrogen requirement. These findings and many other studies imply that agroforestry has a great potential for minimizing nutrient leaching losses through trees. However, fertilizer application in agroforestry crops should be reduced to at least adjacent to the trees. The decline in nutrient leaching losses in AFSs indicates the increased water filtering capacity of the soil (Schmidt et al. 2020).

9.2.4 Soil Erosion Control

Soil erosion, one of the primary causes of soil infertility, is highly influenced by wind and rainfall. Soil erosion, along with other environmental challenges, harms farmers' crops, which are mostly utilized for survival. The energy conveyed from wind or rainfall, or a combination of these factors, sets erosive processes in motion (Hurni et al. 2005). The loss of nitrogen and phosphorus through water and wind erosion, as well as the leaching away of N and K, are the most common physicoclimatic processes that deplete soil nutrients (Amede 2003; Henao and Baanante 2006). In India, a major portion of land (approximately 175 mha) has been degraded through various anthropogenic and natural activities such as different agricultural practices, soil erosion, etc. Loss of nutrients through leaching and erosion is becoming the major cause of land degradation (Mishra and Rai 2013). Soil productivity gets badly impacted by land degradation, and hence, issues like deficiency of food, fodder, cash crops, etc., are generated. Soil quality is not only affected by this process, but it also causes loss of biodiversity and other natural resources. AFSs have been reported to reduce 20% of nitrogen loss through a reduction of 1–10% in soil erosion (Udawatta et al. 2002).

Because all the essential macro and micronutrients are mainly present in the uppermost soil layers, it is the most productive on the planet. Plants require these key nutrients to complete their life cycle. Soil erosion occurs when topsoil is shifted from its original site by several forces, mostly water and wind. Water erosion costs the world roughly 24 billion tons of rich soil each year (Young et al. 2015). Soil erosion releases 1100 MtC into the atmosphere each year, while erosion-induced transportation transports another 300–800 MtC to the ocean (Lal 2011). Due to global climate change, rainfall patterns broadly vary, exacerbating soil erosion. In India, the frequency of high-intensity storms and yearly rainfall is expected to rise in the coming 8–10 years, accelerating erosion and runoff. According to Lee et al. (1996), wind erosion will increase by 15–18% with an increase of 2 °C in temperature. Wind erosion is expected to speed up in arid and semiarid locations unless some better techniques, such as agroforestry, are implemented. Different agroforestry models, such as windbreaks, alley cropping, etc., are all built specifically for wind erosion prevention (Garrett 2009). Therefore, more flexibility will be offered by agroforestry in terms of socioeconomic and environmental services as climatic conditions amend. The kinetic energy of rainfall is reduced by vegetation because of its canopy cover. The amount of energy remaining in raindrops depends on how high the canopy is above the ground. According to one study, a canopy height of four meters reduces kinetic energy by around 80% (Young 1989). Plant litter helps to minimize runoff by boosting soil water holding and infiltration capacity. Plant litter decomposition, root degradation, and rhizosphere exudation increase SOM content and improve soil structure, making it less prone to erosion.

It is important to protect topsoil from erosion to ensure a sustainable food supply and security. Understanding AFSs is one of the ways through which soil erosion can be efficiently controlled. Reports from some developing countries suggest that they have developed AFSs for soil erosion control (Lal 1989; Wiersum 1981, 1984; Pacardo and Montecillo 1983). One such way is alley cropping to reduce soil loss. Here, dense canopy covers, thereby preventing any soil erosion, reduce the kinetic energy in falling rain. In the alley cropping system, approximately the entire energy of rainfall is quenched by the short shrubs, and a barrier to runoff is formed, which takes much time to infiltrate the soil. Therefore, less runoff volume results in lowering the transportation power of the runoff, and consequently, lesser soil erosion is observed (Phoung et al. 2017). A good example of the effectiveness of the alley cropping system comes from Nigeria, where AFS having the combination of *Leucaena* hedge + maize resulted in 76 kg/ha soil loss as compared to a no-till condition resulting in soil loss of 10,737 Kg/ha without *Leucaena* (Lal 1990). In yet another investigation carried out in Dehradun northwestern Himalayan regions, India, very little soil loss was reported in grasses despite high runoff. Other agroforestry methods, like windbreaks and shelterbelts, were found to be very effective in arid regions, especially the regions that are wind erosion-prone. It has been observed that these tree belts increase the growth and crop yield of annual crops and help improve the microclimate. Shelterbelts have also been reported to minimize soil erosion by 50%, according to several researchers (Fanish and Priya 2013).

9.2.5 Water Quality Improvement and Soil Moisture Conservation

Mulching can be a very useful process in AFSs that maintains soil moisture and thus enhances crop yield. The infiltration capacity of soil is increased in AFSs. Kumar et al. (1998) examined the soil water availability effects on barley (*Hordeum vulgare*) yield for several agroforestry models, including *Tecomella undulate, Prosopis cineraria, Azadirachta indica,* and *Acacia albida* in a desert land. *P. cineraria, A. albida, T. undulate*, and *A. indica* were shown to improve crop output by 86%, 57.9%, 48.8%, and 16.8%, respectively, above the control. It has been well proven that agroforestry improves groundwater quality by preventing nutrient leaching from the soil to the groundwater table when compared to the cropping systems where a majority of applied fertilizers drains off, polluting groundwater (Cassman 1999). Agroforestry uses deep-rooted plants to eat surplus nutrients given to the agricultural field. As a result, it serves as a filter and helps in reducing groundwater contamination. Climate change is anticipated to affect 2.7 to 4 billion people owing to reduced water supply (Diop 2008). Water quality in terms of nutrients, sediments, DOC, pesticides, pathogens, and salinity will be affected by climate change (IPCC, 2013). Agroforestry methods will operate as a solution in this changing climatic scenario by improving the microclimate, reducing runoff and evaporation, and increasing groundwater level and soil moisture content, therefore increasing water availability and ensuring food security.

9.3 Agroforestry as Carbon Sink

The role of agroforestry in the carbon cycle and sequestration has been well-known for a long time. It contains an enormous potential to mitigate climate change chiefly by sequestering CO_2 in its soil and various plant species (Albrecht and Kandji 2003; Montagini and Nair 2004). Forests are a huge sink of carbon and play a crucial role in the nutrient cycle, especially carbon (Dixon et al. 1994). Out of two types of carbon cycles (long-term and short-term) occurring in nature, the short-term cycle is of greater importance and occurs in agricultural ecosystems and AFSs over decades and centuries (Nair et al. 2010). Atmospheric CO_2 fixation through the process of photosynthesis and its partial return to the atmosphere in aerobic conditions through respiration by plants, animals, and microbes in the form of CO_2 and anaerobic conditions in the form of CH_4 are a few of the vital processes of short term CO_2 cycle (Fig. 9.4). In contrast, the long-term carbon cycle consists of carbon

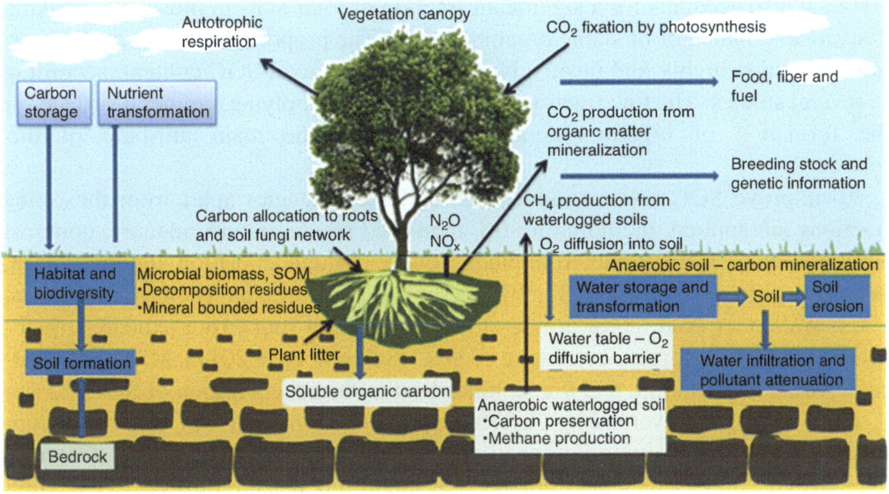

Fig. 9.4 Interrelationship between soil, plant, and carbon and their associated ecosystem services such as photosynthesis, which fixes carbon in the organic forms and increases carbon sink, SOM formation, decomposition and nutrient release, carbon distribution, etc. (Adopted from Berner (2003), Nair et al. (2010))

biogeochemical cycling among various surfaces, namely oceans, the biosphere, the atmosphere, and soil. It manages the atmospheric carbon concentration over more than millions of years of geological timescale (Berner 2003).

9.3.1 Mechanism of Carbon Sequestration

9.3.1.1 Carbon Stabilization in Soil

Soil carbon stabilization is just as critical as carbon fixation. The carbon stabilization process and carbon inputs are the primary factors on which SOC sequestration is mainly dependent. Changes in the stability of SOC can affect carbon release from the soil, and as a result, atmospheric CO2 concentrations increase. The quality and quantity of litter input into the soil that regulates SOC stability can be affected by the mean annual temperature (MAT). Understanding the major parameters that govern SOC stability and specific soil's potential to stabilize more SOC is essential for developing approaches for organic carbon sequestration in soils. Development of organo-mineral compounds having fine-textured soil particles, biochemical recalcitrance, physical protection or inaccessibility, etc., has been proposed as methods for the sequestration of stable SOC (Baldock and Skjemstad 2000; von Lutzow et al. 2006; Dungait et al. 2012; Beare et al. 2014). SOC, which is found in association with fine soil particles, is usually considered extremely stable, having a relatively longer turnover period and delayed response to management changes (Beare et al.

2014). It also accounts for a significant fraction of total SOC in most soils, making it a valuable indicator of stable organic carbon. The proportion of fine mineral particles in soil is highly and positively associated with total SOC content, according to several studies. The fine fractions' significance in supplying a mineral surface for the formation of organo-mineral complexes is the main attribute of this relationship.

To improve SOC content, numerous land-use strategies apart from these fine fractions are applied, including no-till, increased residue return, manure, compost additions, etc. Carbon inputs from plant roots, e.g., rhizodeposition, lignin, suberin, mycorrhizal fungi, and illuviation by leaching are the key sources of SOC (Nguyen 2003; Wallander et al. 2004; Rasse et al. 2005). Plant roots, root exudates, rhizosphere inputs, etc., largely contribute to SOC (Schmidt et al. 2011). Microbial products, partial degradation, and fire residues lead to the accumulation of SOC.

SOC cycling is governed by physical separation from enzymes, electron acceptors, decomposers, sorption/desorption i.e., organometal associations, freezing/thawing, etc., all of which are influenced by environmental circumstances (Schmidt et al. 2011). By proper solubilization and mixing, transit, and following adsorption, surface residual carbon may be absorbed into the soil (Lorenz and Lal 2005). Plant litter and rhizodeposition act as key vectors for the entry of most of the carbon into the SOC pool (Rasse et al. 2005). Consequently, the depth allocation of roots is matched by the light fraction of organic carbon (Schrumpf et al. 2013). Soil, climate, and vegetation types are a few of the factors on which the comparative significance of litter and rhizodeposition vs. other integration processes for SOC allocation and dynamics depend (Rumpel and Kögel-Knabner 2011). However, there is uncertainty in the data sets on root biomass because of uneven sampling, temporal and spatial heterogeneity, and methodological differences among various studies. A large portion of net primary production (NPP) is transferred to soil by fine root turnover, but available estimates on fine root turnover time vary, which can also be a result of different methodologies (Guo et al. 2008). Therefore, the estimates of the belowground carbon inputs to SOC from plant root litter are not certain (Denef and Six 2006). Rhizodeposition expresses the discharge of organic carbon compounds by the roots (Jones et al. 2009). Slight knowledge about carbon rhizodeposition is found, especially in mixed plant communities such as AFSs. The systematic quantification of carbon sequestration in agroforestry soils is mainly hindered by the fact that the amount of rhizodeposition by the trees is nearly unknown (Jones et al. 2009).

Another direct belowground carbon input is dissolved organic carbon (DOC) absorption in the soil, but just a small fraction of soil SOC is represented by DOC as the greater part of the same is eventually returned to the atmosphere in the form of CO_2 (Bolan et al. 2011). In AFSs, the important sources of DOC are stem flow, through fall, newly deposited litter having humus, manure, biosolids, crop residues, etc. DOC retention in subsoils is linked to aluminum (hydroxides) and poorly crystalline iron having a relatively higher surface area. Therefore, the translocation of DOC adds to the mineral-bound SOC formation in the subsoil (Schrumpf et al. 2013). Leaching of DOC from soils to adjoining aquatic ecosystems may also occur (Bolan et al. 2011), and such carbon losses by the leaching process might be predominantly significant for the carbon balance in AFSs (Kindler et al. 2011). As SOC stability depends upon biotic and abiotic environments, it is referred to as ecosystem property (Schmidt et al. 2011). Microbial derivatives play an essential role in SOC stabilization (Kleber et al. 2011).

The SOC turnover appears to be associated with microbial ecology and resource availability within a specified soil environment. Therefore, the processes that are responsible for slowing down mineralization are key centennial-scale stabilization mechanisms for SOC (Sanderman et al. 2010). The decomposition process might be retarded for years by physical protection, while organometallic interactions might be a reason for the enduring stability of most of the non-charred SOC (Kögel-Knabner et al. 2008). Physical protection and biochemical recalcitrance, on the other hand, let SOC remain in the soil for a long time duration, allowing organo-metal compounds to develop (Six et al. 2000). SOC stabilization in AFSs is thus a result of a mix of short- and long-term activities (Nair et al. 2010). Even if SOCs are aged for as many as thousands of years, any interruption in the stabilizing mechanism can cause them to decompose (Ewing et al. 2006). Briefly, the reason how SOC can persist is largely the interactions between SOC and its environment, e.g., the compound chemistry interdependence, climate, soil redox state, soil acidity, water availability, reactive mineral surfaces, and the potential degrader's presence in the direct microenvironment (Schmidt et al. 2011). SOC's relationship with soil minerals is probably the most significant factor in SOC stabilization, irrespective of soil type, vegetation, and land use. The binding site's availability, i.e., the soil's mineral depth and composition, will determine the SOC storage capacity unless additional environmental restrictions prevent decomposition. As a result, long-term SOC storage is narrow (Schrumpf et al. 2013).

9.3.1.2 Increasing Carbon and Reducing its Losses from Soil

By implementing conservation-effective measures that increase biomass production, decrease nutrient and water losses, and prevent SOC losses by enhancing physical, chemical, and biological stabilization mechanisms, biomass carbon inputs to the soil can be improved, and their rate of decomposition can be declined in AFSs (Lal and Follett 2009; Edrisi et al. 2022). Therefore, higher SOC levels are reported in AFSs due to reduced cultivation intensity (Nair et al. 2010). In these systems, SOC sequestration can be improved by providing supplements like composts and fertilizers, irrigation, and mixing organic wastes into the soil, just similar to that of other agricultural systems. Nevertheless, studies on the fertilization of the majority of tree species in tropical areas are insufficient, and no conclusions can be drawn on fertilization effects (Nair et al. 2010). However, compost inputs in agroforestry soils may affect the production and stability of soil aggregates. Furthermore, whereas herbicide treatments have conflicting impacts on SOC levels and aggregation, not much is known about pesticide effects on SOC sequestration (Nair et al. 2010). Conclusively, the impacts of more active and enhanced management approaches on SOC in AFSs vary depending on the location. AFSs with multispecies possess a higher potential to be more productive even than those of the best-performing monocultures. Therefore, it possibly stores additional SOC because of greater interactions belowground, although no conclusion can be made according to the present experimental data (Ong et al. 2004; Rao et al. 2004).

Tree species are selected, and their silvicultural management, e.g., stand density, rotation length, etc., are taken into consideration for SOC sequestration (Nair et al. 2009a). Functionally significant tree species, such as those possessing extensive and deep root systems to improve carbon inputs into the soil, possibly contain a higher potential of enhancing SOC sequestration in AFSs (Kell 2012; Lorenz and Lal 2010). Trees with broad leaves especially have an extensive and deep root system and, as a result, produce higher SOC. More biomass is produced by mixed plantings, including nitrogen-fixing trees, which may further improve SOC sequestration (Nair et al. 2009). To improve SOC sequestration, appropriate tree species must be chosen (Oelbermann et al. 2006). Because of the unavailability of sufficient data regarding manipulative biodiversity having several tree species, it is also uncertain whether mixed tree plantings in AFSs improve SOC sequestration (Scherer-Lorenzen et al. 2005). Conclusively, it is yet to be investigated whether such AFSs can be designed and managed exclusively to improve the belowground carbon sequestration. Direct inputs of organic resources, such as composts, manures, and other recovered organic materials, can further boost inputs. Because of the increased carbon inputs, every management strategy that may boost production from a given area of land should theoretically increase SOC storage. Fertilizer use, increased rotations, improved cultivars, and irrigation are all methods that can result in significant productivity gains. Crop intensification techniques, including double cropping, opportunity cropping, and repeated cropping, can also boost productivity. Emphasis is diverted to figuring out that several of these yield-increasing strategies include the use of irrigation water and fertilizers, both of which take much energy and hence raise CO_2 emissions (Chan 2008). Conservation farming these days is gaining favor as a farming strategy that improves the conservation of resources, e.g., soil and water resources worldwide.

9.4 Carbon Sequestration Potential of AFSs

Carbon sequestration occurs belowground as soil carbon and root biomass and aboveground as standing biomass. The amount of carbon stored in root biomass and soil is far greater than the amount found in crops. Promoting agroforestry is essential to mitigate carbon effects on the atmosphere. The latest study under different AFSs in various ecological circumstances showed that combined tree-crop agricultural systems accumulated comparatively higher carbon in deeper soil layers in the vicinity of the trees than those of tree-less systems. There are different factors, e.g., tree density, species and age, climatic conditions, geographical location, various management practices, etc., that affect carbon sequestration through agroforestry. Higher tree density and species richness lead to higher soil carbon content, and

Table 9.4 Carbon sequestration in different AFSs

Agroforestry system and plant species	Soil depth	Age (year)	Soil C ($Mgha^{-1}$)
Agrisilviculture (*Trifolium subterranean* + *Pseudotsuga menziesii*)	0–45	11	95.89
Agrisilviculture (*Gmelina arborea* + field crops)	10–60	5	27.4
Agrisilviculture (Zea mays + *Gliricidia sepium*)	0–200	10	123
Alley cropping: *Zea mays* + *Phaseolus vulgaris* + *Erythrina poeppigiana*	0–40	19	1.62
Silvopastoral system: (*Arachis pintoi* + *Acacia mangium*)	10–16	0–100	173
Silvopastoral system: (*Brachiaria brizantha* + *Cordia alliodora* + *Guazuma ulmifolia*)	0–100	10–16	132
Tree-based pastures: *Paspalum notatum* (bahiagrass) + *Pinus elliottii* + pature	0–125	8–40	6.9–24.2
Fodder bank (*Pterocarpus lucens* + *P. erinaceus*+ *Gliricidia sepium*)	0–100	6–9	33.4

Nair et al. (2009)

plants having a C3 cycle contributed to more carbon in the silt+clay sized portions, which constitute more steady carbon than plants having a C4 cycle in deeper soil layers. Usually, more carbon is sequestered by tropical humid climates in comparison to arid, semi-arid, and temperate regions. The soils of arid climates and colder regions have, on average, 30 and 800 t ha^{-1} $year^{-1}$ of SOC pool, respectively, below the depth of one meter. Globally, the total land area that has a potential for carbon sequestration is 1023 mha, the same as the area under AFSs worldwide. The amount of carbon sequestered in AFSs depends on environmental conditions and system management strategies. Sequestered carbon trading is a practicable opportunity for financial benefits to the farmers who practice agroforestry, mostly in developing countries. Nevertheless, more precise research outcomes are necessary for an AFS to be used in a worldwide program of carbon sequestration. Soil is the largest source of carbon, approximately five times that of the atmospheric pool. It emits, and stores carbon, and the soil carbon pool is 2500 PgC up to 1 meter depth, and out of this, 1500 PgC is organic. The potential carbon storage in AFSs is reported to be highly variable (Dixon et al. 1994) (Table 9.4 depiction of the carbon sequestration potential of a few AFSs).

The purpose of agroforestry is to increase and improve the livelihood of rural people, but as India is a signatory to all climate change conventions, agroforestry solves the dual purpose of increased production and carbon sequestration. Wastelands and grasslands are being converted into agroforestry zones (Murthey et al. 2013). Abundant data sets are present to suggest that an AFS has a higher overall (biomass) production, soil conservation, soil fertility improvement, nutrient cycling, microclimate improvement, and carbon sequestration potential compared to the monoculture system (Dhyani et al. 2009). As a result, the role of agroforestry as a carbon sequestration strategy has raised substantial expectations.

9.4.1 Aboveground Carbon Sequestration

Due to the complex and intricate relationships between belowground and aboveground carbon sequestration, considering anyone in isolation would be impractical. In brief, the basis of the estimation of aboveground carbon sequestration potential is the supposition that almost 45–50% of the branches and around 30% of the foliage dry weight make carbon great variability is found in these estimates ranging from 0.29–15.21 Mg/ha/year As these estimations are straightly linked to the estimated productivity of the system, their dependency occurs on various factors, including various land-use types, site characteristics, involved species, stand age, and different management practices. Degraded land, semi-arid, and arid AFSs have a lower capacity for carbon sequestration than those of humid lands, whereas tropical AFSs have comparatively higher vegetation carbon sequestration potential than those of temperate ones (Nair et al. 2009). The carbon sequestration capacity of aboveground vegetation can be increased through various land-use practices like reforestation, aforestation, and different AFSs such as agri-silviculture, silvopastoral, etc. (Brown 1996).

9.4.2 Belowground Carbon Sequestrations

There are two ways of carbon sequestration: direct and indirect (SSSA 2001). CO_2 is converted into inorganic carbon compounds such as magnesium and calcium carbonates ($MgCO_3$ and $CaCO_3$) through chemical reactions and indirect carbon sequestration. Indirect carbon sequestration occurs through photosynthesis. All green plants fix CO_2 into sucrose or other organic forms (plant biomass) inside the chloroplast with the help of photosynthetic pigment chlorophyll. Some amount of this plant biomass is sequestered indirectly in the form of SOC during the process of decomposition. The sequestered carbon amounts at a specific location reveal a balance between the release and uptake mechanisms of carbon. It has been estimated that traditional soils account for 58% of the gross CO_2 emission from terrestrial ecosystems into the atmosphere (Lal and Bruce 1999). Examining the literature on soil carbon sequestration, a general trend of rising SOC has been reported in AFSs as compared to other systems of efficient land use (Nair et al. 2009), and various land use systems were ranked in terms of SOC content in the order: forests>agroforests>tree plantations>arable crops (Agro-forests are complex multi-strata systems, which is similar to home-gardens in structural components but bigger). Varying values of sequestered carbon in AFSs reflected the socioeconomic and biophysical characteristics of methodological artifacts or system parameters. Quality and amount of biomass provided by various trees and non-tree components, structural properties of soil, etc., significantly affect agroforestry's impact on soil carbon sequestration. For instance, when the silvopastoral systems are established, certain functional outcomes become inevitable when grass-dominated lands are allowed for tree plantation, such as an open pasture system. These involve modifications in the above and belowground gross productivity of the soil, alterations to root depth and distributions, and litter input quality and quantity. (Jackson et al. 2000; Jobbagy and Jackson 2000). Carbon dynamicity and storage are modified by the above-mentioned changes in soil characteristics, litter components, etc. (Schlesinger et al. 1990). SOC is closely associated with the quality of organic matter and contains different proportions, which have heterogeneous structures and varying decomposability (Agren et al. 1996).

9.4.3 Carbon Sequestration in Agri-Silvicultural Tree Biomass

Maikhuri et al. (2000) assessed the impact of multipurpose trees over five years in the Garhwal district of Himachal Pradesh. Under this assessment, trees of *Alibizzia lebbek* (siris), *Alnus nepalensis* (Utis), *Boehmeria rugulosa* (Genthi), *Celtis australis* (Kharik), *Dalbergia sissoo* (Sesham), *Ficus glomerata* (Gular), *Grewia optiva* (Bhimal), *Prunus cerasoides* (Paiyan), *Pyrus pashia* (Molu), and *Sapium sebiferum* (Charvi) were planted at regular intervals of 3-meters. *A. nepalensis* followed by *D. sissoo,* showed the highest carbon sequestration, but this difference was three times for carbon accumulation, two times for an increase in soil carbon, and 1.5 times for carbon sequestration. Kaushik and Kumar (2001) assessed the effect of Khejri (*Prosopis cineraria*) on crops. All fodder crops were significantly affected in the kharif season only, with the maximum impact on pearl millet and the least impact on baffle grass. The carbon sequestration strategy aims at natural resource management by increasing tree growth, crop biomass, and soil carbon-improving activities. The carbon sequestration rates ranged from 0.85–2.20 t C ha^{-1} $year^{-1}$ for biomass carbon in drought proofing and 0.12–2.61 t C ha^{-1} $year^{-1}$, 0.1–1.97 t C ha^{-1} $year^{-1}$ and 0.1–1.93 t C ha^{-1} $year^{-1}$ for SOC in drought proofing, land development and minor irrigation activities respectively. The total carbon sequestered ranged from 107 to 4.900 Mt. C from all land resource management activities. The total carbon sequestered was 6.96 Mt. C and 25.48 Mt. CO_2, and it is estimated that it will be nearly ten folds by the next decade (Ravindranath and Murthy 2021). Agri-silviculture involves the use of trees and crops in a common space. The trees, being deep-rooted, can be sustained in harsh climates and can also sequester large amounts of carbon compared to crops. The roots form 20–25% of total biomass. Dhyani et al. (2009) studied carbon sequestration and reported 40% less sequestration in tree monocrops and 84% in food crops as compared to agri-silviculture. Eleven-year-old *Dalbergia sissoo* has been reported to accumulate 48–52 t C ha^{-1} of biomass in the silvi-agricultural system (Newaj and Dhyani 2008). The effect of pruning treatment with trees and black gram mustard on the tree biomass ranged from 23.6 to 34.49 t C ha^{-1}. In another study, poplar biomass was reported as 25.2 t C ha^{-1}. In wheat cultivation, a 113.6% increase in net carbon storage was found (Chauhan et al. 2010).

9.4.4 Carbon Sequestration in Silvopastoral Tree Biomass

NRCAF carried out a comparative study on grasslands and silvopastoral system for biomass production. This system was comprised of a few species of woody perennials such as *Dichrostachys cinerea, Albizia amara, etc., grass species, e.g. Chrysopogan fulvus, and legume trees such as Stylosanthes hamata* and *S. scabra.* This study ultimately evaluated that the silvopastoral system carbon sink was found to be twice as that of natural grasslands, and carbon storage of both systems was reported as 6.72 tC/ha/year and 3.14 tC/ha/year, respectively, after the span of 8 years (NRCAF 2007). In six-year-old poplar plantations in Punjab, carbon storage was reported to be 55.43 t/ha in block planting methods and 32.7 t/ha in boundary plantations (Chauhan et al. 2010). Silivi-pastoral systems help in increasing the organic content of the soil. Studies reported around 1.7 to 2.3 times increase with *Cenchrus ciliaris, Leucaenea leucocephala,* and *Stylosanthess hamata* plants compared to a control in AFSs (Kaur et al. 2002; Edrisi et al. 2018). Higher levels of organic carbons were reported in dry and arid lands when crops were integrated with grass species in the silvopastoral system (Venkateswaralu 2010). Similarly, the introduction of *Dalbergia sissoo* and *Prosopis juliflora* tree species with grass has increased carbon flux in NPP (Narain 2008). Singh et al. reported 16,400 tC/year sequestered carbon amount in AFS having tress species such as *Populus deltoides, Anthocephalus chinensis, Eucalyptus sp, etc. (*Singh 2003*).* Rai et al. (2009) studied the effect of the silvopastoral system in natural grasslands in Uttar Pradesh, where Albizia procera species had the highest biomass accumulation of 8.61 t ha^{-1} $year^{-1}$.

9.5 Conclusions and Future Perspectives

Agroforestry has been proven to be a critical candidate for providing a range of ecosystem services, including soil enrichment, carbon sequestration, degraded land restoration, climate change mitigation, etc**.** It is a great way of complete utilization of space and other resources above and below the ground. Its outcomes are further extraordinary in protecting the environment, livelihood security, and soil conservation and sustainability. AFSs on farmlands/croplands ensure food security and considerable economic benefits to the farmer, the society, the region, and the nation as a whole by improving soil fertility and thereby enhancing crop and tree product yields. They also have the potential to provide habitats to the edge species and thereby add to gene pool conservation along with biodiversity enhancement and multiple tree species growth. Moreover, AFSs accumulate a huge amount of carbon and thus mitigate climate change by reducing the effect of GHG emissions, etc. Though our environment is still under continuous threat of climate change and in such a scenario, the adoption of such diversified land use systems (AFSs) by the farming community should be encouraged more since these systems provide stability and sustainability to the farming systems in different regions worldwide where the crops suffer from harsh climatic conditions.

Since AFSs have the potential to improve the efficiency of soil by improving soil quality and preventing land degradation, achieving sustainable agriculture and food security, combating climate change, etc., the future prospects of AFSs are promising. These systems are more profitable compared to sole cropping on a defined area of soil. They have proven themselves to be more sustainable and resilient agricultural systems. However, the adoption of such systems is not very easy as concerted efforts with the Indian government's policymakers are required to overcome the bottlenecks prevailing with these systems. More advertisement and knowledge-sharing events should be organized to educate the farmers on how to overcome the constraints. More adoption of AFSs, particularly in xerophytic conditions, will improve dry land productivity and, thus, the economic well-being of farmers. The government of India adopted a national agroforestry policy in 2014, and this policy is expected to considerably improve farm productivity and livelihood of small and marginal farmers in the future.

References

Agren GI, Bosatta E (1996) Quality: a bridge between theory and experiment in soil organic ° 5 matter studies. Oikos 76:522–528

Albrecht A, Kandji ST (2003) Carbon sequestration in tropical agroforestry systems. Agric Ecosyst Environ 99:15–27. https://doi.org/10.1016/S0167-8809(03)00138-5

Amede (ed) (2003) Natural resource degradation and environmental concerns in the Amhara National Regional State Impact on food security. Ethiopian Soils Sci, Soc, pp 173–183

Andrade SAL, Abreu CA, Abreu MF, Silveira APD (2003) Interac¸a˜o de chumbo, da saturac¸a˜o por bases do solo e de micorrizal arbuscular no crescimento e nutric¸a˜o mineral da soja. R Bras Ci Solo 27:945–954

Annonymous (1987) Annual report on AICRAP. G B Pant University of Agriculture and Technology, Pantnagar

Ayoub AT (1999) Fertilizers and the environment. Nutr Cycl Agroecosyst 55:117–121. https://doi.org/10.1023/A:1009808118692

Bainard LD, Koch AM, Gordon AM, Newmaster SG, Thevathasan NV, Klironomos JN (2011) Influence of trees on the spatial structure of arbuscular mycorrhizal communities in a temperate tree-based intercropping system. Agric Ecosyst Environ 14:3–20

Baldock JA, Skjemstad JO (2000) Role of the soil matrix and minerals in protecting natural organic materials against biological attack. Geochemistry 31:697–710

Beare MH, McNeill SJ, Curtin PDRL, Jones HS, Dodd M, Sharp BJ (2014) Estimating the organic carbon stabilisation capacity and saturation deficit of soils: a New Zealand case study. Biogeochemistry 120:71–87

Berner R (2003) The long-term carbon cycle, fossil fuels and atmospheric composition. Nature 426:323–326. https://doi.org/10.1038/nature02131

Bolan NS (1991) A critical review on the role of mycorrhizal fungi in the uptake of phosphorus by plants. Plant Soil 134:189–207

Bolan NS, Adriano DC, Kunhikrishnan A, James T, McDowell R, Senesi N (2011) Dissolved organic matter: biogeochemistry, dynamics, and environmental significance in soils. Adv Agron 110:1–75. https://doi.org/10.1016/B978-0-12-385531-2.00001-3

Brown S (1996) Present and potential roles of forests in the global climate change debate. Unasylva 185:3–10

Buresh RJ (1995) Nutrient cycling and nutrient supply in agroforestry systems. In: Dudal R, Roy RN (eds) Integrated plant nutrition systems. FAO Fertiizers and Plant Nutrition Bulletin Number 12, Rome, pp 155–164

Carneiro MAC, Siqueira JO, Moreira FMS, Carvalho D, Botelho AS, Saggin-Junior OJ (1998) Micorriza arbuscular em espe´cies arbo´reas e arbustivas nativas de ocorre^ncia no Sudoeste do Brasil. Cerne 41:129–145

Carvalho AMX, Tavares RDC, Cardoso IM, Kuyper TW (2010) Mycorrhizal associations in agroforestry systems. In: Dion P (ed) Soil biology and agricultrue in the tropics. Springer, pp 185–208

Cassman KG (1999) Ecological intensification of cereal production systems: yield potential, soil quality, and precision agriculture. Proc Natl Acad Sci 96:5952–5959

Chan Y (2008) Increasing soil organic carbon of agricultural land. Primefact 735:1–5

Chauhan SK, Sharma SC, Chauhan R, Gupta N, Srivastava R (2010) Accounting poplar and wheat productivity for carbon sequestration in agrisilviculture system. Indian For 136(9):1174–1182

Danjon F, Stokes A, Bakker MR (2013) Root systems of woody plants. In: Eshel A, Beeckman T (eds) Plant roots: the hidden half. CRC Press, Boca Raton

Denef K, Six J (2006) Contributions of incorporated residue and living roots to aggregate-associated and microbial carbon in two soils with different clay mineralogy. Eur J Soil Sci 57:774–786. https://doi.org/10.1111/j.1365-2389.2005.00762

Dhyani SK, Newaj R, Sharma AR (2009) Agroforestry: its relation with agronomy, challenges and opportunities. Indian J Agronomy 54:249–266

Diop S (ed) (2008) Vital water graphics—an overview of the state of the World's fresh and marine waters. United Nations Environmental Programme, Nairobi. https://doi.org/10.1007/s10457-009-9229-7

Dixon RK, Brown S, Houghton RA, Solomon AM, Trexler MC et al (1994) Carbon pools and fluxes of global forest ecosystems. Science 263:185–190

Dungait JAJ, Hopkins DW, Gregory AS, Whitmore AP (2012) Soil organic matter turnover is governed by accessibility not recalcitrance. Glob Chang Biol 18:1781–1796

Edrisi SA, Tripathi V, Abhilash PC (2018) Towards the sustainable restoration of marginal and degraded lands in India. Trop Ecol 59:397–416

Edrisi SA, Tripathi V, Dubey PK, Abhilash PC (2022) Carbon sequestration and harnessing biomaterials from terrestrial plantations for mitigating climate change impacts. In: Biomass, biofuels, biochemicals. Elsevier, pp 299–313

Elsen A, Gervacio D, Swennen R, De Waele D (2008) AMF-induced biocontrol against plant parasitic nematodes in Musa sp.: a systemic effect. Mycorrhiza 18:251–256

Ewing SA, Sandermann J, Baisden WT, Wang Y, Amundson R (2006) Role of large-scale soil structure in organic carbon turnover: evidence from California grassland soils. J Geophys Res 111:G03012. https://doi.org/10.1029/2006JG000174

Fanish SA, Priya RS (2013) Review on benefits of agro forestry system. International J Edu Res 1(1):1–12

Garrett HE (ed) (2009) North American agroforestry: an integrated science and practice, 2nd edn. American Society of Agronomy, Inc, Madison

George TS, Gregory PJ, Robinson JS, Buresh RJ (2002) Changes in phosphorus concentrations and pH in the rhizosphere of some agroforestry and crop species. Plant Soil 246:65

Godfray H, Beddington J, Crute I, Haddad L, Lawrence D, Muir JF, Toulmin C (2010) Food security: the challengeof feeding 9 billion people. Science 327:812–818

Gonçalves CDBQ, Schlindwein MM, Martinelli GDC (2021) Agroforestry systems: a systematic review focusing on traditional indigenous practices, food and nutrition security, economic viability, and the role of women. Sustain For 13(20):11397

Guo D, Li H, Mitchell RJ, Han W, Hendricks JJ, Fahey TJ, Hendrick RL (2008) Fine root heterogeneity by branch order: exploring the discrepancy in root turnover estimates between minirhizotron and carbon isotopic methods. New Phytol 177:443–456. https://doi.org/10.1111/j.1469-8137.2007.02242.x

Hartemink AE, Buresh RJ, Jama B, Janssen BH (1996) Soil nitrate and water dynamics in Sesbania fallow, weed fallows and maize. Soil Sei Soc Am J 60:568–565
Harvey CA, Villanueva C, Villacís J, Chacón M, Munoz D et al (2005) Contribution of live fences to the ecological integrity of agricultural landscapes in Central America. Agric Ecosyst Environ 111(4):200–230
Henao J, Baanante C (2006) Agricultural production and soil nutrient mlning in Africa: implication for resource conservation and policy development. Bull. International Fertilizer Development Center. Muscle Shoals, Al. USA, IFDC Tech
Herridge DF, Peoples MB, Boddey RM (2008) Global inputs of biological nitrogen fixation in agricultural systems. Plant Soil 311:1–18. https://doi.org/10.1007/s11104-008-9668-3
Hinsinger P, Betencourt E, Bernard L, Brauman A, Plassard C, Shen J, Tang X, Zhang F (2011) P for two, sharing a scarce resource: soil phosphorus acquisition in the rhizosphere of intercropped species. Plant Physiol 156:1078–1086
Hoosbeek MR, Remme RP, Rusch GM (2018) Trees enhance soil carbon sequestration and nutrient cycling in a silvopastoral system in South-Western Nicaragua. Agrofor Syst. https://doi.org/10.1007/s10457-016-0049-2
Hurni H, Herweg K, Portner B, Liniger H (2005) Soil erosion and conservation in global agriculture. In: Braimoh AK, Vlek PLG (eds) Land use and soil resources. Springer, Dordrecht. https://doi.org/10.1007/978-1-4020-6778-5_4
Ingleby K, Wilson J, Munro RC (2007) Mycorrhizas in agroforestry: spread and sharing of arbuscular mycorrhizal fungi between trees and crops: complementary use of molecular and microscopic approaches. Plant Soil 294:125–136
Intergovernmental Panel on Climate Change (2007) Assessment report.
IPCC (2013) Intergovernmental panel on climate change. Impacts of Climate Change Special Reports 28 May 2013. http://www.ipcc.ch/ipccreports/sres/regional/index.php?idp=30
Isaac ME, Hinsinger P, Harmand JM (2012) Nitrogen and phosphorus economy of a legume tree-cereal intercropping system. Sci Total Environ 15:434
Jackson RB, Schenk HJ, Jobbagy EG, Canadell J, Colello GD, Dickinson RE, Field CB, Friedlingstein P, Heimann M, Hibbard K, Kicklighter DW, Kleidon A, Neilson RP, Parton WJ, Sala OE, Sykes MT (2000) Belowground consequences of vegetation changeand their treatment in models. Ecol Appl 10:470–483
Jama B, Kwesiga F, Niang A (2006) Agroforestry innovations for soil fertility management in sub-Saharan Africa: Prospects and challenges ahead. World Agroforestry into the. Future 60
Jobbagy EG, Jackson RB (2000) The vertical distribution of soil organic carbon and its ´ 20 relation to climate and vegetation. Ecol Appl 10:423–436
Jones DL, Nguyen C, Finlay RD (2009) Carbon flow in the rhizosphere: carbon trading at the soil-root interface. Plant Soil 321:5–33. https://doi.org/10.1007/s11104-009-9925-0
Kähkölä A-K, Nygren P, Leblanc HA, Pennanen T, Pietikäinen J (2012) Leaf and root litter of a legume tree as nitrogen sources for cacaos with different root colonisation by arbuscular mycorrhizae. Nutr Cycl Agroecosyst 92:51–65
Kang BT, van der Kruijs ACBM, Couper DC (1986) Alley cropping for food crop production in humid and subhumid tropics. Paper presented at the Alley Farming Workshop, Ibadan
Kaur B, Gupta SR, Singh G (2002) Carbon storage and nitrogen cycling in silvipastoral system on sodic soil Northwestern India. Agrofor Syst 54:21–29. https://doi.org/10.1023/A:1014269221934
Kaushik N, Kumar V (2001) Khejri (*Prosopis cineraria*) based agroforestry systems for arid Haryana, India. Journal of Arid Environments 55(2003):433–440
Kell DB (2012) Large-scale sequestration of atmospheric carbon via plant roots in natural and agricultural ecosystems: why and how. Philos Trans R Soc B 367:1589–1597. https://doi.org/10.1098/rstb.2011.0244
Kindler R, Siemens J, Kaiser K, Walmsley DC, Bernhofer C, Buchmann N, Cellier P, Eugster W, Gleixner G, Grûnwald T, Heim A, Ibrom A, Jones SK, Jones M, Klumpp K, Kutsch W, Larsen KS, Lehuger S, Loubet B, McKenzie R, Moor E, Osborne B, Pilegaard K, Rebmann C, Saunders M, Schmidt MWI, Schrumpf M, Seyfferth J, Skiba U, Soussana J-F, Sutton MA, Tefs

C, Vowinckel B, Zeeman MJ, Kaupenjohann M (2011) Dissolved carbon leaching from soil is a crucial component of the net ecosystemcarbon balance. Glob Chang Biol 17:1167–1185

Kleber M, Nico PS, Plante A, Filley T, Kramer M, Swanston C, Sollins P (2011) Old and stable soil organic matter is not necessarily chemically recalcitrant: implications for modeling concepts and temperature sensitivity. Glob Chang Biol 17:1097–1107. https://doi.org/10.1111/j.1365-2486.2010.02278.x

Kögel-Knabner I, Guggenberger G, Kleber M, Kandeler E, Kalbitz K, Scheu S, Eusterhues K, Leinweber P (2008) Organo-mineral associations in temperate soils: integrating biology, mineralogy and organic matter chemistry. J Plant Nutr Soil Sci 171:61–82. https://doi.org/10.1002/jpln.200700048

Kumar BM (2011) Quarter century of agroforestry research in Kerala: an overview. J Trop Agric 49(1–2):1–18

Kumar A, Hooda MS, Bahadur R (1998) Impact of multipurpose trees on productivity of barley in arid ecosystem. Ann Arid Zone 37:153–157

Kumar P, Singh RP, Singh AK, Kumar V (2014) Quantification and distribution of agro forestry-systems and practices at global level. HortFlora Res Spectrum 3(1):1–6

Lal R (1989) Agroforestry systems and soil surface management of a tropical Alfisol. II. Water runoff, soil erosion and nutrient loss. Agrofor Syst 8:97–111

Lal R (1990) Agroforestry systems to control erosion on arable tropical steeplands. IAHS-AISH Publ, No.192

Lal R (2011) Soil carbon sequestration. SOLAW background thematic report- TR04B, food and agriculture Organization of the United Nations, Rome, p 36

Lal R, Bruce J (1999) The potential of world cropland to sequester C and mitigate the greenhouse effect. Environment Science and Policy 2:177–185

Lal R, Follett RF (2009) Soils and climate change. In: Lal R, Follett RF (eds) Soil carbon sequestration and the greenhouse effect, vol 57, 2nd edn. SSSA Special Publication, Madison, pp xxi–xxviii

Lee JJ, Phillips DL, Dodson RF (1996) Sensitivity of the US Corn Belt to climate change and elevated CO2: II. Soil erosion and organic carbon. Agric Syst 52:503–521

Lorenz K, Lal R (2005) The depth distribution of soil organic carbon in relation to land use and management and the potential of carbon sequestration in subsoil horizons. Adv Agron 88:35–66. https://doi.org/10.1016/S0065-2113(05)88002-2

Lorenz K, Lal R (2010) Carbon sequestration in forest ecosystems. J Plant Nutr Soil Sci. https://doi.org/10.1002/jpln.201090008

Maikhuri RK, Semwa RL, Rao KS, Singh K, Saxena KG (2000) Growth and ecological impacts of traditional agroforestry tree species in central Himalaya, India. Agrofor Syst 48:257–272

Middledrop PJM, Briglia M, Salkinoja-Salonen M (1990) Biodegradation of pentachlorophenol in natural polluted soil by inoculated Rhodococcus *chlorophenolicus*. Microb Ecol 20:123–139

Mishra PK, Rai SC (2013) Use of indigenous soil and water conservation practices among farmers in Sikkim Himalaya. Indian J Tradit Knowl 12(3):454–464

Misra PK (2011) Soil fertility management in agroforestry system. Int J Biotechnol Biochem 7(5):637–644

Mommer L, Kirkegaard J, van Ruijven J (2016) Root-root interactions: towards a rhizosphere framework. Trends Plant Sci 21:209–217

Montagnini F, Nair PKR (2004) Carbon sequestration: an underexploited environmental benefit of agroforestry systems. Agrofor Syst 61:281–295

Moser G, Leuschner C, Hertel D, Hölscher D, Köhler M, Leitner D, Michalzik B, Prihastanti E, Tjitrosemito S, Schwendenmann L (2010) Response of cocoa trees (Theobroma cacao) to a 13-month desiccation period in Sulawesi, Indonesia. Agrofor Syst 79:171–187

Munishamappa M, Austin D, Muddumadappa N (2012) Carbon sequestration in litter and soil of coffee based agroforestry systems Central Western Ghats Kodagu district of Karnataka. Environ Ecol 30(3B):985–987

Munroe JW, Soto G, de M. Virginio Filho E, Fulthorpe R, Isaac ME (2015) Soil microbial and nutrient properties in the rhizosphere of coffee under agroforestry management. Appl Soil Ecol 93:40–46
Murthy IK, Gupta M, Tomar S, Munsi M, Tiwari R et al (2013) Carbon sequestration potential of agroforestry Systems in India. J Earth Sci Climate Change 4:131. https://doi.org/10.4172/2157-7617.1000131
Mutuo PK, Cadisch G, Albrecht Palm CA, Verchot L (2005) Potential of agroforestry for carbon sequestration and mitigation of greenhouse gas emissions from soils in the tropics. Nutr Cycl Agroecosyst 71:43–54
Nair PKR (1993) An introduction to agroforestry. Kluwer Academic Publishers, p 499
Nair PKR, Nair VD, Kumar BM, Haile SG (2009) Soil carbon sequestration in tropical agroforestry systems: a feasibility appraisal. Environ Sci Policy 12:1099–1111. https://doi.org/10.1016/j.envsci.2009.01.010
Nair PKR, Nair VD, Kumar BM, Showalter JM (2010) Carbon sequestration in agroforestry systems. Adv Agron 108:237–307
Narain P (2008) Dryland management in arid ecosystem. J Indian Soc Soil Sci 56:337–347
Newaj R, Dhyani SK (2008) Agroforestry for carbon sequestration: scope and present status. Ind J Agroforest 10:1–9
Newsham KK, Fitter AH, Watkinson AR (1995) Multifunctionality and biodiversity in arbuscular mycorrhizas. Trends Ecol Evol 10:407–411
Nguyen C (2003) Rhizodeposition of organic C by plants: mechanisms and controls. Agronomie 23:375–396
NRCAF (2007) Perspective plan vision 2025 National Research Centre for Agroforestry. Jhansi, Uttar Pradesh
Phillips RP, Erlitz Y, Bier R, Bernhardt ES (2008) New approach for capturing soluble root exudates in forest soils. Funct Ecol 22:990–999
Oelbermann M, Voroney RP, Gordon AM, Kass DCL, Schlönvoigt AM, Thevathasan NV (2006) Soil carbon dynamics and residue stabilization in a costa Rican and southern Canadian alley cropping system. Agrofor Syst 68:27–36. https://doi.org/10.1007/s10457-005-5963-7
Olsson PA, Thingstrup I, Jakobsen I, Baath F (1999) Estimation of the biomass of arbuscular mycorrhizal fungi in a linseed field. Soil Biol Biochem 31:1879–1887
Ong CK, Kho RM, Radersma S (2004) Ecological interactions in multispecies agroecosystems: concepts and rules. In: Ong CK, Huxely P (eds) Tree-crop interactions, a physiological approach. CAB International, Wallingford, pp 1–15
Pacardo EP, Montecillo L (1983) Effects of corn/ipil-ipil cropping systems on productivity and stability of upland agro-ecosystems. Annual Report, UPLB PCARRD Res. Project
Pandey DN (2007) Multifunctional agroforestry systems in India. Curr Sci 92(4):455–463. http://www.jstor.org/stable/24097558
Phoung TT, Shrestha RP, Choung HV (2017) Simulation of soil erosion risk in the upstream area of Bo River watershed. In: In redefining diversity and dynamics of natural resources management in Asia, vol 3. Elsevier, pp 87–99
Pragasan LA, Parthasarathy N (2005) Litter production in tropical dry evergreen forests of South India in relation to season, plant life-forms and physiognomic groups. Curr Sci 88(8):1255–1263
Radersma S, Grierson PF (2004) Phosphorus mobilization in agroforestry: organic anions, phosphatase activity and phosphorus fractions in the rhizosphere. Plant Soil 259:209–219
Rai P, Ajit, Chaturvedi PO, Singh R, Singh UP (2009) Biomass production in multipurpose tree species in natural grasslands under semi arid conditions. J Trop For 25:11–16
Rao MR, Schroth G, Williams SE, Namirembe S, Schaller M, Wilson J (2004) Managing below-ground interactions in agroecosystems. In: Ong CK, Huxely P (eds) Tree-crop interactions, a physiological approach. CAB International, Wallingford, pp 309–332
Rasse DP, Rumpel C, Dignac MF (2005) Is soil carbon mostly root carbon? Mechanisms for a specific stabilization. Plant Soil 269:341–356

Ravindranath NH, Murthy IK (2021) Mitigation co-benefits of carbon sequestration from MGNREGS in India. PLoSONE 16(5):e0251825

Rawat N, Nautiyal BP, Nautiyal MC (2009) Litter production pattern and nutrients discharge from decomposing litter in Himalayan alpine ecosystem. NY Sci J 2(6):54–67

Rillig MC, Mummey DL (2006) Mycorrhizas and soil structure. New Phytol 171:41–53

Rillig MC, Wright SF, Nichols KA, Schmidt WF, Torn MS (2001) Large contribution of arbuscular mycorrhizal fungi to soil carbon pools in tropical forest soils. Plant Soil 233:167–177

Roshetko JM, Lasco RD, Angeles MSD (2007) Small holder agroforestry systems for carbon storage. Mitig Adapt Strateg Glob Chang 12:219–242

Rufyikiri GS, Declerck S, Dufey JE, Delvaux B (2000) Arbuscular mycorrhizal fungi might alleviate aluminium toxicity in banana plants. New Phytol 148:343–352

Rumpel C, Kögel-Knabner I (2011) Deep soil organic matter—a key but poorly understood component of terrestrial C cycle. Plant Soil 338:143–158. https://doi.org/10.1007/s11104-010-0391-5

Sanderman J, Farquharson R, Baldock J (2010) Soil carbon sequestration potential: a review for Australian agriculture—a report prepared for Department of Climate Change and Energy Efficiency, CSIRO Land and Water http://www.csiro.au/resources/Soil-Carbon-Sequestration-Potential-Report.html

Sanders IR (2003) Specificity in the arbuscular mycorrhizal symbiosis. In: Van Der Heijden MGA, Sanders I (eds) Mycorrhizal ecology. Springer, Berlin, Heidelberg, pp 415–437

Sarvade S, Singh R, Prasad H, Prasad D (2014) Agroforestry approaches of for improving soil nutrient status. Popular kheti 2:1

Scherer-Lorenzen M, Potvin C, Koricheva J, Schmid B, Hector A, Bornik Z, Reynolds G, Schulze E-D (2005) The design of experimental tree plantations for functional biodiversity research. In: Scherer-Lorenzen M, Korner C, Schulze E-D (eds) Forest diversity and function: temperate and boreal systems ecological studies. Springer-Verlag, Berlin Heidelberg, pp 347–376

Schlesinger WH, Reynolds JF, Cunningham GL, Huenneke LF, Jarrell WM, Virginia RA, Whitford WG (1990) Biologicalfeedbacks in global desertification. Science 247:1043–1048

Schmidt et al (2020) Does temperate agroforestry reduce nutrient leaching losses compared to cropland monocultures? 22nd EGU General Assembly, held online 4–8 May, 2020, id.8528. https://doi.org/10.5194/egusphere-egu2020-8528

Schmidt MWI, Torn MS, Abiven S, Dittmar T, Guggenberger G, Janssens IA, Kleber M, Kögel-Knabner I, Lehmann J, Manning DAC, Nannipieri P, Rasse DP, Weiner S, Trumbore SE (2011) Persistence of soil organic matter as an ecosystem property. Nature 478:49–56. https://doi.org/10.1038/nature10386

Schrumpf M, Kaiser K, Guggenberger G, Persson T, Kögel-Knabner I, Schulze ED (2013) Storage and stability of organic carbon in soils as related to depth, occlusion within aggregates, and attachment to minerals. Biogeosciences 10:1675–1691. https://doi.org/10.5194/bg-10-1675-2013

Sheldrick WF, Syers JK, Lingard J (2002) A conceptual model for conductingnutrient audits at national, regional, and global scales. Nutr Cycl Agroecosyst 62:61–72

Singh TP (2003) Potential of farm forestry in carbon sequestration. Indian Forester 129:839–843

Six J, Elliott ET, Paustian K (2000) Soil macroaggregate turnover and microaggregate formation: a mechanism for C sequestration under no-tillage agriculture. Soil Biol Biochem 32:2099–2103. https://doi.org/10.1016/S0038-0717(00)00179-6

Smith SE, Read DJ (2008) Mycorrhizal symbiosis, 3rd edn. Academic, London, some agroforestry and crop species. Plant Soil 246:53–63

SSSA (2001) Carbon sequestration: position of the Soil Science Society of America (SSSA). Available from: www.soils.org/pdf/pos_paper_carb_seq.pdf

Tavares RC (2007) Efeito da inoculac¸a˜o com fungo micorrı'zico arbuscular e da adubac¸a˜o orga^nica no desenvolvimento de mudas de sabia´ (Mimosa caesalpiniaefolia Benth.). In: sob estresse salino. Magister Scientiae dissertation, Universidade Federal do Ceara´. Fortaleza

Udawatta RP, Krstansky JJ, Henderson GS, Garrett HE (2002) Agroforestry practices, runoff, and nutrient loss: a paired watershed comparison. J. Environ. Qual. 31:1214–1225 under controlled conditions. Sci Total Environ 434:71–78

Upson MA, Burgess PJ (2013) Soil organic carbon and root distribution in a temperate arable agroforestry system. Plant Soil 373:43–58. https://doi.org/10.1007/s11104-013-1733-x

Van der Heijden MGA et al (1998) Mycorrhizal fungal diversity determines plant biodiversity, ecosystem variability and productivity. Nature 396:69–72

Venkateswaralu J (2010) Rainfed agriculture in India: research and development scenario. Indian council of Agricultural Research, New Delhi, p 508

Verchot LV, Noordwijk MV, Kandji S, Tomich T, Ong C (2007) Climate change: linking adaptation and mitigation through agroforestry. Mitig Adapt Strateg Glob Chang 12:901–918

von Lützow M, Kögel-Knabner I, Ekschmitt K, Matzner E, Guggenberger G, Marschner B, Flessa H (2006) Stabilization of organic matter in temperate soils: mechanisms and their relevance under different soil conditions—a review. Eur J Soil Sci 57:426–445. https://doi.org/10.1111/j.1365-2389.2006.00809.x

Wallander H, Goransson H, Rosengren U (2004) Production, standing biomass and natural abundance of 15N and 13C in ectomycorrhizal mycelia collected at different soil depths in two forest types. Oecologia 139:89–97

Wiersum KF (1981) Observations on agroforestry in Java, Indonesia. Agric. Univ. Wageningen, p 131

Wiersum KF (1984) Surface erosion under various tropical agroforestry systems. In: O'Loughlin CL, Pearce AJ (eds) Effects of forest landuse on erosion and slope stability. Univ. of Hawaii, Honululu

Yengwe J, Gebremikael MT, Buchan D, Lungu O, Neve SD (2018) Effects of *Faidherbia albida* canopy and leaf litter on soil microbial communities and nitrogen mineralization in selected Zambian soils. Agrofor Syst. https://doi.org/10.1007/s10457-016-0063-4

Young A (1989) Agroforestry for soil conservation. CAB International, Wallingford, pp101

Young R, Orsini S, Fitzpatrick I (2015) Soil degradation: a major threat to humanity. Sustainable Food Trust. http://assets.fsnforum.fao.org.s3-eu-west1.amazonaws.com/public/discussions/contributions/Soil degradation

Zaidi S, Usmani S, Singh BR, Musarrat J (2008) Significance of *Bacillus subtilis* strains SJ-101 as a bioinoculant for concurrent plant growth promotion and nickel accumulation in *Brassica Juncea*. Chemosphere 64:991–997

Zheng Z (2006) Litter decomposition and nutrient release in a tropical seasonal rain forest of Xishuangbanna. Southwest China Biotropica 38(3):342–347

Zhu YG, Miller RM (2003) Carbon cycling by arbuscular mycorrhizal fungi in soil-plant systems. Trends Plant Sci 8:407–409

Zhuang XL, Chen J, Shim H, Bai Z (2007) New advances in plant growth-promoting rhizobacteria for bioremediation. Environ Int 33:406–413

Chapter 10
Impact of Various Agronomic Approaches on Nutrient Use Efficiency of Crops

Sagarika Parida

Abstract Agricultural soils are invariably deficient or imbalanced in one or more essential nutrients. As a result, the yield levels of the plant are declining. This issue encourages the need to increase nutrient use efficiency by different fertilization strategies according to the crop requirements. Proper functioning of the soil significantly depends on soil organic matter (SOM), water-holding capacity, and nutrient contents, which results in higher growth and vigor with greater productivity in natural environments and agronomical practices. This will ensure the increase in food security for the growing population as well as maintain the ground and surface water level, thereby reducing the negative impacts on the ecosystem. This chapter is aimed at harnessing and maintaining the nutrients in the agricultural system through different agronomic approaches such as crop rotations, proper tillage operations, and residue management integrated with nutrient management. The above mentioned approaches fundamentally involves use of organic fertilizers, viz. application of farm yard manure, compost and vermicompost in the soil, foliar fortification with macro and micronutrients spraying of the crop at the juvenile stages of crop growth. It can help ensure the soil health sustainability, efficient nutrient uptake and increased photosynthetic rates of the crops resulting in increased soil organic carbon (SOC). Integrated Nutrient Management (INM) to enhance crop productivity is also the part of discussion in this chapter. It will focus on the increase in nutrient uptake and the growth of different crops in water stress conditions by bioinoculation of different rhizobacterial strains in soil.

Keywords Conventional agricultural practices · Nutrient use efficiency · Agronomic approaches · Integrated nutrient management · Nano formulations

S. Parida (✉)
Department of Botany, School of Applied Sciences, Centurion University of Technology and Management, Bhubaneswar, Odisha, India
e-mail: sagarika.parida@cutm.ac.in

S. A. Edrisi et al. (eds.), *Sustainable Plant Nutrition and Soil Carbon Sequestration*, Sustainable Plant Nutrition in a Changing World,
https://doi.org/10.1007/978-3-031-53590-1_10

10.1 Introduction

The global population will reach around 9 billion by 2050. Global grain production of about 70% should have to increase to meet the food demand of the world population (FAO 2009; Alexandratos and Bruinsma 2012). Demand for food is the main cause of the depletion of natural resources, leading to environmental degradation (Tilman et al. 2002; Cassman et al. 2003). High nutrient use efficiency in relation to higher crop productivity by the crop plants is a challenge for a sustainable environment. Soil fertility will have to be preserved to feed the growing population. All the materials for fertilizers depend on geological resources. Nitrogen fertilizer depends on fossil fuels, whereas potassium and phosphates are derived through mining. Mineral nutrients are still needed in the soil for plant growth. Soil nutrients have been removed after each crop production, and a significant decrease of the soil nutrients has been observed and has to be maintained by soil fortification with different macro and micronutrients (Manning 2015). The nutritional content and microbiological status of the soil were taken as indicators of soil quality necessary for the efficient use of nutrients for plant growth. Organic farming and innovative agricultural practices are also important in maintaining soil health and strengthening carbon sequestration by storing carbon in sinks like oceans, forests, and crop fields (Lal 2004; Sarkar et al. 2021). Now, innovations in agricultural technology have been given importance that can reduce resource use production costs and increase production while maintaining environmental quality (Gupta et al. 2002). The important challenge is to build up a possible system that is energy- and labor-efficient and also simultaneously helps to maintain soil health and a better environment, maximizing crop yield with less cost (Gathala et al. 2011). It was reported that poor fertilizer response (FR) in maize resulted because of imbalanced soil fertilization (Vanlauwe et al. 2019). FR is determined by the increase in crop yield independent of the type and amount of fertilizers used. FR is used to identify the 'non-responsive soil' where the effect of fertilization is not observed (Ichami et al. 2019; Njoroge et al. 2017). In Kenya, it was observed that soil with a pH below 5.2, less rainfall, and less silt have lower FRs, resulting in low agronomic nitrogen use efficiency (Kihara et al. 2017). The aim of this chapter is to gather information on various modified agronomical methods for the cultivation of crops with the fortification of soil with microbes, different types of manures, fertilizers, use of fertilizers in nano forms, and foliar spraying of nutrients to restore soil health to increase the efficiency of nutrient uptake of the plants for growth and yield.

10.2 Agricultural Practices

Carbon sequestration is the most important action plan to lessen climate change and enhance the soil fertility of agricultural soils. Cultivated land requires a certain amount of organic matter called SOM, which is required for carbon sequestration.

Fig. 10.1 Agronomic approaches for nutrient use efficiency of crops. (Edrisi et al. 2023; Majumdar et al. 2023)

Organic manures, the addition of harvested crop remains or residues, peat soils, and various types of compost, along with good cultivation practices, are needed to enhance carbon sequestration. Agricultural practices like farm management, no-tillage techniques, mulching, crop rotations, residue management, nutrient management, and soil biota management are used for carbon sequestration (Fig. 10.1).

10.2.1 Farm Management

Upgrading the farm management practices supports in two ways, i.e., food production and maintaining soil health, which in turn will help for a sustainable environment. Best management practices have been implemented by researchers, and through the implementation of these agricultural practices, farmers achieve optimum production along with the safekeeping of the environment (Griffith and Murphy 1991). The fertilizer industry has framed out the Global 4R Nutrient Stewardship Framework that encourages nutrient application by utilizing the right source at the right rate, at the right time, and at the right place (IFA 2009; Phillips et al. 2009; Bruulsema et al. 2016). This strategy is established to manage the fertilizer application to ensure the economic, social, and environmental goals. Soil parameters, prevailing climatic conditions, and types of cropping should be considered before selecting the fertilizers for the best agronomic management systems for getting maximum yield from the produce (Dubey et al. 2020, 2021a). Each farm and farm operating system has to be adapted to implement improved nutrient management practices. Major challenges for smallholder farmers are the availability of labor, sources of fertilizer nutrients, types of crops cultivated in the particular area, multiple cropping systems, and availability of analytical laboratories. Therefore, ways should be found for the maximum use of nutrients for enhancing the yield and making a profit while keeping the environment safe.

10.2.2 Zero Tillage

Brazilian farmers began zero tillage in the early 1970s. This system captured the motivation with the discovery of systemic non-selective and selective herbicides and guaranteed the control of weeds with greater effectiveness. Zero tillage refers to the agricultural practice where soil disturbances are not being caused by plowing (Govaerts et al. 2005). In the conventional system, the tillage practice system was operated where six plowings using the country plow and 12 laddering for leveling the soil were employed for wheat (Saunders 1994). The evolution of zero tillage has arisen from the cultivation of wheat in a rice-wheat cropping system that enables an ideal shifting of the conventional approach of cultivation to a zero tillage system. Zero tillage is a one-time operation where the seed and fertilizer are applied in a pit in the undisturbed seedbed. The pit has been packed to preserve sufficient surface residues to prevent soil erosion. Seed drill designs proved to produce higher yield, greater residue production, and ensure maximum soil protection. In conservation tillage, previous leftover residues of crop plants remain on the field to increase SOM (Holland 2004). Therefore, it offers less soil erosion from wind and water, less soil compaction, and less evaporation of moisture, and the soil becomes more fertile and resilient because of the mulch cover of previous crops.

Potential benefits of zero tillage extend to early planting, substantial productivity, lower maintenance costs, better time management, and a less muddy environment with greater satisfaction. It has been reported that a drastic reduction of diesel fuel of about 50–70% with a proportionate reduction of GHG emissions (Gentil 1995). Essentially, zero tillage has a major effect in the reduction of CO_2 gas emissions by immobilizing carbon in the soil surface residues and SOM in comparison to the conventional system of tillage practice method (Derpsch 1998). It has been reported from the literature that there is a gradual decline in crop yield (Duxbury 2001; Ladha et al. 2003). Cultivation practices in rice-wheat systems deteriorate the soil, water holding capacity, and water resources (Ali and Byerlee 2002; Gupta et al. 2003; Ladha et al. 2003; Chauhan et al. 2012). Effects of five tillage systems (conventional tillage, reduced tillage, raised bed planting, rotavator tillage, and zero tillage) for wheat cultivation were investigated in Indo-Gangetic plains of South Asia for energy and economic use efficiency (Kumar et al. 2013; Saharawat et al. 2012; Pathak et al. 2011). It has been suggested that minimum tillage

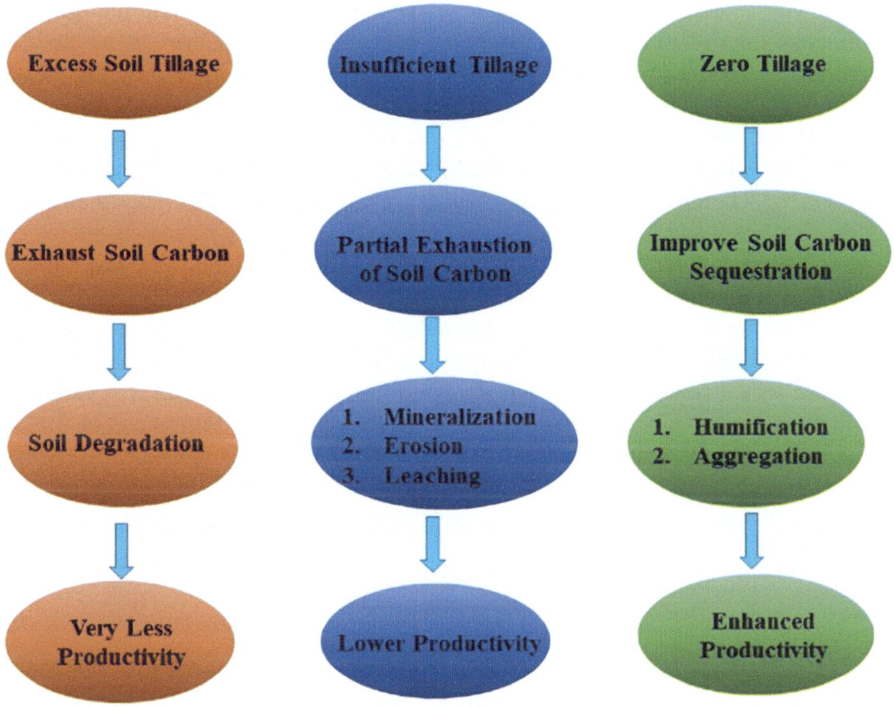

Fig. 10.2 Different types of tillage practices adopted in the agricultural fields and their relative productivities. (Adopted from Kumar et al. 2013; Saharawat et al. 2012; Pathak et al. 2011)

implementation leads to improved operational field capacity. Technologies based on conservation agriculture increased energy efficiency to enhance crop yield (Fig. 10.2).

Energy use, seedling efficiency, and wheat yield were studied, and it reported that the operational field capacity had been increased by 18%, followed by 17% and 13% with respect to specific energy and energy use efficiency, respectively, in comparison with conventional tillage cropping system by implementing zero tillage cultivation practices (Kumar et al. 2013). In this study, it was also stated that the maximum income in zero tillage is about 33%, followed by reduced tillage with 20% in comparison with the third method. These results revealed that conservation agriculture-based crop farming practices are the suitable cultivation approaches for nutrient use efficiency to get higher yield and profitability. Crop establishment in rice-wheat cropping systems requires a 25–30% price value of the cost of total wheat production in South Asia and results in a reduced benefit-cost ratio (Saharawat et al. 2012; Pathak et al. 2011).

The Gross Primary Production by the plants in the terrestrial systems and the microbial activity (respiration) in the soil determines the fate of SOM, thereby the regulating the SOC pool. It is controlled by both abiotic and biotic factors that include climatic factors, parent materials, biological diversity, and spatiotemporal factors. These factors fundamentally regulate the production or assimilation and decomposition (by microbes) of SOM (Singh et al. 2020; Edrisi et al. 2022). Therefore, the soil carbon dynamics is the function of soil formation by microbial activities, soil water retention, and biomass production (i.e. Net Primary Production (NPP). Hence, the conservation tillage or zero tillage practices help reduce the loss of labile carbon and soil moisture and ultimately the microbial respiration (Dubey et al. 2021b). Therefore, reducing the respiration rate can eventually helps maintaining the SOC pool and its dynamics.

Zero tillage with retention of crop residue enhanced crop yield over conventional tillage practice (Nandan et al. 2019). In India, zero tillage started, and it was widespread in Haryana in the North-Western Indo-Gangetic plains (Malik et al. 2005). Zero tillage helped the farmers in Haryana to overcome two hindrances, viz. late sowing of wheat because of late maturing and harvesting of Basmati rice varieties and high incidence of weed *Phalaris minor* (Harrington 1992; Fujisaka et al. 1994, Malik et al. 2005). Wheat yield gain of 15.4% was reported in farm trials of Haryana and reported to be decomposed into 9.4% because of timeliness and 6% increased fertilizer and water use efficiency along with a reduction in weed population (Mehla et al. 2000).

10.2.3 Mulching

Mulching is a practice of covering the soil surface to make the conditions appropriate for crop growth as mulching protects the soil from hot sunny rays, prevents drying from winds, protects the soil moisture, maintains coolness, protects from heavy rainfall and soil erosion, and prevents weed growth. Mulching helps to maintain soil fertility as mainly the soil surface is covered with rice or wheat straw or with detached plant parts to secure excessive evaporation, withstand cold stress, and simultaneously upgrade the SOM contents in the soil (Mulumba and Lal 2008). Thereby, the physical and chemical properties can be improved, and in turn, the soil can sequester more carbon, hence, mitigating climate change effects (Farooqi et al. 2018).

10.2.4 Crop Rotations and Cover Crops

SOM is the reservoir that holds plant nutrients. These nutrient reservoirs are not sufficient for the cultivated crops; therefore, different crop residues are incorporated by crop rotation. The nutritional needs of the crops are different for different crops. Most of the crops consume soil nutrients during their growth. Crop rotation has its own advantages in maintaining soil fertility. These nutrients are present in the harvested farm products, and the rest remain on the field as crop residues. Detached vegetative crop residues are knowingly left after harvesting over the field to decompose by the soil microorganisms, facilitating improvement in soil aggregation that improves water infiltration, aeration, and soil's water holding capacity, inducing root growth by facilitating nutrient use (Rangarajan et al. 2009). Crop rotation is considered a good agronomical practice to increase soil fertility, make better use of nitrogen, improve soil structure, and increase crop yields (Sainju et al. 2020). Cover crops are grown to get the benefit of the soil for maintaining the nutritional balance but not for that crop yield.

10.3 Nutrient Management

Nutrient management plays a significant role. Over or under-application of nutrients has a great impact on agricultural productivity. Nutrient use efficiency by the crop plants can be maintained by managing the soil biota, use of organic fertilizers viz. farm yard manure, compost and vermicompost, soil carbon sequestration, nitrogen fertilization, foliar fertilization, nutrient stewardship management, nanomaterial, nano fertilizers use and foliar fertilization to increase the nutrient uptake efficiency by the crops.

10.3.1 Nutrient Stewardship and Adaptive Management

Availability of food in the future depends upon adequate incorporation of nutrients for efficient use by the crops. The increase in demand for food, fiber, and feed crop products is severely punishing the world's soil resources. Soil health influences food security, aiming for a clear link between the quality of the soil and crop production. Chemical, physical, and biological hindrances of world soils affect future crop productivity and require integrated agronomical approaches for nutrient use efficiency of plants to enhance productivity. Nutrient management is associated with crops, soil, irrigation, and weather factors (Havlin et al. 1990).

Intercropping systems of cultivations were also found to increase the grain yield by 13.9% and 45.1%. Phosphorus use efficiency in maize under field conditions in regosol soil was significantly higher as compared to the mono-cropping system without any additional supply of phosphorus. The primary application of Olsen-P was applied at the rate of 12.6 mg kg^{-1} followed by phosphorus at the rate of 35 kg ha^{-1} $year^{-1}$. Sufficient P supply slows down the leaf senescence process by lessening the chlorophyll degradation and remobilization of P, influencing leaf longevity and thereby increasing the length of the photosynthesis process and maintaining the root growth. It is revealed that translocated sucrose in the roots contributes to root growth and enhances the uptake of P to fulfill the nutrient requirements of growing shoots (Zhou et al. 2021).

10.3.2 Organic Fertilizers

Organic cropping systems include animal compost, vermin compost, organic fertilizers, and farmyard manure. Organic fertilizers are derived from animal and plant residues. Differences in concentrations of different elements like C, N, P, K, Ca, and Mg were noticed in different types of animal compost (Ueno and Yamamuro 2001; Stockdale et al. 2001).

10.3.3 Soil Carbon Sequestration and Nitrogen Fertilization

Carbon is the most important organic matter of soil and is responsible for the water-holding capacity, structure, and fertility of soil. SOC dynamics is determined by the input and output balance of carbon in the soil. Carbon sequestration occurs when carbon input exceeds carbon output. Therefore, SOC is influenced by soil types, climatic condition of the cultivated area, and other chemical and biological aspects such as availability of minerals, microbial community, agronomic practices like crop rotations, zero tillage system, residual management or mulching and covering of soil with crop residues, and proper utilization of fertilizers. Soil carbon maintains the fertility and productivity of the soil. Conventional tillage practices significantly reduce the particulate organic carbon from the soil with the flow of runoff water deposited in low-lying areas and influence SOC dynamics. Soil carbon is an energy source for soil flora and fauna that is responsible for the degradation of SOC and its conversion to CO_2. Soil fauna fragment the plant residues and other farm wastes to make them suitable for microbial decomposition and thereby play a major role in carbon dynamics. Soil CO_2 is lost by the plants through the respiration pathway and revert into the atmosphere. SOC is also reduced by leaching, oxidation, and erosion by wind and runaway water on the surface of the soil. Conservation tillage can sequester soil carbon and contribute to mitigating GHGs emissions (Hati et al. 2020). Plant residue incorporation with the addition of nutrients may reduce SOC loss. The soil with sugarcane residue with the addition of nutrients was associated with soil carbon changes. The fertility of the soil was changed with nutrient addition and reduced total soil carbon depletion (Coonan et al. 2020). An intensive cropping system also disturbs the soil, leading to SOC depletion. Crop rotation also favors maintaining the soil C. Topsoil comprising a depth from 0 to 30 cm of soil profile has been focused on soil carbon sequestration. Cereal-based cropping systems, for example, double rice and forestry (*Madhuca longifolia* (L.) J.F. Macbr. and *Diospyros melanoxylon* Roxb.) or horticulture (*Psidium guajava* L.) system to conserve organic carbon in soil which would help to increase the food and timber production (Badole et al. 2020).

SOC has been known to increase by input of crop residues. For high grain yield, the soil has been fertilized with nitrogen fertilizers, which are known to sequester SOC. The effect of N fertilization on the production of crop residues, organic carbon content, and total soil nitrogen in non-till dry land cropping systems was evaluated, and it observed that an increased amount of crop residue resulted in increased SOC and total soil nitrogen in 0–7.5 cm soil depth after 11 crops. Soil bulk density decreases with an increase in SOC. Nitrogen fertilization promotes improvement in soil quality productivity and increases carbon sequestration in the soil (Halvorson et al. 1999). This concept is contradictory when considering the hundred years' data of SOC content from the oldest experimental site of Morrow Plots cultivated continuously with corn (*Zea mays* L.) cultivation. Application of higher doses of synthetic fertilizer for about 40–50 years showed a net decline of soil carbon and maximum loss in *Glycine max* (L.) Merr., and *Avena sativa* L., despite the incorporation of carbon residue and synthetic nitrogen fertilization for about 40–50 years (Khan et al. 2007).

10.3.4 Foliar Fertilization

The application of fertilizer by spraying nutrients on the leaves of crop plants has been proven to be an effective method to improve the efficient use of nutrients for crop growth. An increase in yield, disease, pest resistance, and crop quality has been observed in foliar fortification. The response of plants to foliar fertilization depends upon species, form of fertilizers, their concentration, application frequencies, spraying timing, and the crop growth stage (Farooqi et al. 2018). Foliar fortification can also be used to recover the plants from transplant shock and extreme weather. Fertilizer quantity can be adjusted according to the nutrient composition of the soil and specific stages of crop growth. Nutrient absorption is 8–20 times more efficient than ground application (Anonymous 1985; Parida et al. 2020). Nutrient sprays are generally diluted in water and sprayed. The ionic and polar solutes penetrate through the leaf surface. This method is proven to be environmentally friendly because the nutrients are delivered directly to the targeted crops in optimum amounts. This ensures and reduces the environmental impact related to soil fertilization.

Soil application of essential plant nutrients has been practiced to enhance plant growth and vigor and obtain maximum yields. Application of fertilizers on soil is based on soil tests, and foliar spraying of nutrients is done on the basis of visible symptoms. However, under certain specific conditions, foliar spraying of nutrients is more economical and potent for correcting nutrient deficiency. For sufficient absorption of nutrients, foliar nutrient supplement requires a higher leaf area index, and to combat the nutrient deficiency, more than one application may be required depending upon the intensity of nutrient deficiency. Foliar fortification of micronutrients facilitates quick use of nutrients and foliar application of 1%. Humic acid liquid fertilizer and 2% mono ammonium phosphate (MAP) revealed the highest nodulation in green gram in comparison to control conditions in variable proportions of soil and compost (Nayak et al. 2020). The effect of micronutrient foliar spray was studied in two villages under Krishi Vigyan Kendra, Angul, Odisha. In this study, there is an increase in growth, number of nodules per plant, and yield of

Vigna radiata L. with an increase in a net return of 88.56% was reported in foliar application of Oregon 80 with other plant protection measures in the improved variety IPM 02-03 (Das and Parida 2020).

Foliar application of 1% of DAP and $CaNO_3$, 2% KCl, and $ZnSO_3$ in black gram variety PU35 facilitated nutrient use efficiency and growth of black gram (*Vigna mungo* L.) in relation to height, leaf area, number of root nodules, pods per plant as well as number of seeds per pod and maximum yield of 4.6 to 5.5 q/ha over the existing variety "Khuntibiri" was reported (Parida et al. 2020). Therefore, insecticides, herbicides, and fungicides can be applied by mixing with foliar fertilization to increase the yield and management of insects, pests, and diseases. This will reduce the application cost (Fageria et al. 2009).

10.3.5 Applications and Aspects of Nanomaterial for Nutrient Use Efficiency

Nanoparticles (NPs) have distinctive physicochemical properties like high surface area, particle morphology, and high reactivity. The use of optimum nutrients has been proven to be a major nutrient management practice in agriculture for sustainable crop production, and priority has been given to agricultural research. Therefore, nanonutrition based upon nano-carrier mediated fertilizer and plant growth boosting nanoparticles (NPs) enhanced nutrient use efficiency because of delivering nano-sized nutrients for sustainable crop production. Nanoparticles, nanomaterials, and nano-carrier-based fertilizers have been shown to increase crop growth and yield with unclear mechanisms.

Nano fertilizers (NFs) are nanomaterials that provide macro and micronutrients to plants. NFs also help as carriers for efficient nutrient utilization because of their unique mechanisms of action and reduced nutrient loss over conventional chemical fertilizers. The absorption of NFs increased because of its small size, larger surface area, and reduced N loss due to leaching in comparison to conventional fertilizers (Fig. 10.3). Because of the small size, the energy consumption is reduced during the transport of the nutrients (Brackhage et al. 2013, DeRosa et al. 2010, Mastronardi et al. 2015).

Past literature data concentrated on the nanotechnology potential in fertilizer and other plant protection products to increase the potential of the crop field (Gogos et al. 2012; Hong et al. 2013; Nair et al. 2010). Data revealed that nano-sized fertilizer particles, when mixed with water, could enter through the stomata and cuticular surface of the leaves and could enter into the root through ion channels (Eichert et al. 2008; Rico et al. 2011). Nano fertilizers appeared to target the delivery of nutrients better, enhancing the nutrient use efficiency of the crops (Solanki et al. 2015).

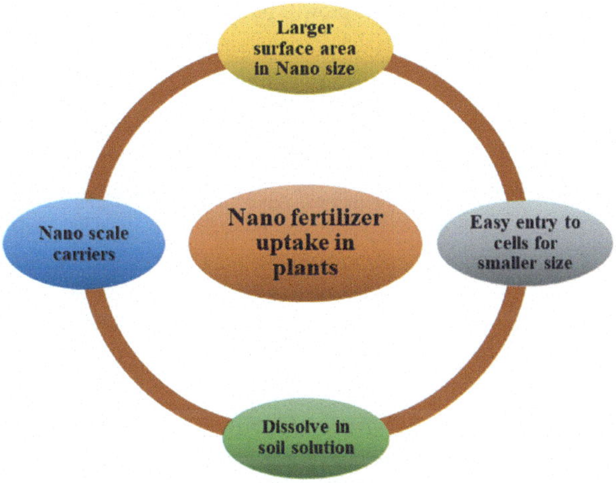

Fig. 10.3 Mechanisms of nano-fertilizers uptake for better nutrient use efficiency by crops

10.4 Soil Biota Management

Soil biota depends on diverse organisms present in the soil and is known to contribute to the sustenance of the natural ecosystem. Soil organisms influence land productivity, directly affecting crop production immediately. They also contribute to participating in different nutrient cycles, modifications of soil structure and food chain, and food web interaction of the ecosystem, increasing the use of nutrients by crop plants (Barrios 2007). Different organisms are considered to nurture the soil biota for the survival of life and contribute to ecosystem services. Different organisms like micro symbionts, micro regulators, decomposers, soil-born pests, and diseases are known to maintain the soil biota and are responsible for nutrient mobilization, transformation, and increase in fertilizer use efficiency in the management of sustainable agricultural ecosystems. *Bacillus* species are also employed to promote plant growth in different crops (Sansinenea 2019). Bioinoculation of plant growth-promoting rhizobacteria (PGPR) such as *Cupriavidusnecator* 1C2 and *Pseudomonas fluorescens* S3X was anylyzed (Pereira et al. 2020). The ratios of 103 cells gm^{-1} and 106 cells gm^{-1} dry weight of soil was assessed on different paramenters. These were the crop growth, nutrient uptake, and use in maize crops in different levels of irrigation in greenhouse condition and reported to increase in the productivity in water stress condition (Pereira et al. 2020). Crops with optimum micronutrient contents are needed to combat the micronutrient deficiency. Therefore, plant growth promoting microorganisms is known to enrich the micro and macronutrients in food crops, and potential microorganisms, along with the application of mineral fertilizers, can enhance the mineral elements uptake, leading to growth and yield. Different microorganisms and their impact on the growth and yield of a few crops are depicted in Table 10.1.

Table 10.1 Effect of plant growth promoting bacteria on nutrient use efficiency, crop growth, and yield

Crop	Plant Growth Promoting Bacteria (PGPB)	Growth or grain yield (Kg ha^{-1})	References
Zea mays L.	*Azospirillumbrasilense* Sp245	3702 ± 133 over control with 2580 ± 280	Martins et al. (2018)
	A. brasilense AbV5 + AbV6	3815 ± 158 over control with 2580 ± 280	Martins et al. (2018)
	Serratiaodorifera KC425221	Maximum photosynthesis was observed with algal biochar and seeds of maize treated with *S. odorifera* KC425221 in water stress conditions.	Ullah et al. (2020)
Triticumaestivum L. cultivar Buck Suren$^{(R)}$	Inoculation with bacterial combinations such as *Bacillus* sp., *Agrobacterium tumefaciens, Klebsiella pneumonia, K. variicola, Pseudomonas* sp.,	This resulted in an increase in phosphorus, potassium, and alkali-hydrolyzed nitrogen in the soil by 49.46, 99.51, and 19.38% respectively.	Wang et al. (2020)
	*Strenotrophomona s*spp.	Significant increases in N in shoot and root are 76% and 36%, respectively.	Majeed et al. (2015)
	Acetobacter pasteurianus	Significant increases in N in shoot and root are 76% and 36%, respectively.	Majeed et al. (2015)
	Acinetobacter guillouiae EU-B2RT.R1	Plant growth promoter of wheat cultivar PBW 343 in pot experiment under greenhouse conditions and can be used as biofertilizers.	Rana et al. (2020)
Capsicum annum L.	*Bacillus megaterium* CDK25	Seeds of *C. annum* L. treated showed better germination percentage, the plant length, girth, number of branches, and fresh and dry weight of shoot and root after 60 days with maximum number of fruits and maximum Zn content in fruits was estimated after 120 days.	Bhatt and Maheshwari (2020)

(continued)

Table 10.1 (continued)

Crop	Plant Growth Promoting Bacteria (PGPB)	Growth or grain yield (Kg ha^{-1})	References
Oryza sativa L.	*Acidovorax delafieldii*	Combined with half the recommended dose of inorganic fertilizers, it increased the root and shoot fresh and dry weight and yield of the PSB Rc23 rice variety.	Cavite et al. (2021)
	Serratia spp. BRM 32114	Increase in stomatal conductance by nearly 20%; N, Ca, and Mg contents with 7, 11, and 9% and number of grain per plant by 17% and mass of 1000 grains by 2% and yield of grains by 22% in upland rice cultivar	Nascente et al. (2019)

10.4.1 Rhizosphere Management

Rhizosphere, or root management, is now gaining importance for modern agriculture susceptibility. It was reported that the production of cereal grain using chemical fertilizers increased by 512% from 1980 to 2010 (Zhang et al. 2012). It was also mentioned that the application of chemical fertilizers resulted in low nutrient use efficiency for over the last 20 years in Chinese intensive farming systems. Overuse of chemical fertilizers is the most important factor for low nutrient use efficiency. The rhizosphere is the zone where interaction between plants and soil takes place, and the rhizosphere has a major impact on the transformation mobilization of soil nutrients to be used efficiently by the plants through the presence of rhizospheric microorganisms (Kour et al. 2019). Root zone nutrient supply at an optimal quantity of nutrients according to the crop demand in the rhizosphere region plays a significant role in maximizing the efficiency of the rhizosphere in nutrient mobilization and acquisition (Shen et al. 2013). Rhizosphere management has been proven to be the most effective approach to enhance nutrient use efficiency. This approach can be attained by manipulating the growth of the root, strengthening acidification and exudation of organic acids, carboxylate, phosphatases, some signaling substances, proton release, and redox changes in the rhizosphere region and balancing root zone nutrient supply according to crop demand (Zhang et al. 2010, Chen et al. 2011, Marschner 2012). Reduction in nitrous oxide (N_2O) emission and increase in nitrogen use efficiency are needed to increase agricultural production.

Rhizosphere soils and roots of *Robinia pseudoacacia* were studied for root nitrogen and phosphorus resorption efficiencies at different stages of successions on the Loess Plateau in China and reported that the initial increase in diversity of soil microorganisms and soil nutrients in the rhizosphere followed by gradual declining

(Xu et al. 2021). The result indicated that the nutrient contents in the soil surrounding the rhizosphere were significantly associated with microbial diversity and correlated with nutrient use efficiency and soil nutrients of the rhizosphere (Xu et al. 2021). Management of the rhizosphere is an effective method to enhance nutrient use efficiency (Marschner 2012). Excess use of fertilizers may result in the deposition of high nutrient concentrations in the rhizospheric region, inhibiting the growth of the root (Li et al. 2009; Mi et al. 2010).

10.5 Conclusions

Nutrient use efficiency of the crop plants can be achieved by integrated nutrient management, foliar fertilization, nano-carrier based fertilizers, and plant growth boosting nanoparticles (NPs), crop rotations, proper tillage operations, residue management and use of organic fertilizers viz. farm yard manure, compost, and vermicompost, use of plant growth promoting bacteria to increase the crop growth and yield. The important challenge is to build up a possible system that is energy- and labor-efficient and simultaneously helps maintain soil health and a better environment. Zero tillage is cost-effective over the conventional tillage system with respect to low energy and fuel input than the conventional tillage system. Minimum tillage is known to increase root density, length, and weight and enhance nutrient and water use efficiency. Likewise, management of the rhizosphere is essential for the efficient use of nutrients by crops, enhancing crop growth.

The efficient use of fertilizer is affected by factors such as the properties of soil, crop efficiency, persistent climatic conditions, fertilizer type, soil microbes, and other factors. Therefore, it can be concluded that the efficient nutrient uptake from the soil by the crops to enhance the growth, vigor, and yield with balanced nutrients in the produce depends on various attributes. These attributes include soil health and quality, its water holding capacity, pH, soil biota especially presence of specific microbes in the region of the rhizosphere is the most important aspects for solubilization of nutrients to make them available for plants. Healthy soil makes the soil responsive to fertilizer solubilization and enhances its efficient uptake by the crops, resulting in maximum yield. Further research will clarify and contribute to confirming different agronomical approaches and conservation practices for different crops to maintain the rhizospheric biota to gain soil carbon content to enhance the nutrient use efficiency of crops.

References

Alexandratos N, Bruinsma J (2012). World agriculture towards 2030/2050: the 2012 revision

Ali M, Byerlee D (2002) Productivity growth and resource degradation in Pakistan's Punjab: a decomposition analysis. Econ Dev Cult Chang 50(4):839–863

Anonymous TNA (1985) Principles of foliar feeding

Badole S, Datta A, Krishna Chaitanya A, Majumder SP, Mandal B (2020) Soil carbon dynamics under different land-use and management systems. In: Carbon management in tropical and sub-tropical terrestrial systems. Springer, Singapore, pp 103–121

Barrios E (2007) Soil biota, ecosystem services and land productivity. Ecol Econ 64(2):269–285

Bhatt K, Maheshwari DK (2020) Zinc solubilizing bacteria (Bacillus megaterium) with multifarious plant growth promoting activities alleviates growth in Capsicum annuum L. *3*. Biotech 10(2):1–10

Brackhage C, Schaller J, Bäucker E, Dudel EG (2013) Silicon availability affects the stoichiometry and content of calcium and micro nutrients in the leaves of common reed. SILICON 5(3):199–204

Bruulsema TW, Fixen PE, Sulewski GD (2016) 4R plant nutrition manual: a manual for improving the management of plant nutrition, North American version (revised). International Plant Nutrition Institute, Norcross

Cassman KG, Dobermann A, Walters DT, Yang H (2003) Meeting cereal demand while protecting natural resources and improving environmental quality. Annu Rev Environ Resour 28(1):315–358

Cavite HJM, Mactal AG, Evangelista EV, Cruz JA (2021) Biochemical characteristics and inoculation effects of multi-trait plant growth-promoting rhizobacteria on upland rice (Oryza sativa L. cv PSB Rc23) seedling growth. Arch Microbiol 203(6):3533–3540

Chauhan BS, Mahajan G, Sardana V, Timsina J, Jat ML (2012) Productivity and sustainability of the rice–wheat cropping system in the Indo-Gangetic Plains of the Indian subcontinent: problems, opportunities, and strategies. Adv Agron 117:315–369

Chen XP, Cui ZL, Vitousek PM, Cassman KG, Matson PA, Bai JS et al (2011) Integrated soil–crop system management for food security. Proc Natl Acad Sci 108(16):6399–6404

Coonan EC, Richardson AE, Kirkby CA, Kirkegaard JA, Amidy MR, Strong CL (2020) Soil fertility and nutrients mediate soil carbon dynamics following residue incorporation. Nutr Cycl Agroecosyst 116(2):205–221

Das A, Parida S (2020) Integrated plant protection measures and micronutrient foliar spray on growth and yield of green gram (*Vigna radiata* L.). Indian J Nat Sci 10(59):18269–18276

DeRosa MC, Monreal C, Schnitzer M, Walsh R, Sultan Y (2010) Nanotechnology in fertilizers. Nat Nanotechnol 5(2):91–91

Derpsch R (1998, March) Historical review of no-tillage cultivation of crops. In: Conservation tillage for sustainable agriculture. Proceedings from an international workshop, Harare, pp 22–27

Dubey PK, Singh GS, Abhilash PC (2020) Adaptive agricultural practices: building resilience under changing climate. Springer Nature Switzerland, Springer brief in environmental science, ISBN 978–3–030-15518-6. https://doi.org/10.1007/978-3-030-15519-3

Dubey PK, Singh A, Chaurasia R, Pandey KK, Bundela AK, Dubey RK, Abhilash PC (2021a) Planet friendly agriculture: farming for people and the planet. Curr Res Environ Sustain 3:100041

Dubey RK, Dubey PK, Chaurasia R, Rao CS, Abhilash PC (2021b) Impact of integrated agronomic practices on soil fertility and respiration on the Indo-Gangetic Plain of North India. Agronomy 11:402. https://doi.org/10.3390/agronomy11020402

Duxbury JM (2001) Long-term yield trends in the rice-wheat cropping system: results from experiments and Northwest India. J Crop Prod 3(2):27–52

Edrisi SA, Tripathi V, Dubey PK, Abhilash PC (2022) Carbon sequestration and harnessing biomaterials from terrestrial plantations for mitigating climate change impacts. In: Biomass, biofuels, biochemicals. Elsevier, pp 299–313

Edrisi SA, Singh A, Dubey PK, Abhilash PC (2023) *Tectona grandis* Lf. mediated restoration of marginal lands in Eastern Uttar Pradesh, North India. Land Degrad Dev 34(4):929–942

Eichert T, Kurtz A, Steiner U, Goldbach HE (2008) Size exclusion limits and lateral heterogeneity of the stomatal foliar uptake pathway for aqueous solutes and water-suspended nanoparticles. Physiol Plant 134(1):151–160

Fageria NK, Filho MB, Moreira A, Guimarães CM (2009) Foliar fertilization of crop plants. J Plant Nutr 32(6):1044–1064
Farooqi ZUR, Sabir M, Zeeshan N, Naveed K, Hussain MM (2018) Enhancing carbon sequestration using organic amendments and agricultural practices. In: Carbon capture, utilization and sequestration, p 17
Food and Agriculture Organization (2009) How to feed the World in 2050. In: Executive summary-proceedings of the expert meeting on how to feed the World in 2050. Food and Agriculture Organization, Rome
Fujisaka S, Harrington L, Hobbs P (1994) Rice-wheat in South Asia: systems and long-term priorities established through diagnostic research. Agric Syst 46(2):169–187
Gathala MK, Ladha JK, Kumar V, Saharawat YS, Kumar V, Sharma PK et al (2011) Tillage and crop establishment affects sustainability of South Asian rice–wheat system. Agron J 103(4):961–971
Gentil LV (1995) Aspectos econômicos do plantio direto. I Seminário Internacional do Sistema Plantio Direto. Resumos. EMBRAPA-CNPT, Ponta Grossa, RS. Agosto, 09-12
Gogos A, Knauer K, Bucheli TD (2012) Nanomaterials in plant protection and fertilization: current state, foreseen applications, and research priorities. J Agric Food Chem 60(39):9781–9792
Govaerts B, Sayre KD, Deckers J (2005) Stable high yields with zero tillage and permanent bed planting? Field Crop Res 94(1):33–42
Griffith WK, Murphy LS (1991) The development of crop production systems using best management practices. Potash & Phosphate Institute, Norcross
Gupta RK, Naresh RK, Hobbs PR, Ladha JK (2002) Adopting conservation agriculture in the rice-wheat system of the Indo-Gangetic Plains: new opportunities for saving water. In: Water wise rice production. Proceedings of the international workshop on water wise rice production. International Rice Research Institute, Los Banos, pp 207–222
Gupta RK, Naresh RK, Hobbs PR, Jiaguo Z, Ladha JK (2003) Sustainability of post-green revolution agriculture: the rice–wheat cropping systems of the Indo-Gangetic Plains and China. In: Improving the productivity and sustainability of rice-wheat systems: issues and impacts, vol 65, pp 1–25
Halvorson AD, Reule CA, Follett RF (1999) Nitrogen fertilization effects on soil carbon and nitrogen in a dryland cropping system. Soil Sci Soc Am J 63(4):912–917
Harrington LW (1992) Wheat and rice in Karnal and Kurukshetra Districts, Haryana, India: farmers' practices, problems, and an agenda for action. CIMMYT
Hati KM, Biswas AK, Somasundaram J, Mohanty M, Singh RK, Sinha NK, Chaudhary RS (2020) Soil organic carbon dynamics and carbon sequestration under conservation tillage in tropical vertisols. In: Carbon management in tropical and sub-tropical terrestrial systems. Springer, Singapore, pp 201–212
Havlin JL, Kissel DE, Maddux LD, Claassen MM, Long JH (1990) Crop rotation and tillage effects on soil organic carbon and nitrogen. Soil Sci Soc Am J 54(2):448–452
Holland JM (2004) The environmental consequences of adopting conservation tillage in Europe: reviewing the evidence. Agric Ecosyst Environ 103(1):1–25
Hong J, Peralta-Videa JR, Gardea-Torresdey JL (2013) Nanomaterials in agricultural production: benefits and possible threats. Sustainable nanotechnology and the environment: advances and achievements, 73–90.
Ichami SM, Shepherd KD, Sila AM, Stoorvogel JJ, Hoffland E (2019) Fertilizer response and nitrogen use efficiency in African smallholder maize farms. Nutr Cycl Agroecosyst 113(1):1–19
International Fertilizer Association (2009) The global "4R"nutrient stewardship framework: developing fertilizer best management practices for delivering economic, social and environmental benefits. International Fertilizer Industry Association (IFA), Paris, 10 pp
Khan SA, Mulvaney RL, Ellsworth TR, Boast CW (2007) The myth of nitrogen fertilization for soil carbon sequestration. J Environ Qual 36(6):1821–1832

Kihara J, Sileshi GW, Nziguheba G, Kinyua M, Zingore S, Sommer R (2017) Application of secondary nutrients and micronutrients increases crop yields in sub-Saharan Africa. Agron Sustain Dev 37(4):1–14

Kour D, Rana KL, Yadav N, Yadav AN, Kumar A, Meena VS et al (2019) Rhizospheric microbiomes: biodiversity, mechanisms of plant growth promotion, and biotechnological applications for sustainable agriculture. In: Plant growth promoting rhizobacteria for agricultural sustainability. Springer, Singapore, pp 19–65

Kumar V, Saharawat YS, Gathala MK, Jat AS, Singh SK, Chaudhary N, Jat ML (2013) Effect of different tillage and seeding methods on energy use efficiency and productivity of wheat in the Indo-Gangetic Plains. Field Crop Res 142:1–8

Ladha JK, Dawe D, Pathak H, Padre AT, Yadav RL, Singh B et al (2003) How extensive are yield declines in long-term rice–wheat experiments in Asia? Field Crop Res 81(2–3):159–180

Lal R (2004) Soil carbon sequestration impacts on global climate change and food security. Science 304(5677):1623–1627

Li SX, Wang ZH, Malhi SS, Li SQ, Gao YJ, Tian XH (2009) Nutrient and water management effects on crop production, and nutrient and water use efficiency in dryland areas of China. Advances in Agronomy 102:223–265

Majeed A, Abbasi MK, Hameed S, Imran A, Rahim N (2015) Isolation and characterization of plant growth-promoting rhizobacteria from wheat rhizosphere and their effect on plant growth promotion. Front Microbiol 6:198

Majumdar A, Dubey PK, Upadhyay MK, Giri B, Srivastava AK, Roychowdhury T (2023) Effects of tillage and no-tillage agronomic practices on soil carbon content, nutrient dynamics and microbial community change: field study to genomics. In: Goldschmidt 2023 conference. GOLDSCHMIDT

Malik RK, Yadav A, Gupta RK, Singh S, Hobbs PR, Bellinder RR (2005) Introduction and success of zero tillage in wheat under rice-wheat cropping system in Haryana, India-related stories. In: Accelerating the adoption of resource conservation technologies in rice wheat system of the Indo-Gangetic Plains. Proceedings of the project workshop June 1–2, pp 1–17

Manning DA (2015) How will minerals feed the world in 2050? Proc Geol Assoc 126(1):14–17

Marschner P (2012) Mineral nutrition of higher plants. Academic, San Diego, p 651

Martins MR, Jantalia CP, Reis VM, Döwich I, Polidoro JC, Alves BJR et al (2018) Impact of plant growth-promoting bacteria on grain yield, protein content, and urea-15 N recovery by maize in a Cerrado Oxisol. Plant Soil 422(1):239–250

Mastronardi E, Tsae P, Zhang X, Monreal C, DeRosa MC (2015) Strategic role of nanotechnology in fertilizers: potential and limitations. In: Nanotechnologies in food and agriculture. Springer, Cham, pp 25–67

Mehla RS, Verma JK, Hobbs PR, Gupta RK (2000) Stagnation in the productivity of wheat in the Indo-Gangetic Plains: zero-till-seed-cum-fertilizer drill as an integrated solution. Facilitation Unit, Rice-Wheat Consortium for the Indo-Gangetic Plains

Mi G, Chen F, Wu Q, Lai N, Yuan L, Zhang F (2010) Ideotype root architecture for efficient nitrogen acquisition by maize in intensive cropping systems. Science China Life Sciences 53:1369–1373

Mulumba LN, Lal R (2008) Mulching effects on selected soil physical properties. Soil Tillage Res 98(1):106–111

Nair R, Varghese SH, Nair BG, Maekawa T, Yoshida Y, Kumar DS (2010) Nanoparticulate material delivery to plants. Plant Sci 179(3):154–163

Nandan R, Singh V, Singh SS, Kumar V, Hazra KK, Nath CP, Poonia S, Malik RK, Bhattacharyya R, McDonald A (2019) Impact of conservation tillage in rice–based cropping systems on soil aggregation, carbon pools and nutrients. Geoderma 340:104–114

Nascente AS, Lanna AC, de Sousa TP, Chaibub AA, de Souza ACA, de Filippi MCC (2019) N fertilizer dose-dependent efficiency of Serratia spp. for improving growth and yield of upland rice (Oryza sativa L.). Int J Plant Prod 13(3):217–226

Nayak S, Nayak D, Parida S (2020) Micronutrient foliar spray on growth performance of green gram (*Vigna radiata* L.). Asian J Biol Life Sci 9(2):235

Njoroge R, Otinga AN, Okalebo JR, Pepela M, Merckx R (2017) Occurrence of poorly responsive soils in western Kenya and associated nutrient imbalances in maize (Zea mays L.). Field Crop Res 210:162–174

Parida S, Mishra A, Das A (2020) Impact of foliar application of fertilizers on growth and seed yield of Vigna mungo L. Int J Modern Agric 9(4):296–307

Pathak H, Saharawat YS, Gathala M, Ladha JK (2011) Impact of resource-conserving technologies on productivity and greenhouse gas emissions in the rice-wheat system. Greenhouse Gases Sci Technol 1(3):261–277

Pereira SIA, Abreu D, Moreira H, Vega A, Castro PML (2020) Plant growth-promoting rhizobacteria (PGPR) improve the growth and nutrient use efficiency in maize (Zea mays L.) under water deficit conditions. Heliyon 6(10):e05106

Phillips SB, Camberato JJ, Leikam D (2009) Selecting the right fertilizer rate: a component of 4R nutrient stewardship. Crops Soils 42(4):14–18

Rana KL, Kour D, Kaur T, Sheikh I, Yadav AN, Kumar V et al (2020) Endophytic microbes from diverse wheat genotypes and their potential biotechnological applications in plant growth promotion and nutrient uptake. Proc Natl Acad Sci India Sect B Biol Sci 90(5):969–979

Rangarajan A, Mohler CL, Johnson SE (2009) Crop rotation effects on soil fertility and plant nutrition. In: Crop rotation on organic farms. University of Maryland: NRAES

Rico CM, Majumdar S, Duarte-Gardea M, Peralta-Videa JR, Gardea-Torresdey JL (2011) Interaction of nanoparticles with edible plants and their possible implications in the food chain. J Agric Food Chem 59(8):3485–3498

Saharawat, Y. S., Ladha, J. K., Pathak, H., Gathala, M. K., Chaudhary, N., & Jat, M. L. (2012). Simulation of resource-conserving technologies on productivity, income and greenhouse gas GHG emission in rice-wheat system

Sainju UM, Ghimire R, Mishra U, Jagadamma S (2020) Reducing nitrous oxide emissions and optimizing nitrogen-use efficiency in dryland crop rotations with different nitrogen rates. Nutr Cycl Agroecosyst 116(3):381–395

Sansinenea E (2019) Bacillus spp.: as plant growth-promoting bacteria. In: Secondary metabolites of plant growth promoting rhizomicroorganisms, pp 225–237

Sarkar D, Dubey PK, Chaurasia R, Sankar A, Shikha CN, Ganguly S, Meena VS, Meena SK, Parewa HP, Rakshit A (2021) Organic interventions conferring stress tolerance and crop quality in agroecosystems during UN decade on ecosystem restoration. Land Degrad Dev 32(17):4797–4816. https://doi.org/10.1002/ldr.4094

Saunders DA (1994) Wheat in heat-stressed environments: irrigated, dry areas, and rice-wheat farming systems: proceedings of the international conferences: wheat in hot, dry, irrigated environments, wad Medani, Sudan, 1–4 February 1993, wheat in warm area, Rice-wheat farming systems, Dinajpur, Bangladesh, 13–15 Feburary [sic] 1993. CIMMYT

Shen J, Li C, Mi G, Li L, Yuan L, Jiang R, Zhang F (2013) Maximizing root/rhizosphere efficiency to improve crop productivity and nutrient use efficiency in intensive agriculture of China. J Exp Bot 64(5):1181–1192

Singh AP, Singh SK, Rai S, Kumar M (2020) Soil carbon dynamics in relation to soil surface management and cropping system. In: Carbon management in tropical and sub-tropical terrestrial systems. Springer, Singapore, pp 159–172

Solanki P, Bhargava A, Chhipa H, Jain N, Panwar J (2015) Nano-fertilizers and their smart delivery system. In: Nanotechnologies in food and agriculture. Springer, Cham, pp 81–101

Stockdale EA, Lampkin NH, Hovi M, Keatinge R, Lennartsson EKM, Macdonald DW et al (2001) Agronomic and environmental implications of organic farming systems. Adv Agron 70:261–327

Tilman D, Cassman KG, Matson PA, Naylor R, Polasky S (2002) Agricultural sustainability and intensive production practices. Nature 418(6898):671–677

Ueno H, Yamamuro S (2001) Fate of nitrogen derived from 15N-labeled plant residues and composts in rice-planted paddy soil. Soil Sci Plant Nutr 47(4):747–754

Ullah N, Ditta A, Khalid A, Mehmood S, Rizwan MS, Ashraf M et al (2020) Integrated effect of algal biochar and plant growth promoting rhizobacteria on physiology and growth of maize under deficit irrigations. J Soil Sci Plant Nutr 20(2):346–356

Vanlauwe B, Coe RIC, Giller KE (2019) Beyond averages: new approaches to understand heterogeneity and risk of technology success or failure in smallholder farming. Exp Agric 55(S1):84–106

Wang J, Li R, Zhang H, Wei G, Li Z (2020) Beneficial bacteria activate nutrients and promote wheat growth under conditions of reduced fertilizer application. BMC Microbiol 20(1):1–12

Xu M, Jian J, Wang J, Zhang Z, Yang G, Han X, Ren C (2021) Response of root nutrient resorption strategies to rhizosphere soil microbial nutrient utilization along Robinia pseudoacacia plantation chronosequence. For Ecol Manag 489:119053

Zhang F, Shen J, Zhang J, Zuo Y, Li L, Chen X (2010) Rhizosphere processes and management for improving nutrient use efficiency and crop productivity: implications for China. Adv Agron 107:1–32

Zhang F, Cui Z, Chen X, Ju X, Shen J, Chen Q et al (2012) Integrated nutrient management for food security and environmental quality in China. Adv Agron 116:1–40

Zhou T, Wang L, Sun X, Wang X, Pu T, Yang H et al (2021) Improved post-silking light interception increases yield and P-use efficiency of maize in maize/soybean relay strip intercropping. Field Crop Res 262:108054

Chapter 11
Recycling of Biodegradable Waste in Agriculture: Transformation, Use, and Effects on Soil Quality

Gabriella Rossi and Claudio Beni

Abstract The objective of this chapter is to highlight the importance of the recovery and recycling of biodegradable waste in agriculture to improve the management of organic matter and nutrients and to promote the economic transition towards the circularity of resources. The advantages of the recovery and recycling of biodegradable waste deriving from livestock and agro-industrial origin in agriculture are discussed. The optimal management of these biomasses transforms them from waste to resource, reducing the amount of waste disposed in landfills and their impact on the environment. Bio-waste contains organic matter and nutrients, especially nitrogen and phosphorus, which can be recovered to improve soil fertility and crop nutrition. The recycling of bio-waste in agriculture reduces energy consumption by increasing the production of renewable energy and reducing the energy demand for the production of mineral fertilizers. Furthermore, fertilization with bio-based fertilizers administers organic matter to the soil, mitigating SOC loss with positive effects on soil quality. The processes of biodegradable waste transformation in bio-based fertilizers are illustrated: composting and anaerobic digestion. The results of some studies on the effects of fertilization with compost or digestate on the chemical-physical, mechanical, and biological properties of the soil are reported.

Keywords Biomass · Agriculture · Biodegradable waste · Anaerobic digestion · Carbon sequestration

---

G. Rossi (✉)
Consiglio per la ricerca in agricoltura e l'analisi dell'economia agraria—CREA, Centro di ricerca Agricoltura e Ambiente—Rome (Italy). Council for Agricultural Research and Economics, Research Center for Agriculture and Environment, Rome, Italy
e-mail: gabriella.rossi@crea.gov.it

C. Beni
Consiglio per la ricerca in agricoltura e l'analisi dell'economia agraria—CREA Centro di ricerca Ingegneria e Trasformazioni agroalimentari—Monterotondo, Rome (Italy). Council for Agricultural Research and Economics, Research Center for Engineering and Agro-food processing—Monterotondo, Rome, Italy

S. A. Edrisi et al. (eds.), *Sustainable Plant Nutrition and Soil Carbon Sequestration*, Sustainable Plant Nutrition in a Changing World,
https://doi.org/10.1007/978-3-031-53590-1_11

11.1 Introduction

Promoting biodegradable waste recycling and recovery techniques is an urgent priority. Optimizing the management of these biomasses can contribute to the replacement of chemical fertilizers with bio-based products fertilizers (Chojnacka et al. 2020; Chew et al. 2019; Tur-Cardona et al. 2018). The interest of researchers and stakeholders is aimed at evaluating the suitability in fertilization practices of bio-based fertilizers derived from the transformation of biodegradable waste of zootechnical and agro-industrial activities. One of the main constraints in their use in agriculture is the possible presence of chemicals and biologicals that are potentially polluting the environment and toxic to human health (Rossi et al. 2002; Goss et al. 2013; Mattana et al. 2014; Mohapatra et al. 2016; Urra et al. 2019). However, biodegradable waste contains nutrients (primarily nitrogen and phosphorus) and organic matter that could be recovered rather than disposed of. Fertilization practices involving the use of bio-based fertilizers could improve soil quality. In fact, the organic fraction of these products administered to soil could counteract the phenomenon of soil carbon decline due to climatic conditions and the adverse effects on soil texture, soil drainage, land use, and vegetation type and management (FAO 2015). The optimization of biodegradable waste transformation processes and the correct management of bio-based fertilizers contribute to the reduction of the consumption of chemical fertilizers (Chojnacka et al. 2020; Chew et al. 2019; Tur-Cardona et al. 2018).

European regulation (Regulation (EU) 2019/1009) defines the criteria according to which a biodegradable waste can be marketed as fertilizer if it complies with strict conformation criteria (End of Waste). These can be summarized as requirements for (1) product quality, (2) input materials, (3) processes and treatment techniques, (4) provision of information, (5) quality management procedures, and (6) reduction of the quantity of waste produced. This approach was adopted by the European Commission, which established the principles for the sustainable management of biodegradable waste in compliance with the circular economy model (Circular Economy Package-European Commission 2016). In this context, treating the organic fraction of biodegradable waste through composting or anaerobic digestion techniques links organic matter and nutrients with energy recovery. The current priority is to optimize processes/products for the agricultural utilization of biodegradable waste derived from livestock and by-products of agro-industrial activity (Bernal et al. 2017). The recovery and recycling of the organic and mineral fraction of biodegradable waste derived from animal farming, agriculture, and agro-industry by-products favor the transition towards the circular economy. They are actions that transform waste into a resource, enhancing its use and reducing the impact of waste products on the environment. Circular agriculture development requires a pilot promotion structure to act as a hub of technology and science, especially in districts with a large volume of agricultural waste generation (Fig. 11.1).

In these districts, introducing sustainable agricultural technology and practices is in high demand (Atinkut et al. 2020). For example, on-farm composting is a

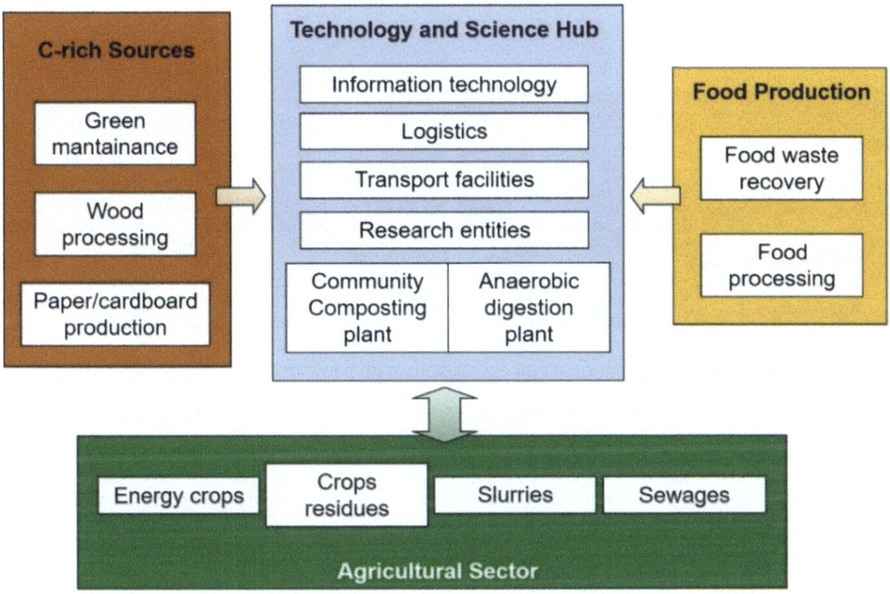

Fig. 11.1 Association of the technology and science to manage agricultural biodegradable waste

low-cost procedure that reduces the disposal of agricultural crop residues. High-quality compost represents a driving force in connecting food, waste, the economy, and the environment (Duque-Acevedo et al. 2020; Razza et al. 2018).

11.2 Soil Organic Carbon Decline and Organic Amendments

According to Lal (2008), over 2700 gigatons (Gt) of carbon are stored in soils. The report issued by FAO indicates that the global loss of the SOC pool since 1850 is estimated at 66 billion tons, mainly caused by the change in land use and climate (FAO 2015). Agricultural soils contain 25% to 75% less SOC than their counterparts in undisturbed or natural ecosystems (Lal 2010). SOC loss is particularly evident in Mediterranean agricultural areas. Low SOC contents characterize these areas. They are often prone to soil degradation and SOC depletion due to the changes in extensive land use in recent decades (Fig. 11.2). These same areas are also highly vulnerable to climate change. Mediterranean climatic conditions threaten the soil's fertility due to the high mineralization of SOM (Francaviglia et al. 2017; Scharlemann et al. 2014; Jones et al. 2012).

Agronomic systems that adopt organic amendment and crop rotation practices can increase SOC levels by improving soil fertility and agronomic productivity (Doyeni et al. 2021; Kauer et al. 2015). They do this by acting as a carbon and nitrogen source and by contributing to the mitigation of and adaptation to climate change.

Fig. 11.2 Effect of erosive phenomena on clayey soils with a low organic carbon content (Pictured by Ulderico Neri—CREA-AA)

Table 11.1 Processes, factors, and causes affecting SOC sequestration

Processes (mechanisms)	Factors (environment)	Causes (anthropogenic activities)
Photosynthesis	Climate	Land use
Humification	Soil	Farming systems
Aggregation	Landscape	Tillage
Illuviation	Vegetation	Water management
		Nutrient management
		Pest management

Modified by Lal (2017)

Sustainable agricultural soil management attracts organic amendments as recommended management practices (RMPs) (Table 11.1). This soil fertility management practice can also contribute to "carbon leakage" (Amelung et al. 2020; Lal 2017; Diacono and Montemurro 2015).

A survey published in 2013 (Koch et al. 2013) compared 24 tests in the Mediterranean basin between biological and non-biological systems. The former managed to capture 3559.9 kg of CO_2 per hectare every year. The study found that if the system were deployed worldwide, the seized SOC would rise to 17.4 Gt annually. The Rodale Institute in Pennsylvania, USA, has conducted long-running comparisons of organic and conventional cropping systems for over 30 years. The Farming Systems Trial manuring organic plots showed that CO_2 was captured in the soil at the rate of 3596.6 kg of CO_2 per hectare per year. When extrapolated globally across agricultural lands, this would capture 17.5 Gt of CO_2 in soil per year (La Salle and Hepperly 2008). Some authors have found that organic amendment contributes more substantially to soil carbon sequestration than the cover crop technique alone (Vincente-Vincente et al. 2016). The slower nitrification of N and the high cation exchange capacity of organic amendments stabilize soil chemical properties by increasing soil buffering capacity and slowly releasing essential nutrients for more sustainable plant growth (Scotti et al. 2015; Case et al. 2019).

11.3 Biodegradable Waste Transformation into Bio-Based Fertilizers: Composting and Anaerobic Digestion

The residual biodegradable waste of crops, agro-industrial transformations, livestock farms, and bio-energy productions can, therefore, play a prominent role in improving the environmental sustainability of the rural territory. In some cases, they can even generate by-products with considerable nutritional value. Biowastes such as agricultural waste, green and garden waste, livestock waste, etc., are processed through composting or anaerobic digestion techniques in order to stabilize their organic fraction and produce bio-based fertilizers. The nature of the raw materials and their chemical composition, particularly nitrogen concentration, C/N ratio, and moisture content (Table 11.2), determine the qualitative value and the potential use of the end product (compost or anaerobic digestate). The controlled transformation of biodegradable waste into compost or digestate induces chemical changes in by-products (Table 11.3) with effects on the organic component and the C/N ratio (Long Li et al. 2019; Mortier et al. 2016; Makan et al. 2012).

Table 11.2 Total N content, C/N ratio, and moisture content of the main biodegradable waste

Material	N (% d.m.)	C/N (w/w)	Moisture (% f.m.)
Apple pomace	1.1	48	88
Cocoa shells	2.3	22	8
Corn stalks	0.7	66	12
Fruit wastes	1.4	40	80
Olive husks	1.3	33	9
Tomato processing wastes	4.5	11	62
Vegetables	2.7	19	87
Fish wastes	10.6	3.6	76
Poultry carcasses	2.4	5	65
Paunch manure	1.8	25	83
Swine manure	3.1	14	69
Garbage (food waste)	2.4	15	69
Paper from domestic refuse	0.2	152	19
Corrugated cardboard	0.1	563	8
Paper pulp	0.6	90	82
Sawdust	0.2	442	39
Hardwoods	0,1	560	–
Grass clippings	3.4	17	82
Leaves	0.9	54	38
Seaweeds	1.9	17	53
Tree trimmings	3.1	16	70
Shrub trimmings	1.0	53	15

Source: Rynk, R. (1992). Appendix A, Table A.1. d.m.: dry matter; w/w: weight overweight; f.m.: fresh matter

Table 11.3 Average fertilizing value of compost and digestate in percentage of dry matter (w/w)

Nutritive element	Liquid digestate	Solid digestate	Compost
Total N	4.0	3.5	4.0
Soluble N	2.0	0.7	0.1
P_2O_5	1.5	1.7	1.7
K_2O	4.1	2.8	3.6
Mg	0.9	1.5	2.1
Ca	5.4	25.5	22.8
S	0.3	0.4	0.5

Source: National Directive of Switzerland on compost and digestate quality (Fuchs 2015)

The agricultural utilization of bio-based fertilizers can perform an essential function in the turnover of the SOM soil with a positive residual effect on soil fertility. It also presents a plant nutritional effect variable depending on the biomass composition and waste treatment method. The different processing of biodegradable waste influences the C/N ratio of bio-based fertilizer (compost or digestate) with different effects on the SOM cycle, soil microbiological activity, nutrient bioavailability, and crop productivity. Through composting digestion, the final product of composting or anaerobic digestion may be more suitable for supporting the mineral nutrition of crops in a short time and preserving the SOM content for a long time with a positive effect on the overall soil fertility (Nascimento et al. 2020; Rossi and Beni 2018; Rossi et al. 2020). For example, Rossi (Rossi et al. 2020) conducted a fertilization test using three zootechnical digestates characterized by different C/N ratios. The results showed that the one with the lowest C/N ratio (digested pig slurry as such) in the soil behaved as a nitrogen fertilizer with readily available nitrogen. For this reason, it should be used with particular attention in nitrogen fertilization practices in soils vulnerable to nitrogen leaching risk (Council Directive 91/676/EEC). The one with the highest C/N ratio (digested pig slurry-solid fraction) showed a behavior similar to an organic amendment. The large-scale production and use of compost and digestates are still limited due to the difficulties encountered in the following phases: matrices supply and analysis, logistics and transport management, farmers, technical operators, and professionals training (Table 11.4).

11.3.1 Composting

Composting is a bio-oxidative process that transforms undecomposed biodegradable waste into stable organic matter through the activity of microflora and mesofauna. To proceed efficiently, the process requires optimal conditions regarding the aeration, humidity, and chemical composition of the starting matrices (Epstein 1997). Compost is a soil amendment whose agronomic use requires an adequate chemical-physical characterization regarding pH, electrical conductivity, nutrient content, and unwanted substances and elements. In addition to these parameters, the

Table 11.4 Limitations of the application of composting and anaerobic digestion process and use

Composting	Anaerobic digestion
Process limitations	
Availability of matrices	Availability/cost of digesting microorganisms
Energy costs for temperature/humidity control	Sizing of the plant
Matrices transportation costs	Seasonality of the availability of waste
Use limitations	
Stakeholder formation and information	Costs of investment and management of machinery
Distance of the plants	Insufficient agricultural area for spreading
Compost transportation costs	Impossibility of storage for liquid phase
Impossibility of storage	Lack of information on the chemical composition
Lack of adequate machinery	Lack of safety information
Lack of information on the chemical composition	
Lack of safety information	

characterization of biological properties is also becoming increasingly important, especially regarding gas exchange rates and enzymatic activity. Furthermore, the growing diffusion of molecular biology analytical methods permits the definition of an important qualitative parameter: the microbiological fingerprinting of the amendment obtained through the bio-oxidation process. In order to ensure the quality of the soil amendment, waste mixtures must be characterized by a C/N ratio between 25 and 30, a water content between 55% and 65%, and a degree of aeration between 5% and 15%. Knowledge of the chemical composition of the biodegradable waste, at least regarding the concentration of C and N, is necessary in order to obtain suitable mixtures. The steps of shredding, mixing, turning over, and wetting the biodegradable waste heaps must also be carried out rationally through process parameters control, temperature, humidity, and pH (Ayilara et al. 2020; Sayara et al. 2020). The addition of bio-accelerators to waste is becoming an increasingly widespread practice to accelerate compost's stabilization and maturation processes. Among these organisms, earthworms, insect larvae (*Cetonia* spp., *Hermetia* spp.), thermophilic microorganisms, and mycorrhizae find an ever-wider application (Cremeniac and Boclaci 2014; Lalander et al. 2020; Faesal et al. 2020).

A further quality parameter of the resulting compost is the degree of sanitation obtained by controlling the temperature during stabilization (thermophilic phase) and the composition of the microbial flora present in the waste (Manyi-Loh et al. 2016). The temperatures reached during stabilization, up to 55–60 °C, induce biomass sanitization in the oxidizing process, thus killing most of the microorganisms harmful to cultivated plants. The presence of competitive microorganisms toward plant parasites represents a further biological control tool for pests probably present in the original matrices (Hadar 2011).

pH variations and the abiotic factors' synthesis of allelopathic toxins or lytic action molecules show the greatest efficacy. At the same time, competition for nutrients and antibiosis is the main biotic factor of biocontrol. Many pathogens and parasites of plants show sensitivity to the effects described above (Alternaria, Fusarium, Monilia, Sclerotinia, Verticillium, etc.) (Pane et al. 2014). Compost can be easily stored (even for a long period), transported, and distributed with conventional agricultural machinery, such as manure spreaders (Khater 2012; Gondek et al. 2014). Compost production in Europe concerns 30.5 tons per year of fresh bio-waste derived from maintaining public and private green areas and wet residues (Siebert 2019). Bio-oxidation provides 3.5 tons per year of mature compost based on dry matter. A percentage of 50% of the production is destined directly for open-field agriculture, 15% for producing growing media for nurseries, 15% for green maintenance, 14% for hobby gardening, and 6% for other uses.

11.3.2 Anaerobic Digestion

Anaerobic digestion is a technology used to treat the organic waste stream, primarily due to its capacity to produce methane as renewable energy. Primary feedstocks for anaerobic digestion, both mono-fermented and co-digested, are animal manures (Neshat et al. 2017), food industry wastes (Banks et al. 2018), municipal solid wastes (Seruga 2021), crop residues (Sukhesh and Polisetty 2018) and energy crops (Herrmann 2016). Anaerobic digestion is a complex process through which biomass, in the absence of oxygen, is converted into biogas, consisting mainly of methane and CO_2. The percentage of methane varies according to the type of waste used and the process adopted, from 50% to 80% (Lohani and Havukainen 2018). Furthermore, the by-product of anaerobic digestion is a nitrogen-rich fertilizer product without risk to the environment when the production process is properly controlled and regulated (Banks et al. 2018). Anaerobic digestion is carried out by microorganisms capable of degrading waste into acetic acid, CO_2, and hydrogen, which methanogenic bacteria use to synthesize methane. Anaerobic microorganisms have low growth reproduction rates and reaction rates, so process control is essential. The temperature must be kept around 35 °C if using mesophilic bacteria, 55 °C with thermophilic bacteria, and the pH must be close to 7. The process in the reactor can include some preliminary steps, such as moisture reduction and hydrolysis of biomass. In contrast, the reactor delivery can be continuous or batch (Lamb 2020; Kirk and Gould 2020).

Digestate is a by-product of the anaerobic digestion process and has fertilizing properties due to the partially degraded organic matter content and mineral nutrients (Campos et al. 2019). However, its agricultural use requires a careful evaluation of nutritional and potentially toxic elements (Pecorini et al. 2020). In Italy, the use of digestate in agriculture is regulated by the Ministerial Decree n. 5046-25/02/2016, which implements the previous European Directives and national legislation. The Ministerial Decree has the objective of regulating the agricultural use of zootechnical digestate and digestate from agro-industrial sources, considering the nitrogen fraction that derives from digestate in accordance with the Nitrates Directive for vulnerable zones and ordinary areas and soil requirements. In line with the circular economy principles, energy crops, and agricultural processing waste provide carbon to supplement animal manure and obtain an optimal blend for anaerobic digestion and maximize conversion to biogas (Cuetos et al. 2011; Adebayo et al. 2014; Wey et al. 2015). To be used in agriculture, the anaerobic digestate must have minimum quality requirements, such as the content of organic matter, total nitrogen, total phosphorus, and absence *of Salmonella* spp. In fact, in the case of using mesophilic bacteria, the process temperatures do not allow the sanitation of the waste during digestion. Digestate utilization has been studied to evaluate the improving and fertilizing effect on soil, considering soil microbiology and ecosystem aspects (Muscolo et al. 2017; Risberg et al. 2017).

11.4 Effects of Bio-Based Fertilizers on Soil Quality

The different chemical characteristics of bio-based fertilizers applied to the soil influence carbon and nitrogen dynamics in the soil-plant system and influence yield productivity, nitrogen uptake, nitrogen leaching, or volatilization (Fig. 11.3a, b).

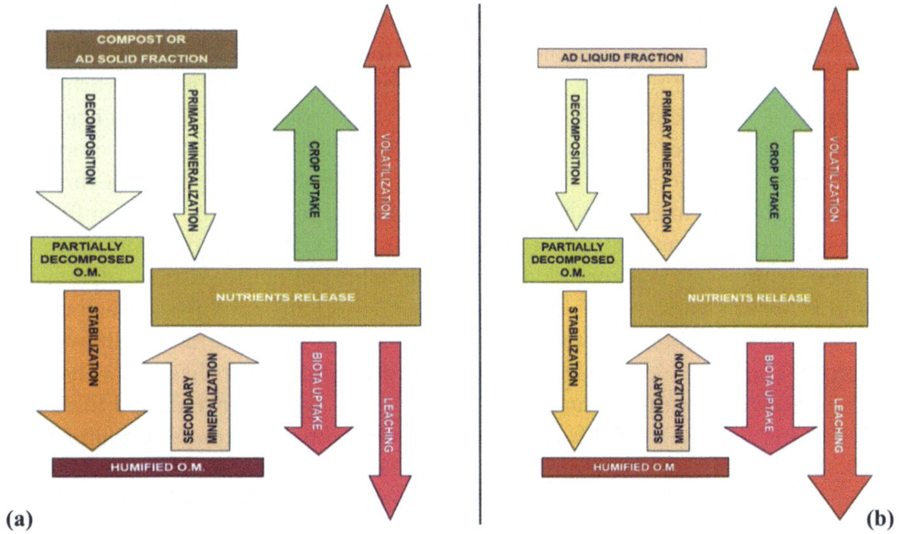

Fig. 11.3 Changes in soil organic matter (SOM) cycle following the amendment with (**a**) compost or solid fraction of anaerobic digestate. The different behavior between biosolids (compost or solid fraction of anaerobic digestates) and the liquid fraction of anaerobic digestate is shown through the different thicknesses of the arrows that indicate the various phases of the organic matter cycle in the soil. (**b**) a liquid fraction of anaerobic digestate—The different behavior between biosolids (compost or solid fraction of anaerobic digestates) and the liquid fraction of anaerobic digestate is shown through the different thickness of the arrows that indicate the various phases of the organic matter cycle in the soil

The administration to the soil of biomasses of different compositions and C/N ratio adds to the soil variable quantities of available nitrogen and differently affects the humification/mineralization process. Mature compost contains humified organic matter, which induces a quantitative and qualitative improvement of the colloidal status of the soil (Zeng et al. 2015). Compost contributes to the restoration of the chemical, physical, and biological fertility of soils. In fact, it favors the bioavailability of nutrients, the structural stability of the soil, water retention, animal and microbial diversity (worms, larvae, bacteria, and *fungi*, including mycorrhizae), and yields crop stability (Agegnehu et al. 2016; FAO 2020; Aytenew 2021). In a recent study on loamy sand soil, compost was added in two doses, 10 and 20 g kg^{-1} d.m. of soil (Strachel et al. 2017). After two weeks from the soil amendment, organotrophic, saprotrophic, and oligotrophic bacteria, as well as actinomycetes, increased in relation to the compost dose. Also, several species of fungi increased with the amendment. Arslan (Arslan et al. 2008) studied the effect of kitchen waste compost on sandy loam soil (10-20-30%). After 28 days, total bacteria increased in the major doses (20–30%). Actinomycetes grew for compost at 10% and 20% but decreased for the dose at 30%.

Compost use provides essential ecosystem services such as soil erosion reduction, soil water retention, soil fertility, and increased soil biodiversity. The soil amendment with compost derived from biodegradable waste increases carbon sequestration in the soil and reduces GHG emissions from agricultural soils with a beneficial effect on the environment. These goals are highly desirable in food security (FAO 2017). Compost amendment plans (the percentage of administration to the soil, the timing of addition, and the method of administration) are established on the basis of the qualitative characteristics of the compost itself, the chemical-physical properties of the soil, and the nutritional requirements of the crops. In organic agriculture, compost is used as it is, often without integration with other fertilizers. Even in integrated agriculture, in which compost participates in formulating the supply of nutrients and synthetic fertilizers, knowledge of its properties is required. In this way, it is possible to assess the supply of nutrients, the variations in buffering capacity, and the influence on microbial diversity and activity, water retention, structural stability, trafficability, and workability. This information allows you to prepare correct guidelines for fertilization with compost. Fertilization with compost derived from waste biomass from agricultural, agro-industrial, and livestock activities improves soil quality, plant nutrition, and crop productivity (Ebraimi et al. 2019; Pane et al. 2016; Papafilippaki and Nikolaidis 2020; Wolka and Melaku 2015). Research on the fertilizing value of compost from biodegradable waste has been encouraged in recent decades. Sullivan et al. (2002) reported that N mineralized from food waste compost increased the grass yield of tall fescue during the second and third seasons after application. Furthermore, mineralizable N and available P increased by 1.8 and 1.6 times, respectively (Zaman et al. 2002). The application of brewery sludge to the soil at the rate of up to 120 t ha^{-1} has enhanced the SOC, exchangeable cations, soluble cations and anions, and NH_4^+ and NO_3^--N content of the amended soil (Erdem and Ok 2002). In a study carried out on consumer vegetable fertilization with compost derived from wine-making residues, compared to manure or mineral fertilization, the authors showed that the use of distiller residue compost positively affected the crop's productivity. They reached a higher yield than the other treatments due to the greater availability of nutrients for the plant. Nevertheless, the amendment with manure induced a residual soil organic fertility enhancement. (Rossi et al. 2008, 2016). Other authors found that organic compost at planting positively affected the production and physiological quality of zucchini seeds. The sulfur fertilization at topdressing improves seeds' physiological quality and, in the presence of organic compost at planting, its production (Bardiviesso et al. 2020).

Francaviglia (Francaviglia et al. 2020) conducted a study to identify the benefits due to crop associations and alternative agricultural practices in four selected regions of Europe (Atlantic, Boreal, Northern Mediterranean, and Southern Mediterranean) in terms of crop production (CP). The conversion from traditional monocropping systems with intensive tillage and mineral fertilization to diversified alternative methods using crop rotations, no-tillage, and organic fertilization showed a crop performance improvement. These results were observed especially when longer crop rotations ($\geq$3 years) and no-tillage were adopted, particularly on medium-textured soils and in dry climates. Organic fertilization with manure, slurry, or compost allowed the enhancement of crop productivity, especially in Mediterranean areas with conditions on medium soil textures.

Soil fertilization with digestate has short-term positive effects on soil microbial activity, microbial biomass, and community composition by promoting soil microbial diversity. The mineralization of organic matter applied by digestate provides an available carbon source for Gram+ bacteria. Also, a major proliferation in fungal communities is observed as a consequence of carbon supply by digestate (García-Sánchez et al. 2015; Sapp et al. 2015; Goodall et al. 2019) microbial biomass. Digestate does not affect soil earthworms' content in the short term, even if the structure stability is ameliorated by earthworms' activity and dynamics. Digestate added to soil covered with green mulch, once incorporated into the topsoil, also improves the earthworm content (Frøseth et al. 2014).

The impact on the soil enzymatic activity differs in relation to the type of soil, crop, and digested biodegradable waste type. Different experimental conditions measured the increase or decrease in alkaline phosphatase, β-glucosidase, urease, and protease activities (Möller 2015). In a three-year trial on sandy loam soil, household compost (17 t ha^{-1}) and sewage sludge anaerobic digestate (4.2 t ha^{-1}) were added to evaluate fluorescein diacetate hydrolysis activity, biomass carbon, extracellular polysaccharides and phospholipid fatty acids dynamic (Debosz et al. 2002). No significant effect was observed despite the slight increase of these parameters in the amended soils.

The administration of digestates to soil induces an increase in SOM mineralization processes with a reduction in the soil C/N ratio (Möller 2015). In this condition, a decrease in the availability of nutrients for crops may occur due to competition from telluric microorganisms (Kuzyakov and Xu 2013). Therefore, fertilization plans based on anaerobic digestates must replace carbon by administering biomass with a higher C/N ratio. For this reason, many anaerobic digestion plants in Europe are equipping themselves with composting plants to work in parallel with digestion. Composting the digestate improves the C/N ratio, lowers the load of impurities and contaminants, and improves its fertilizer value (Knoop et al. 2018). For the same reason, digestates can be conveniently used to fertilize soils on which cereals have been grown, leaving a large number of high-carbon residues.

Digestates find an optimal application as fertilizers or soil amendments, even if most of them are not suitable for direct administration to the soil. Their composition might cause environmental impacts due to ammonia emissions, nitrate leaching, and nutrient overdoses, requiring complementary treatments to remove or recover nutrient surplus. One of the simplest digestate post-treatment methods is the separation of the solid and the liquid fraction. With the solid/liquid separation of the digestate, a high percentage of nitrogen is concentrated in the clarified fraction in ammonia form, which can be readily assimilated by crops. Organic nitrogen with slower mineralization is concentrated in the solid fraction, which makes this material more suitable for amending use. Fertilizer values of liquid digestates are comprised of livestock manures and inorganic fertilizers (Rossi et al. 2020; Nkoa 2014). The digestate finds valuable agronomic reuse as such or can subsequently be composted to be more easily stored and transported. The separation of the solid and the liquid phases is frequently applied when transporting solid fractions over longer distances because of the reduction of water content and solid material volume. There are many advantages of the recovery and recycling of digestate in agriculture. These advantages are biowaste sanitization and deodorization, improved biosecurity from pathogen reduction, supply of organic matter and NPK fertilizers, reduction of mineral fertilizers use and a decrease of raw material and energy input for their production. In addition, it also help in better control of ammonia emissions, better management of agricultural soil nitrogen, and improvement of physical-chemical soil functions (Holm-Nielsen et al. 2009; Grigatti et al. 2011; Beni et al. 2012; Zangh et al. 2014; Koszel and Lorencowicz 2015; Battista et al. 2019; Barlòg et al. 2020; Nascimento et al. 2020).

A recent study illustrated the positive effects of applying digestate derived from sugar beet pulp to achieve economic and environmental-friendly performance (Suhartini et al. 2018). The positive effects of the use of digestate as a bio-based fertilizer on the quality of the soil, on the plant nutrition, and on crop productivity depend on the choice of the correct dose of administration, time of application to the soil, suitable soil tillage and chemical-physical characteristics of the soil itself (Gilbert et al. 2020).

11.5 Short Case Studies

The most relevant results of more extensive studies conducted by the authors are reported.

11.5.1 Compost Amendment

The research aimed at evaluating the effects of different fertilization techniques on the physical-mechanical properties of loamy-sandy soil. Triennial use of mixed composted amendments in different doses, both as they are and mixed with mineral fertilizers, was compared to the treatment with mineral fertilizers and untreated control. A multi-year experimental trial (2012–2015) was set up in a greenhouse horticultural rotation, "indeterminate tomato—green bean—lettuce," according to an experimental design with randomized blocks. The treatments were: compost 15 t ha^{-1} (C15); 50% compost 15 t ha^{-1} + 50% mineral fertilizer 40 kg N ha^{-1} (C15 + N/2); compost 30 t ha^{-1} (agronomic dose- C30); compost 45 t ha^{-1} (C45); mineral fertilizer 80 kg N ha^{-1} (Min); control (CNT).

In the third and fourth years of the trial, the main physical-mechanical parameters of the soil were analyzed. The following parameters were determined: shear resistance and infiltration rate in the field, colloid content, bulk density, moisture, uniaxial compression resistance, and crackability in the laboratory. The best performance was observed in soils treated with the highest dose of mixed compost (C45). The infiltration rate value was the highest, while the cracking index value was the lowest (Figs. 11.4 and 11.5).

11.5.2 Amendment with a Solid Phase of Anaerobic Digestate

A field test was carried out on clayey alluvial soil cultivated for three years (2016–2018) with romaine lettuce. The effects of the application of distillery residues on the soil's chemical and physical-mechanical properties, both as such and

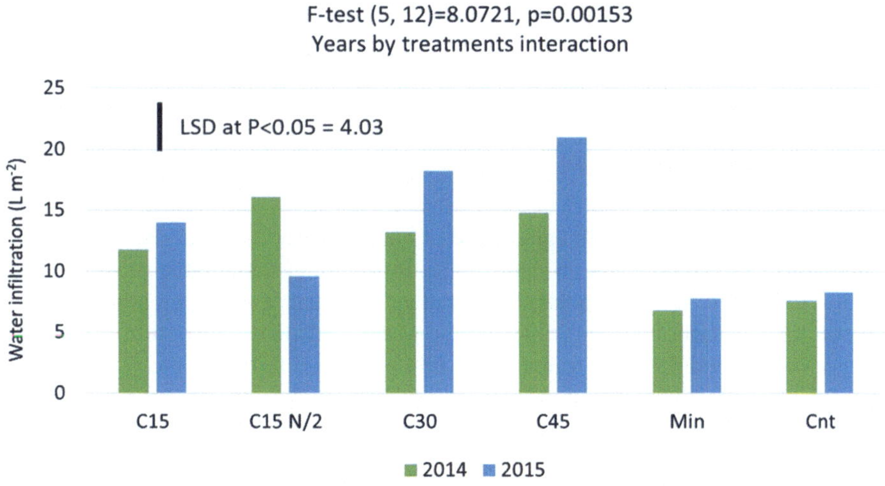

Fig. 11.4 Water infiltration per surface unit in two hours field test (L m^{-2})

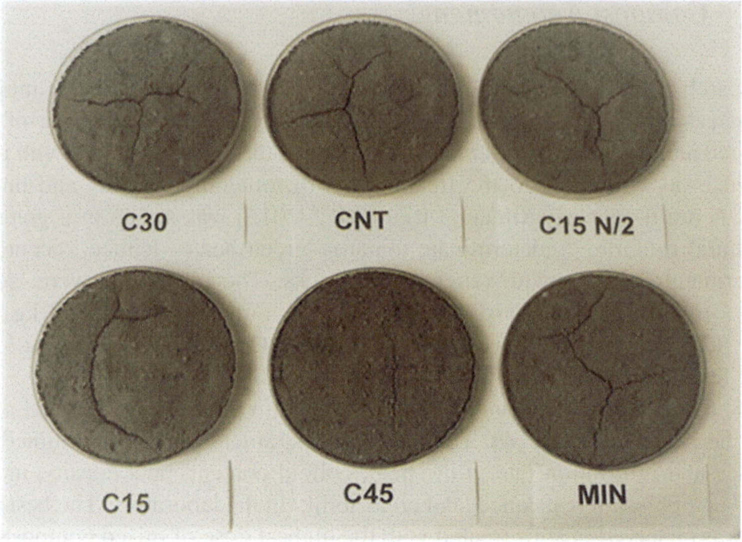

Fig. 11.5 Soil samples of the different treatments at the end of the cracking test

anaerobically digested, were determined. The soil amendments were compared with mineral fertilization and untreated control. The amendments were distillation residues 15 t ha^{-1} (DR), digestate from distillation residues 30 t ha^{-1} (AD), mineral fertilizer 45 kg N ha^{-1} (Min), and control (C).

The soil's main chemical and physical-mechanical parameters, both in the field (water infiltration rate) and in the laboratory (colloid content, apparent density, humidity, oedometric deformation), were determined at the end of the third year of testing. The application of distillery residues has improved the availability of P compared to other treatments (not reported data). The administration of anaerobic digestate improved the resistance to soil compression (Fig. 11.6), penetration, and soil infiltration rate (Fig. 11.7) compared to other treatments.

11.6 Conclusions

The recovery and recycling in agriculture of bio-based fertilizers derived from biodegradable waste represents an opportunity to combine sustainable practices and circular economy. It also represents a strategy to create a positive carbon budget (carbon input > carbon output) contributing to soil carbon sequestration. The recycling of biodegradable waste transformed into bio-based fertilizer allows the recovery of valuable fertilizer components. Biomass recovery supports soil fertility and prevents SOC decline, soil degradation, and desertification. The agricultural valorization of biodegradable waste constitutes a benefit to the development of the circular economy and of eco-sustainable management practices that enhance ecosystem

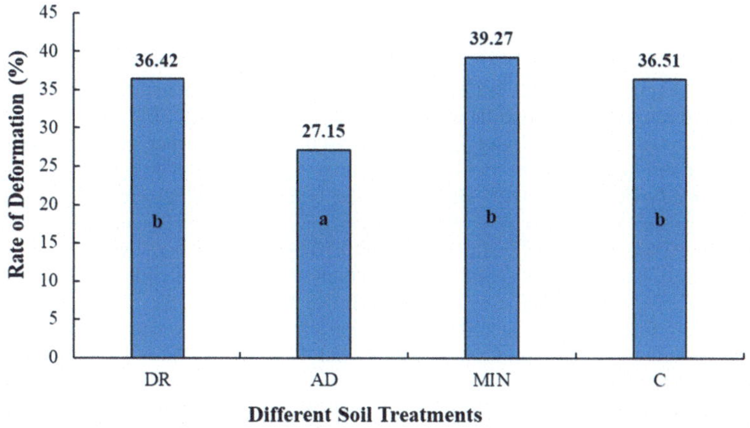

Fig. 11.6 Oedometric deformation detected with a load of 200 daN. Bars with different letters were significant ($P < 0.05$) using Duncan's multiple range test. (Note: *DR* distillation residues, *AD* digestate from distillation residues, *MIN* mineral fertilizer, *C* control)

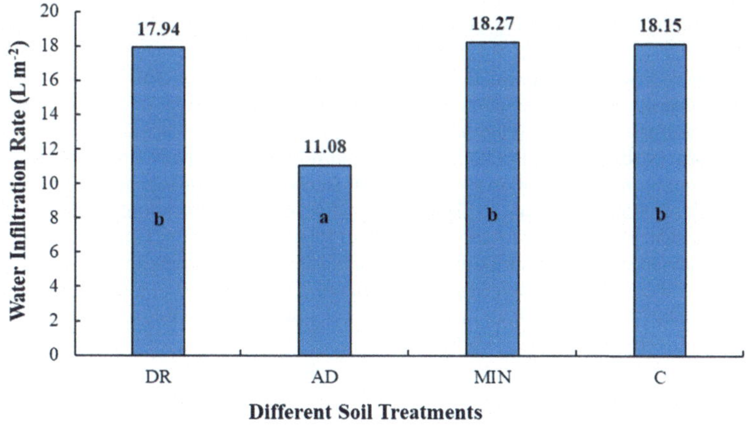

Fig. 11.7 Water infiltration per surface unit in two hours field test (L m^{-2}). Bars with different letters were significant ($P < 0.05$) using Duncan's multiple range test (Note: *DR* distillation residues, *AD* digestate from distillation residues, *MIN* mineral fertilizer, *C* control)

services. It is necessary to promote waste recycling practices and the optimization of by-product transformation processes, considering that some types of bio-based fertilizers are by-products of the production process of renewable energy from agro-zootechnical sources. At the local level, actions aimed at creating small processing plants and supporting non-profit investments allow us to achieve better environmental performance.

Acknowledgments This work was supported by the Italian Ministry of Agriculture (MiPAAF) under the AGROENER project (D.D. n. 26329, 1st April 2016).

References

Adebayo AO, Jekayinfa SO, Linke B (2014) Effect of co-digestion on anaerobic digestion of pig slurry with maize cob at mesophilic temperature. J Nat Sci Res 4(22):66–73

Agegnehu G, Bass AM, Paul N. Nelson PN et al. (2016) Benefits of biochar, compost and biochar–compost for soil quality, maize yield and greenhouse gas emissions in a tropical agricultural soil. Sci Total Environ 543(A):295–306. https://doi.org/10.1016/j.scitotenv.2015.11.054

Amelung W, Bossio D, de Vries W et al (2020) Towards a global-scale soil climate mitigation strategy. Nat Commun 11:5427. https://doi.org/10.1038/s41467-020-18887-7

Arslan EI, Öbek E, Kirbag S et al (2008) Determination of the effect of compost on soil microorganisms. International Journal of Science & Technology 3(2):151–159

Atinkut HB, Yan T, Zhang F et al (2020) Cognition of agriculture waste and payments for a circular agriculture model in Central China. Sci Rep 10:10826. https://doi.org/10.1038/s41598-020-67358-y

Ayilara MS, Olanrewaju OS, Babalola OO et al (2020) Waste management through composting: challenges and potentials. Sustain For 12(11):4456. https://doi.org/10.3390/su12114456

Aytenew M (2021) Soil biodiversity as a key sponsor of regenerative agriculture. Biodiversity of Ecosystems. IntechOpen, London, In. https://doi.org/10.5772/intechopen.99716

Banks CJ, Heaven S, Zhang Y et al (2018) Food waste digestion: anaerobic digestion of food waste for a circular economy. In: Murphy JD (ed) IEA Bioenergy Task, vol 37, Paris, p 12

Bardiviesso EM, Lanna NBL, Aguilar AS et al (2020) Production and quality of zucchini seeds after applying Sulphur as top dressing and organic compost at planting. AJCS 14(07):1064–1071. https://doi.org/10.21475/ajcs.20.14.07.p2049

Barłóg P, Hlisnikovský L, Kunzová E (2020) Effect of Digestate on soil organic carbon and plant-available nutrient content compared to cattle slurry and mineral fertilization. Agronomy 10:379. https://doi.org/10.3390/agronomy10030379

Battista F, Frison N, Bolzonella D (2019) Energy and nutrients' recovery in anaerobic digestion of agricultural biomass: an Italian perspective for future applications. Energies 12(17):3287

Beni C, Servadio P, Marconi S et al (2012) Anaerobic digestate administration: effect on soil physical and mechanical behavior. Commun in Soil Sci Plant Anal 43(5):821–834

Bernal MP, Sommer SG, Chadwick D et al (2017) Chapter three—current approaches and future-banks trends in compost quality criteria for agronomic, environmental, and human health benefits. In: Sparks DL (ed) Advances in agronomy, vol 144. Academic Press, pp 143–233. https://doi.org/10.1016/bs.agron.2017.03.002

Campos JL, Crutchik D, Franchi Ó et al (2019) Nitrogen and phosphorus recovery from anaerobically pretreated agro-food wastes: a review. Front Sustain Food Syst 2:91. https://doi.org/10.3389/fsufs.2018.00091

Case SDC, Jensen LS (2019) Nitrogen and phosphorus release from organic wastes and suitability as bio-based fertilizers in a circular economy. Environ Technol 40(6):701–715. https://doi.org/10.1080/09593330.2017.1404136

Chew KW, Chia SR, Yen HW et al (2019) Transformation of biomass waste into sustainable organic fertilizers. Sustain For 11(8):2266. https://doi.org/10.3390/su11082266

Chojnacka K, Moustakas K, Witek-Krowiak A (2020) Bio-based fertilizers: a practical approach towards circular economy. Bioresour Technol 295:122223. https://doi.org/10.1016/j.biortech.2019.122223

Circular economy package-European Commission (2016) Proposal for a regulation of the European Parliament and of the Council laying down rules on the making available on the market of CE marked fertilising products and amending Regulations (EC) No 1069/2009 and (EC) No 1107/2009 https://eur-lex.europa.eu/legal-content/EN/ALL/?uri=CELEX:52016PC0157. Accessed 13 Oct 2021

Cremeniac L, Boclaci T (2014) Worm compost to improve soil fertility. In: Dent D (ed) Soil as world heritage. Springer, Dordrecht. https://doi.org/10.1007/978-94-007-6187-2_41

Cuetos MJ, Fernández C, Gómez X et al (2011) Anaerobic co-digestion of swine manure with energy crop residues. Biotechnol Bioprocess Eng 16(5):1044
Debosz K, Petersen OS, Kure LK et al (2002) Evaluating effects of sewage sludge and household compost on soil physical, chemical and microbiological properties. Appl Soil Ecol 19(2002):237–248
Diacono M, Montemurro F (2015) Effectiveness of organic wastes as fertilizers and amendments in salt-affected soils. Agriculture 5(2):221–230
Doyeni MO, Stulpinaite U, Baksinskaite A, Suproniene S, Tilvikiene V et al (2021) The effectiveness of Digestate use for fertilization in an agricultural cropping system. Plants 10(8):1734. https://doi.org/10.3390/plants10081734
Duque-Acevedo M, Belmonte-Ureña LJ, Plaza-Úbeda JA et al (2020) The management of agricultural waste biomass in the framework of circular economy and bioeconomy: an opportunity for greenhouse agriculture in Southeast Spain. Agronomy 10(4):489
Ebrahimi E, Asadi G, von Fragstein und Niemsdorff P (2019) A field study on the effect of organic soil conditioners with different placements on dry matter and yield of tomato (Lycopersicon esculentum L). Int J Recycl Org Waste Agric 8(1):59–66
Epstein E (1997) The science of composting, 1st edn. CRC Press, Boca Raton. https://doi.org/10.1201/9780203736005
Erdem N, Ok SS (2002) Effect of brewery sludge amendments on some chemical properties of acid soil in pot experiments. Bioresour Technol 84(3):271–273
Faesal, Djaenuddin N, Syafruddin (2020) In: IOP Conference Series: Earth and Environment Science. 484:012053. https://doi.org/10.1088/1755-1315/484/1/012053
FAO (2017) Unlocking the potential of soil organic carbon. Outcome document. Global symposyium on soil organic matter, Food and Agriculture Organization of the United Nations, Rome
FAO and ITPS (2015) Status of the World's soil resources (SWSR)—Main report. Chapter 6 Global soil status, processes and trends, 116–118. In: Food and Agriculture Organization of the United Nations and Intergovernmental Technical Panel on Soils, Rome
FAO, GSP (2020) Soil biodiversity: a nature-based solution. In: SoilBiodiversity/20/Report. Food and Agriculture Organization of the United Nations and Global Soil Partnership, Rome
Francaviglia R, Renzi G, Ledda L et al (2017) Organic carbon pools and soil biological fertility are affected by land use intensity in Mediterranean ecosystems of Sardinia, Italy. Sci Total Environ 599–600:789–796. https://doi.org/10.1016/j.scitotenv.2017.05.021
Francaviglia R, Álvaro-Fuentes J, Di Bene C et al (2020) Diversification and management practices in selected European regions. A data analysis of arable crops production. Agronomy 10:297
Frøseth RB, Bakken AK, Azzaroli Bleken M et al (2014) Effects of green manure herbage management and its digestate from biogas production on barley yield, N recovery, soil structure and earthworm populations. Eur J Agron 52(B):90–102. https://doi.org/10.1016/j.eja.2013.10.006
Fuchs JG (2015) Compost and digestate, valuable products for farmers. FIBL—Research Institute of Organic Agriculture, Lausanne. https://www4.ti.ch/fileadmin/DFE/DE-SA/comunicati/orti/2015/orti-profi_compost_1.pdf. Accessed 13 Oct 2021
García-Sánchez M, Siles JA, Cajthaml T et al (2015) Effect of digestate and fly ash applications on soil functional properties and microbial communities. Eur J Soil Biol 71:1–12. https://doi.org/10.1016/j.ejsobi.2015.08.004
Gilbert J, Ricci-Jurgensen M, Ramola A (2020) Benefits of compost and anaerobic digestate. ISWA—International Solid Waste Association, Rotterdam. https://app.livestorm.co/iswa/benefits-of-compost-and-anaerobic-digestate. Accessed 13 Oct 2021
Gondek K, Kopeć M, Mierzwa M et al (2014) Chemical and biological properties of composts produced from organic waste. J Elem 19(2):377–390. https://doi.org/10.5601/jelem.2014.19.2.670
Goodall TI, Griffiths RI, Blaud A et al (2019) Soil biological properties from a digestate experiment on winter wheat at North Wyke and Henfaes Farm. NERC Environmental Information Data Centre, Lancaster. https://doi.org/10.5285/391c0294-07f1-4856-b592-428bd44055ca
Goss MJ, Tubeileh A, Goorahoo D (2013) A review of the use of organic amendments and the risk to human health. Adv Agron 120:275–379

Grigatti M, Di Girolamo G, Chincarini R et al (2011) Potential nitrogen mineralization, plant utilization efficiency and soil CO_2 emissions following the addition of anaerobic digested slurries. Biomass Bioenergy 35:4619–4629

Hadar Y (2011) Suppressive compost: when plant pathology met microbial ecology. Phytoparasitica 39:311–314. https://doi.org/10.1007/s12600-011-0177-1

Herrmann C, Idler C, Heiermann M (2016) Biogas crops grown in energy crop rotations: linking chemical composition and methane production characteristics. Bioresour Technol 206:23–35. https://doi.org/10.1016/j.biortech.2016.01.058

Holm-Nielsen JB, Al Seadi T, Oleskowicz-Popiel P (2009) The future of anaerobic digestion and biogas utilization. Bioresour Technol 100:5478–5484

Jones A, Panagos P, Barcelo S et al (2012) The state of soil in Europe. JRC 68418: Publications Office of the European Union, Luxembourg. https://doi.org/10.2788/77361

Kauer K, Tein B, Sanchez de Cima D et al (2015) Soil carbon dynamics estimation and dependence on farming system in a temperate climate. Soil Tillage Res 154:53–63. https://doi.org/10.1016/j.still.2015.06.010

Khater E (2012) Chemical and physical properties of compost. Misr J Agric Eng 29(4):1567–1582. https://doi.org/10.21608/mjae.2012.101382

Kirk DM, Gould MC (2020) Bioenergy and anaerobic digestion, editor(s): Anju Dahiya, bioenergy (second edition). Academic, Cambridge, pp 335–360. https://doi.org/10.1016/B978-0-12-815497-7.00017-8

Knoop C, Dornack C, Raab T (2018) Effect of drying, composting and subsequent impurity removal by sieving on the properties of digestates from municipal organic waste. Waste Manag 72:168–177. https://doi.org/10.1016/j.wasman.2017.11.022

Koch A, McBratney A, Adams M et al (2013) Soil security: solving the global soil crisis. Global Pol 4(4):434–441. https://doi.org/10.1111/1758-5899.12096

Koszel M, Lorencowicz E (2015) Agricultural use of biogas digestate as a replacement fertilizers. Agric Sci Procedia 7:119–124

Kuzyakov Y, Xu XL (2013) Competition between roots and microorganisms for nitrogen: mechanisms and ecological relevance. New Phytol 198(3):656–669

Lal R (2008) Sequestration of atmospheric CO_2 in global carbon pools. Energy Environ Sci 1:86–100

Lal R (2010) Managing soils and ecosystems for mitigating anthropogenic carbon emissions and advancing global food security. Bioscience 60(9):708–721

Lal R (2017) Soil organic carbon sequestration: importance and state of science. Proceedings of the Global Symposium on Soil Organic Carbon. Food and Agriculture Organization of the United Nations. Rome

Lalander C, Ermolaev E, Wiklicky V et al (2020) Process efficiency and ventilation requirement in black soldier fly larvae composting of substrates with high water content. Sci Total Environ 729:138968. https://doi.org/10.1016/j.scitotenv.2020.138968

Lamb JJ (2020) Anaerobic digestion—from biomass to biogas. SCIO Publishing, Copenaghen, p 462. https://doi.org/10.48216/9788269203325

LaSalle T, Hepperly P (2008) Regenerative organic farming: a solution to global warming. The Rodale Institute, Kutztown

Lohani SP, Havukainen J (2018) Anaerobic digestion: factors affecting anaerobic digestion process. In: Varjani S, Gnansounou E, Gurunathan B, Pant D, Zakaria Z (eds) Waste bioremediation. Energy, environment, and sustainability. Springer, Singapore. https://doi.org/10.1007/978-981-10-7413-4_18

Long L, Fuqing X, Xumeng G et al (2019) Biological treatment of organic materials for energy and nutrients production—anaerobic digestion and composting. In: Li Y, Ge X (eds) Advances in Bioenergy. Elsevier, Cambridge. https://doi.org/10.1016/bs.aibe.2019.04.002

Makan A, Mountadar M (2012) Effect of C/N ratio on the in-vessel composting under air pressure of organic fraction of municipal solid waste in Morocco. J Mater Cycles Waste Manag 14:241–249. https://doi.org/10.1007/s10163-012-0062-0. Accessed 13 Oct 2021

Manyi-Loh CE, Mamphweli SN, Meyer EL et al (2016) An overview of the control of bacterial pathogens in cattle manure. Int J Environ Res Public Health 13:843. https://doi.org/10.3390/ijerph13090843

Mattana S, Petrovicová B, Landi L et al (2014) Sewage sludge processing determines its impact on soil microbial community structure and function. Appl Soil Ecol 75:150–161. https://doi.org/10.1016/j.apsoil.2013.11.007

Mohapatra DP, Cledón M, Brar SK et al (2016) Application of wastewater and biosolids in soil: occurrence and fate of emerging contaminants. Water Air Soil Pollut 227:1–14. https://doi.org/10.1007/s11270-016-2768-4

Möller K (2015) Effects of anaerobic digestion on soil carbon and nitrogen turnover, N emissions, and soil biological activity. A review. Agronomy for Sustainable Development, Springer Verlag/EDPSciences/INRA 35(3):1021–1041. https://doi.org/10.1007/s13593-015-0284-3. hal-01284312

Mortier N, Velghe F, Verstichel S (2016) Organic recycling of agricultural waste today: composting and anaerobic digestion. In: Poltronieri P, D'Urso OF (eds) Biotransformation of agricultural waste and by-products in the 4F economy: the food, feed, fiber, fuel (4F) economy. Elsevier, Oxford. https://doi.org/10.1016/C2015-0-00053-3

Muscolo A, Settineri G, Papalia T et al (2017) Anaerobic co-digestion of recalcitrant agricultural wastes: characterizing of biochemical parameters of digestate and its impacts on soil ecosystem. Sci Total Environ 586:746–752

Nascimento AL, De Souza AJ, Oliveira FC et al (2020) Chemical attributes of sewage sludges: relationships to sources and treatments, and implications for sludge usage in agriculture. J Clean Prod 258:120746. https://doi.org/10.1016/j.jclepro.2020.120746

Neshat SA, Mohammadi M, Najafpour GD et al (2017) Anaerobic co-digestion of animal manures and lignocellulosic residues as a potent approach for sustainable biogas production. Renew Sust Energ Rev 79:308–322. https://doi.org/10.1016/j.rser.2017.05.137

Nkoa R (2014) Agricultural benefits and environmental risks of soil fertilization with anaerobic digestates: a review. Agron Sustain Dev 34:473–492

Pane C, Celano G, Zaccardelli M (2014) Metabolic patterns of bacterial communities in aerobic compost teas associated with potential biocontrol of soilborne plant diseases. Phytopathol Mediterr 53(2):75. https://doi.org/10.14601/Phytopathol_Mediterr-13363

Pane C, Palese AM, Spaccini R et al (2016) Enhancing sustainability of a processing tomato cultivation system by using bioactive compost teas. Sci Hortic 202:117–124

Papafilippaki A, Nikolaidis NP (2020) Comparative study of wild and cultivated populations of Cichorium spinosum: the influence of soil and organic matter addition. Sci Hortic 261:108942

Pecorini I, Peruzzi E, Albini E et al (2020) Evaluation of MSW compost and digestate mixtures for a circular economy application. Sustain For 12(7):3042. https://doi.org/10.3390/su12073042

Razza F, D'Avino L, L'Abate G et al (2018) The role of compost in bio-waste management and circular economy. In: Benetto E, Gericke K, Guiton M (eds) Designing sustainable technologies, products and policies. Springer Cham, New York. https://doi.org/10.1007/978-3-319-66981-6_16

Regulation (Eu) 2019/1009 of the European Parliament and of the Council of 5 June 2019 laying down rules on the making available on the market of EU fertilising products and amending Regulations (EC) No 1069/2009 and (EC) No 1107/2009 and repealing Regulation (EC) No 2003/2003. https://eur-lex.europa.eu/legal-content/EN/TXT/HTML/?uri=CELEX:32019R1009&from=IT

Risberg K, Cederlund H, Pell M et al (2017) Comparative characterization of digestate versus pig slurry and cow manure—chemical composition and effects on soil microbial activity. Waste Manag 61:29–538. https://doi.org/10.1016/j.wasman.2016.12.016

Rossi G, Beni C (2018) Effects of medium-term amendment with diversely processed sewage sludge on soil Humification—mineralization processes and on Cu, Pb, Ni, and Zn bioavailability. Plan Theory 7(1):16

Rossi G, Pennelli B, Socciarelli S et al (2002) Effects of medium-term amendment with sewage sludges on heavy metal distribution in soil. In: Violante A, Huang PM, Bollag JM, Gianfreda

L (eds) Soil mineral-organic matter-microorganism interactions and ecosystem health. Developments in soil science, vol 28A. Elsevier Science B.V, Amsterdam, pp 99–107
Rossi G, Beni C, Socciarelli S (2008) Fertilization with distiller's residue compost: effects on nutrients accumulation and yield in consumer vegetables. Agrochimica 52(5):325–336
Rossi G, Socciarelli S, Aromolo R et al (2016) The effects of mineral and organic fertilizers on zucchini (Cucurbita pepo L) investigated by the multiparameters analyses and magnetic resonance imaging (MRI). Agrochimica 60(4):275–287
Rossi G, Neri U, Felici B et al (2020) Effects of different zootechnical digestates on fertilization and nitrogen leaching. Agrochimica LXIV(3):239–251. https://doi.org/10.12871/00021857201917
Rynk R (1992) On-farm composting handbook. Northeast regional agricultural engineering service. Cooperative Extension, New York. https://hdl.handle.net/1813/67142. Accessed 13 Oct 2021
Sapp M, Harrison M, Hany U et al (2015) Comparing the effect of digestate and chemical fertiliser on soil bacteria. Appl Soil Ecol 86:1–9. https://doi.org/10.1016/j.apsoil.2014.10.004
Sayara T, Basheer-Salimia R, Hawamde F et al (2020) Recycling of organic wastes through composting: process performance and compost application in agriculture. Agronomy 10:1838
Scharlemann JPW, Tanner EVJ, Hiederer R et al (2014) Global soil carbon: understanding and managing the largest terrestrial carbon pool. Carbon Manag 5(1):81–91. https://doi.org/10.4155/cmt.13.77
Scotti R, Bonanomi G, Scelza R et al (2015) Organic amendments as sustainable tool to recovery fertility in intensive agricultural systems. J Soil Sci Plant Nutr 15(2):333–352. https://doi.org/10.4067/S0718-95162015005000031
Seruga P (2021) The municipal solid waste management system with anaerobic digestion. Energies 14:2067. https://doi.org/10.3390/en14082067
Siebert S (2019) Compost production in Europe. In Proceeding of ISWA World Congress, October 7–9, Bilbao. https://www.compostnetwork.info/download/201910_iswa2019_compost-production-in-europe/. Accessed 13 Oct 2021
Strachel R, Wyszkowska J, Baćmaga M (2017) The role of compost in stabilizing the microbiological and biochemical properties of zinc-stressed soil. Water Air Soil Pollution 228:349. https://doi.org/10.1007/s11270-017-3539-6
Suhartini S, Heaven S, Banks CJ (2018) IOP Conferences Series: Earth and Environmental Science 131:012048. https://doi.org/10.1088/1755-1315/131/1/012048
Sukhesh MJ, Polisetty VR (2018) Anaerobic digestion of crop residues: technological developments and environmental impact in the Indian context. Biocatal Agric Biotechnol 16:513–528. https://doi.org/10.1016/j.bcab.2018.08.007
Sullivan DM, Bary AI, Thomas DR et al (2002) Food waste compost effects on fertilizer nitrogen efficiency, available nitrogen, and tall fescue yield. Soil Sci Soc Am J 66(1):154–161
Tur-Cardona J, Bonnichsen O, Speelman S et al (2018) Farmers' reasons to accept bio-based fertilizers: a choice experiment in seven different European countries. J Clean Prod 197(1):406–416. https://doi.org/10.1016/j.jclepro.2018.06.172
Urra J, Alkorta I, Garbisu C (2019) Potential benefits and risks for soil health derived from the use of organic amendments in agriculture. Agronomy 9(9):542. https://doi.org/10.3390/agronomy9090542
Vicente-Vicente JL, García-Ruiz R, Francaviglia R et al (2016) Soil carbon sequestration rates under Mediterranean woody crops using recommended management practices: a meta-analysis. Agric Ecosyst Environ 235:204–214. https://doi.org/10.1016/j.agee.2016.10.024
Wei Y, Li X, Yu L et al (2015) Mesophilic anaerobic co-digestion of cattle manure and corn Stover with biological and chemical pretreatment. Bioresour Technol 198:431–436
Wolka K, Melaku B (2015) Exploring selected plant nutrient in compost prepared from food waste and cattle manure and its effect on soil properties and maize yield at Wondo genet. Ethiopia Environ Syst Res 4:9. https://doi.org/10.1186/s40068-015-0035-0

Zaman M, Di HJ, Samamoto K et al (2002) Effects of sewage sludge compost and chemical fertilizer application on microbial biomass and N mineralization rates. Soil Sci Plant Nutr 48(2):195–201

Zeng G, Wu H, Liang J et al (2015) Efficiency of biochar and compost (or composting) combined amendments for reducing Cd, Cu, Zn and Pb bioavailability, mobility and ecological risk in wetland soil. RSC Adv 5:34543–34548

Zhang L, Xu C, Champagne P et al (2014) Overview of current biological and thermo-chemical treatment technologies for sustainable sludge management. Waste Manag Res 32(7):586–600. https://doi.org/10.1177/734242X14538303

Chapter 12
Role of Biochar Technology in Carbon Sequestration and Agro-Environmental Sustainability

Abhilasha Tripathi, Pradeep Kumar Dubey, Munish Kumar Upadhyay, and Purnendu Bose

Abstract Globally, soil-based emissions such as forest fires (as occurred in Australia and Brazil in the recent past) and biomass burning adversely affect the environment's biodiversity and cause warming effects. In coalition, anthropogenic emissions also simultaneously augment global temperature. The Paris Climate Change Convention (Nov 2016) held a consensus to limit the rise in global temperature by below 2 °C until this century's end. Against this backdrop, sequestering more carbon in soil has become an urgent priority. Agroecosystem soil, being both a source and sink of carbon, is therefore intervened with greater effort by the national and global agencies. Sustainable agricultural practices have shown promise in the direction of increasing organic carbon stock and restoring degraded/contaminated soils. Yet, advanced technology such as biochar use and application, progress in the method of preparation (pyrolysis), and cost-effectiveness need attention. Pyrolysis tends to convert biomass rapidly into a recalcitrant form known as biochar. This needs further developments, and newer agro-residue-based substrates need to be exploited for large-scale effective biochar preparation and their recycling in agriculture. Biochar addition in agricultural soil sequesters carbon (mainly as Soil Organic Carbon (SOC)) and, in the long run, manages various soil quality indicators, *viz.* C:N ratio, microbial load, ion conductivity. These harmonize the soil-water-pant-nutrient nexus and sustain crop yield. The soil's physical property is improved (such as reduced bulk density), thereby increasing aeration, rhizospheric expansion, dense root system, and limiting soil erosion. The liming action of biochar in acidic soil maintains soil pH, which is one major driver for plant nutrition.

A. Tripathi · M. K. Upadhyay · P. Bose (✉)
Department of Civil Engineering, Indian Institute of Technology, Kanpur, India
e-mail: pbose@iitk.ac.in

P. K. Dubey
Institute of Environment and Sustainable Development, Banaras Hindu University, Varanasi, Uttar Pradesh, India

WRI India, New Delhi, India

S. A. Edrisi et al. (eds.), *Sustainable Plant Nutrition and Soil Carbon Sequestration*, Sustainable Plant Nutrition in a Changing World,
https://doi.org/10.1007/978-3-031-53590-1_12

Overall, biochar technology needs to be exploited to mitigate GHG emissions, reclaim soil health fertility, and enhance plant growth/yield. Therefore, the present chapter highlights strategies to advance the biochar technological and application-based interventions for enhancing carbon capture and managing the soil ecosystem, thereby meeting some of the important UN-SDGs (12-responsible consumption and production and 13-climate action) by the year 2030.

Keywords Carbon sink · Climate change · GHGs emissions · Plant productivity · Pyrolysis · Soil quality

12.1 Climate Change and Carbon Imbalance: A Global Scenario

Carbon dioxide (CO_2) is the major GHG emitted by anthropogenic activity such as fossil fuel burning, forest fires, etc. (IPCC 2005). Electricity generation and heating (25%) contributes to the largest source of GHG emissions emission, followed by agriculture, forestry and land use (24%), industries (21%), transportation (14%), buildings (10%), and other sources (10% of global GHG emissions during 2010) (EPA 2020). The year 2020 tied with the year 2016 to be the warmest year (Brown 2021), and 2019 was the second warmest year between 1880 and 2020, with global land and ocean temperature rise of 1.02 and 0.95 °C above average temperature. Five out of ten warmest years between 1880 and 2019 have been observed since 2015, which shows a very alarming situation (NOAA 2019). Due to changes in the trends of heat waves due to the climate crisis, India alone encountered 30,000 forest fires in 2019 (mongabay.com).

Globally, frequent forest fires are also observed in Canada (2014, 2017, 2018), Siberia (2003, 2012, 2019), the Mediterranean (2017, 2018), Australia (2009, 2013, 2019, 2020), California (2017, 2018, 2020), Brazil (2020) (futureearth.org; Burak Bir 2020; Greenpeace 2020). Forests are major territorial carbon sinks; during forest fires, the legacy carbon reserves present in the top layer of soil also vent out and intensify the effect of GHGs in the atmosphere (Walker et al. 2019). To cope with the current climate crisis, it is indispensable to control carbon by multiple approaches through removal and reduction. Currently, techniques like Carbon farming, Carbon trading (Nath et al. 2015), Direct air capture (DAC) (Marcucci et al. 2017), bioenergy with carbon capture and storage (BECCS) (Vergragt et al. 2011), and CO_2 storage in the ocean (Herzog 2001) are being implemented for carbon sequestration.

Carbon sequestration is the process of transforming carbon into a relatively stable form (or "sink") that is immune to decay or release compared to its original form that can be released readily in the atmosphere (Lal 2008). The carbon is naturally consumed by the plants from the atmosphere through photosynthesis and converted into carbohydrates. But as the plants die, the carbon is released back into the atmosphere, making the cycle of photosynthesis and decay a carbon-neutral process

(Godbout et al. 2010; Anderson 2020). Carbon can be sequestered naturally by afforestation, reforestation, and by adding carbon-rich materials to the soil to increase the SOC content (Bier 2020). A three-year-long field study by (Dubey et al. 2022) shows the increase in soil SOC upon biochar application.

Agriculture is a major activity associated with soil. Food security can be aided by the selection of suitable practices based on three perspectives, i.e., landscape, farm or field, and crop/species. The soil can be treated sustainably to maintain a balance between these levels (Dubey et al. 2020a, b). Many carbon sequestration techniques noted above are energy intensive, but using soil as a carbon sink can be an energy-producing, sustainable option. Historical evidence of soil carbon enrichment already exists. The traditional approach adopted between 7000 and 500 cal yr. BP to build soil carbon within the Amazon region can be employed in the current scenario for carbon storage. Carbon-rich, fertile soil in this region, called *terra preta* (also known as Amazonian Dark Earth), is surrounded by infertile Acrisols, Lixisols, Ferralsols, and Arenosols. The high fertility of *terra preta,* with nearly three times higher Soil Organic Matter (SOM), microbial activity, nutrient holding capacity, and productivity compared to surrounding soil, attracted researchers to decode its formation and composition. Several scientific studies conclude that soil is found to be 70% richer in charred biomass compared to the surrounding soil (Lombardo et al. 2021). Charring of an organic matter concentrates its carbon content (Czimczik et al. 2002) in a fractal porous structure. Biochar (solid product from charring/ pyrolysis) amendment to soil enhances the soil health index.

Soil health improvement with biochar can be a pivotal solution to several of the 17 Sustainable Development Goals targeted by United Nations member states for 2030 (UNDP.org). Healthy soil can lead to zero hunger by enhanced crop productivity. Good health can be attained by healthy food produced by healthy soil. Climate action is taken by using soil as a carbon sink and fertilizing the soil with sustainable and natural fertilizers. Saving biodiversity begins by preventing desertification of soil, etc. Soil health improvement to achieve the targets needs to follow a sustainable approach. Dubey et al. (2019) states that the adaption of sustainable agricultural practices can positively impact the climate.

The burning of crop residue annually in India, China, and Africa contributes heavily to air pollution, but clean pyrolysis of this biomass has the potential to control CO_2 emissions. The replacement of slash and burn practice by slash and char can offset up to 12% of total anthropogenic carbon emission by land-use change, i.e., 0.21 Pg C globally (Lehmann et al. 2006). Yang et al. (2021) states that conversion of 1 t crop residue can sequester nearly 920 kg CO_2-equivalent. Moreover, incorporation of 1 t crop residue biochar into the soil will retain 599.22 kg CO_2-equivalent stable carbon and facilitate additional capture of 18.50 kg CO_2-equivalent by promoting plant growth through assimilation of more atmospheric CO_2 compared to the soil without biochar treatment. Biochar and Energy Technology (BC & E) technology can remove up to 9.2 Gt CO_2 from the atmosphere per year by 2050 (Anderson 2020). Biochar production generates syngas and pyrolytic oil, which can be further used as renewable fuels and make the system self-sustaining in the long run.

Pyrolysis is the process of disintegrating biomass by heat in the absence of oxygen. The biomass is primarily disbanded into three useful coproducts: biochar, syngas (non-condensable gas that contains predominantly carbon and hydrogen), and pyrolytic oil (condensed fraction of syngas). In the process, 50% of the biomass is transformed into recalcitrant carbon that takes several years to mineralize and is not easily degradable by microbes when added to soil. This biochar forms a long-lasting carbon sink in soil, whereas open-air burning and normal biological degradation retain only 3% and <10–20% carbon, respectively, after 10 years (Lehmann et al. 2006).

Biochar is a carbon-negative material (i) as it contains the carbon of parent biomass in a recalcitrant form that would have been released primarily as CH_4 and CO_2 under natural degradation processes, (ii) as the rate of release of CO_2 from the char is less than the rate of drawdown by the parent biomass, and (iii) when added to soil, it also pumps in atmospheric CO_2 in the form of increased plant biomass (Glaser et al. 2009). With pyrolysis, half of the carbon is retained in the char, while volatiles released from the biomass can be captured as syngas and pyrolytic oil, reducing emissions even further (Das et al. 2014). Biochar provides additional benefits to the soil, such as holding water, providing microbial habitat, increasing nutrient retention, decreasing bulk density, increasing aggregate formation, and capturing toxins and heavy metals (Fig. 12.1).

The bio-oil and syngas formed along with the biochar can replace fossil fuels sustainably. One tonne of dry agricultural biomass can become the energy source that would otherwise require two oil barrels and add another 0.86 tons of CO_2 to the atmosphere. BC & E reduces atmospheric CO_2 (1 tonne of dry biomass transformed into biochar removes 0.6 tonne of atmospheric CO_2 permanently) (Anderson 2020).

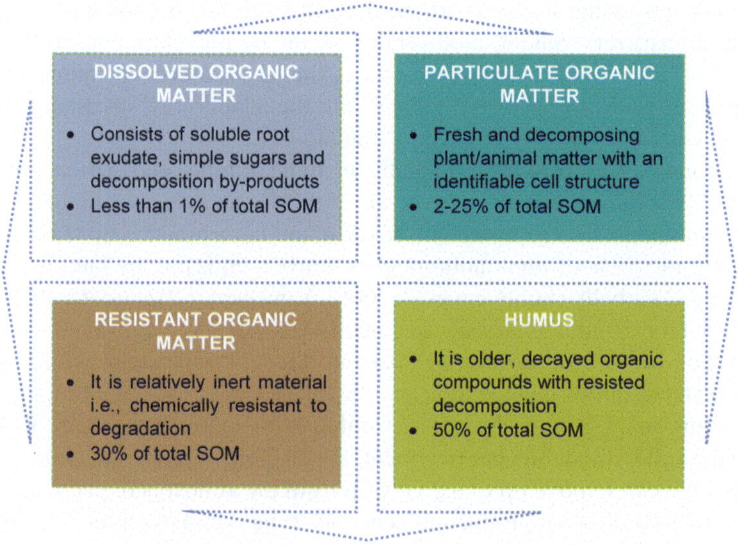

Fig. 12.1 Different fractions of soil organic matter (SOM) in a healthy soil system

12.2 Soil carbon Sequestration and Climate Change Mitigation

12.2.1 Carbon Sequestration in Agroecosystems

The soil can benefit from the incorporation of carbon and minerals through several agricultural practices. Organic amendments such as straw/poultry/farmyard manure incorporation in soil (Zhang et al. 2012) enhance the organic carbon, mineral, and nutrient content of soil but make an insignificant contribution to carbon biomass gain compared to fertilizer treatment.

No-tillage practice is an effective way to retain carbon in the soil. Arai et al. (2014) state that no-tillage with weed mulch management retains more carbon in the soil compared to conventional tillage practice. No-tillage allows more soil aggregate formation, preserving soil carbon and avoiding the damage to soil structure by tillage. Minimum tillage with agricultural residue (MTR) and only minimum tillage (MT) practices retained higher organic carbon compared to the conventional tillage practice (CT) (Ghuman and Sur 2001). However, in some cases, subsoiling (mechanical agitation, turning, digging, etc.) may be necessary to break the hard pan of soil and rejuvenate the bulk density and porosity of the soil. The opportunistic addition of biochar/ straw/ organic matter during tillage can reduce the necessity of future tillage. The further possibility of compaction can also be reduced by biochar addition during tillage as the bulk density will be reduced.

Humus, an organic residue obtained from the decomposition of plant and animal residue, is a complex organic compound less prone to degradation by microorganisms. It can resist degradation by adhering to the minerals with the help of functional groups, thus forming a greasy protective layer around minerals. This reduces the chances of oxidation; thus, an increase in humus content in soil can also contribute to carbon sequestration (Li et al. 2015). The addition of biochar increases the humic content of the soil (Zhang et al. 2017a).

12.2.2 Balancing Carbon Stock in Forest Soils and Peatlands

The incomplete combustion of plants and trees during wildfires generates heterogeneous pyrogenic carbon that has the potential to add soil carbon in recalcitrant form. Due to the wide variety of biomass and combustion conditions, there is a huge variation in the characteristics of this pyrogenic carbon. With low temperature (<250 °C) fires, charring intensity is reduced, and due to the formation of polycyclic aromatic structures, the initial tendency to degradation is reduced, but the resulting pyrogenic carbon remains water soluble and degradable (Santín et al. 2016). Long-term carbon sequestration of pyrogenic carbon from landscape fires is not accounted for in global carbon cycle models (Jones et al. 2019). Nevertheless, crop residue burning does not contribute significantly to pyrogenic carbon; instead, it releases aerosols, GHGs, and pollutants and damages soil microbial biomass.

12.2.2.1 Soil Organic Carbon

Carbon is stored in the terrestrial reserves in the form of SOC (~1530 PgC), soil inorganic carbon [carbonates (~940 PgC), and bicarbonates (~1404 PgC)]. In topsoil, these collectively contribute to the soil health index (Curtis Monger et al. 2015).

SOC is the carbon fraction of soil organic matter (SOM) (composed of Carbon, Hydrogen, Oxygen, and other minerals). SOM is further divided into four types: Dissolved Organic matter, Particulate Organic matter, Humus, and Resistant Organic matter. SOC improves the physical, chemical, and biological characteristics of the soil. SOC regulates several functions within the soil to maintain the nutrient supply for plants, water retention, alleviate soil aggregation, and support the biotic component of soil (Bationo et al. 2007). Thus, higher SOC content enhances the fertility of the soil, increasing the crop yield and yield stability (Pan et al. 2009). Soil inorganic carbon is stored in the form of paedogenic $CaCO_3$ in dryland soils. In humid regions, the dissolution of limestone ($CaCO_3$) releases bicarbonates, which are then transported to groundwater and act as temporary storage for carbon (Monger et al. 2015).

12.2.2.2 Labile and Stable Carbon in Soil

Char consists of two major carbon fractions: Stable–C and labile–C. The labile and stable carbon fractions are nearly 3% and 97%, with a mean residence time of nearly 108 days and 556 years, respectively, with the stable carbon fraction directly contributing to carbon sequestration (Wang et al. 2016). The fractions of the aforementioned forms of carbon are determined through diverse factors, as highlighted in Table 12.1.

Table 12.1 Factors affecting the fraction of labile and stable-C of biochar during pyrolysis (Crombie and Mašek 2015)

Pyrolysis conditions	Labile carbon	Stable carbon
HTT (Highest treatment temperature)	Labile carbon content is heavily dropped upto <0.1 wt.% at high HTT (650 °C)	Stable carbon fraction from 57–65% at 350 °C is raised to 95–100% at higher HTT (650 °C)
Residence time	Independent	Independent
Feedstock type	Significant at lower HTT	Relatively more prominent for higher HTT (650 °C) over lower HTT (350 °C)
Carrier gas flow rate	No significant effect	The yield and stability of stable carbon is increased at a decreased flow rate of carrier gas (N_2) at lower HTT

Labile SOM, which consists of light fraction (LF) and Particulate organic matter (POM), contributes to the short-term carbon and nitrogen cycling in the terrestrial ecosystem (Gulde et al. 2008). The LF is a major source of N for soil and comes mostly from the decomposition of leaves and roots (Ford et al. 1968). The POM is a fraction with relatively low degradability in the form of aggregates (Besnard et al. 1996). Short-lived labile substances (e.g., condensable volatiles) trigger microbial activity as soon as the biochar is applied to the soil because it is the readily available form for microbes (Lehmann et al. 2011; Crombie and Mašek 2015; El-Naggar et al. 2018).

Stable-C is the most significant player as a C-sink for the soil. Carbon must be included in micro aggregates (the building blocks of macroaggregates) to remain stable. Microaggregates, with long-term carbon storage potential despite low carbon storage capacity, and macroaggregates, with relatively higher carbon storage capacity, regulate the soil's carbon storage and stability together (Gulde et al. 2008).

To cite an example of how amending soil with biochar compares with unpyrolyzed organic matter, Ma et al. (2020) state that the application of cotton stalk (ST)/ cotton stack biochar (BC) along with N fertilizer results in SOC 23.7–59.2% in the case of BC and 13.4–36.6% in ST higher than control. The ST treatment enhanced the labile carbon and enzyme activity. At the same time, BC decreased the DOC (Dissolved Organic Carbon): TOC (Total Organic Carbon) EOC (Easily oxidizable organic carbon): TOC ratios with higher soil non-LOC (non-labile organic C): TOC ratio. BC also attained higher microbial biomass carbon (MBC) compared to the ST. A meta-analysis (Liu et al. 2016) also shows that biochar addition to soil increases the MBC content by 18%. The microbial population increases in the beginning in biochar-amended soil due to the presence of labile carbon, but due to the presence of aromatic carbon in aged biochar, wide diversity in microbial species is observed with time (Li et al. 2020). A meta-analysis by Dai et al. (2020) shows that most of the biochar analyzed in the study is alkaline ($pH > 7$) in nature and aids soil productivity. The soil MBC and SOC content was observed to be highest for the biochar with pH 8.1–9 (Liu et al. 2016).

Application of cotton stalk and biochar increases the soil base respiration, β-glucosidase activity, and N-acetyl-glucosaminidase activity in the soil that is proportional to EOC and MBS content. With the higher activity of β-glucosidase and N-acetyl-glucosaminidase on the application of cotton stalks than biochar, the degradation rate of soil organic matter is higher in cotton stalk treatment (Ma et al. 2020). Similarly, low soil respiration was observed in biochar-treated soil compared to the biomass-treated soil during 237 days of incubation (Hernandez-Soriano et al. 2016). As carbon amendment to the soil in the form of compost, biosolids, or any organic waste may degrade briskly and release CO_2, biochar is a preferred option as its aromaticity resists biological degradation. Hence, some of the carbon will be stored for several thousand years when added to soil in the form of biochar (El-Naggar et al. 2015; Mohan et al. 2018). This chapter ponders upon how the inclusion of biochar technology in agricultural practices (from residue management to soil additive) can be one of the meaningful approaches to head towards SDGs. It also provides insights into how biochar application can sequester carbon and GHG emissions.

12.3 Biochar Technology: Fostering Carbon Capture in Agricultural Soil

12.3.1 Effect of Biochar on Soil quality Indicators

With a wide range of physicochemical properties, biochar has multidimensional impacts on soil properties. The direct impact of Biochar amendment has been observed to be fruitful for soil in terms of a rise in yield and biomass gain, i.e., directly proportional to the soil health index. The liming property of biochar increases the pH of highly weathered soil (Jien and Wang 2013). The biochar ash consists of alkaline earth metals that dissolve in water, and the adsorption of active protons by oxygen-containing functional groups of biochar neutralizes the soil acidity (Haghighatjou 2020). The feedstock also affects the liming potential of biochar; e.g., the biochar prepared from leguminous biomass residue has higher alkalinity than non-leguminous as the legume plant biomass accumulates more alkaline minerals during the growth phase (Yuan and Xu 2011).

The biochar amendment increases the fertility of saline-alkaline soils by adjusting the pH, SAR (sodium adsorption ratio), SOC, water holding capacity (WHC), and hydraulic conductivity of soil (K_s) (Amini et al. 2016; Edrisi et al. 2021). The electric conductivity (EC) for soil is adjusted in saline soils, and sodic soils benefit from increased total and aggregate associated SOC, WHC, microbial biomass carbon, and K_s when treated with biochar (Fu et al. 2019; Edrisi et al. 2021). Increased soil microbial biomass is positively correlated with the soil pH (a result of biochar addition). The highly porous structure of biochar contains macro and micropores, i.e., suitable habitat for the microorganisms (Jien and Wang 2013). During pyrolysis, the carboxylic group in the biochar is increased, responsible for increased CEC (Cation exchange capacity) (Chen et al. 2011; Jien and Wang 2013). The low-density biochar particles, when added to soil, cause physical dilution; hence, the bulk density of soil is reduced; for sandy soil, the bulk density was 9% less than the control during the column study (Basso et al. 2013). The porous structure of biochar added improves the hydraulic conductivity of the soil (Jien and Wang 2013).

The addition of biochar increases the soil aggregate stability due to the presence of hydroxyl and carboxyl groups, which help bind the clay and soil particles (Wuddivira et al. 2009; Jien and Wang 2013). Water holding capacity is also increased for sandy (Basso et al. 2013) and loamy sand soil (Yu et al. 2013). Further, biochar amendment in agricultural soil may also improve the microbe's related properties. It includes a variety of characteristic processes, viz. soil enzyme activities, soil respiration, soil C:N ratio, and microbial biomass (Dubey et al. 2020a, b). Agricultural plant growth, GHG emission, yield, and quality significantly depend on these properties (Dubey et al. 2016). Thus, biochar amendment to soil is suitable in terms of climate-smart agricultural productivity along with management against soil erosion.

12.3.2 Fate of Biochar in Sustaining Plant productivity and Yield

The addition of biochar increases above- and below-ground carbon sequestration in the form of enhanced biomass growth. Soil properties are significantly improved by biochar addition, as discussed in the previous section, thus facilitating the biomass gain (Table 12.2). A meta-analysis by Biederman and Harpole (2013) shows that the addition of biochar increased soil productivity. The dark color of biochar aids germination by amending the soil's thermal dynamics compared to the control (treatment without any biochar). The water holding capacity, soil nutrient condition, and bulk density altered by the presence of biochar support the biomass gain. The liming property of biochar increases the bioavailability of P, which is otherwise bound to Al or iron oxide particles. The exchangeable Al (a prime cause of soil acidity) is converted to Al-OH (as precipitate) due to increased pH. Hence, phosphate availability to plants increases (Lee et al. 2010; Yuan and Xu 2011). The phytotoxic elements like Al and Cd are also reduced due to increased soil pH.

As the bulk density of soil with biochar is less, the uprooting of groundnut plants for harvest is easier with less pod loss than control. Moreover, more pods per plant were observed during harvest (https://leisaindia.org/biochar-for-healthy-soils/). Dubey et al. (2020a, b) observed a linear relationship between sustainable yield index (SYI) and major soil microbial enzyme (dehydrogenase), several indicators of carbon in soil (total organic carbon, microbial biomass carbon, and SOC), and critical nutrients for plant growth (NPK) and SYI of biochar was found to be higher than

Table 12.2 Alteration of soil properties with biochar amendments and their benefits on soil and plant nutrient dynamics

Increase (↑) or decrease (↓) in soil property due to biochar amendments		Benefits for soil/plant nutrient dynamics
Bulk Density	↓	Beneficial for available water and air capacity in soil
Porosity	↑	Useful in root penetration and water movements
Water Holding Capacity	↑	Increases irrigation use efficiency
Soil Aggregate Stability	↑	Protection against soil erosion
GHGs Emissions	↓	Reduced CH_4, N_2O, CO_2 emissions form soil
Hydraulic Conductivity (K_{sat})	↑	Increase the permeability of soil
pH	↑	Liming has positive effect on soil biota and nutrient availability
Electrical Conductivity	↑	Regulates the microbial activity in soil
Nutrient Holding Capacity	↑	Increases the nutrient use efficiency of plant and prevent nutrient leaching
Cation Exchange Capacity	↑	Increases nutrient availability for plants
Soil Organic carbon (SOC)	↑	Improve physical, chemical, and biological properties of soil
Microbial Biomass	↑	Aid the decomposition of dead plants/leaves and increases soil fertility

vermicompost and control. The addition of biochar has a positive impact on the above-stated parameters, which, in turn, increases primary productivity (Jeffery et al. 2017; Payen et al. 2020). With enhanced CEC, soil treated with biochar retains nutrients in the rhizosphere and exhibits increased nutrient use efficiency (Prapagdee and Tawinteung 2017). Liu et al. (2020) state that a net increase in root volume, length, diameter, and number of roots, along with the yield and potassium uptake, is observed in biochar-treated soil with maize plantation. In a study by Agegnehu et al. (2016), the biochar treatment increased the grain yield, total biomass yield, harvest index, average chlorophyll content, leaf water content, and reduced nutrient leaching. Biochar amendment in soil also reduces the Phyto availability of organic contaminants (Hurtado et al. 2017) and heavy metal content in the plants (Chen et al. 2018) and makes crops safe for human consumption (Nie et al. 2018).

The addition of biochar can reduce overall fertilizer consumption in the soil. The per annum requirement of N, P, and K fertilizer can be reduced by 0.1 kg, 0.02 kg, and 0.01 kg, respectively, by a one-time application of 337.83 kg biochar obtained from 1 t biomass (Yang et al. 2021). The effect of biochar addition to constructed wetlands (CW) is shown in a study by Ji et al. (2021) where compared to the control, the biochar-treated CWs had abundant *mcr*A and *nos*Z genes and higher ratios of *nos*Z/(*nir*K + *nir*S) and *pmo*A/*mcr*A that resulted in reduced CH_4 and N_2O emissions that can reduce global warming potential by 55.8%. The increased activity of nitrifiers facilitated Total Nitrogen removal by 41.3%.

12.3.3 Role of Biochar as an Efficient Plant Nutrient

Several research studies have been conducted in the past few decades, which showed the positive impact of biochar in terms of essential nutrient availability to agricultural crops. Van Zwieten et al. (2010) showed that biochar application significantly improved exchangeable Ca^{+2} and total carbon content in Ferrosol soil. Similarly, they also showed significant improvement in exchangeable K^+ and carbon content along with other essential nutrients in *Medicago sativa* L. under the influence of Calcarosol soil type (Zhang et al. 2019). Biochar and compost mixed formulations also gaining wide attention. During the composting process, the generation of oxygen-containing functional groups is mainly responsible for improved plant growth and soil health, which eventually leads to improved nutrient-holding capacity (Schulz and Glaser 2012). Biochar is also able to improve the CEC of agricultural soil through its own oxidation process under the presence of water and oxygen. It helps in binding the essential cationic nutrients over the surface of clay particles, humus, and biochar itself (Laird et al. 2010). However, the extent of variation in the ion exchange capacity of biochar depends on different things, including raw material used and pyrolysis conditions (Cheng et al. 2014). Thus, the application of biochar to soil can significantly influence fertilizer-use efficiency in terms of lesser dependency on multiple fertilizer application cycles.

12.4 Biochar Preparation and Their Mechanistic Interventions

12.4.1 Pyrolysis and Its Type

Based on the heat rate, there are broadly two types of pyrolysis: *Fast* pyrolysis and *Slow* pyrolysis. In *fast* pyrolysis, the biomass is heated at the rate of 1000–10,000 °C s^{-1}, and pyrolytic oil is the significant fraction produced, while in *slow* pyrolysis, the biomass is heated at the rate of 1–20 °C min^{-1} and syngas and biochar are the major yields (Brewer et al. 2009; Basu 2010). Thus, the heat rate is a system design parameter selected based on the final products required (Table 12.3).

12.4.2 Factors Affecting Carbon Sequestration Potential of Biochar

The carbon sequestration potential of all the biochar cannot be uniform. It depends on several factors ranging from feedstock used to prepare the biochar to the structural orientation of the biochar. The biochar's carbon content is primarily affected by the *feedstock*; e.g., the wood biochar has higher carbon content compared to the crop residues. The carbon stability differs for the different feedstock and biomass sizes chosen. The higher lignin content of wood feedstock gives a higher stable carbon fraction to the biochar (Leng and Huang 2018).

The carbon sequestration potential is also dependent on the *soil type*. Soils with lower nutrient, organic matter, and clay content, like sandy soils, have low microbial activity; thus, there is a lower tendency of CO_2 efflux from biochar addition to sandy soil compared to sandy, loamy soil with conditions conducive for soil microbial activity (El-Naggar et al. 2018).

Table 12.3 Common ways exploited for biochar preparation (You et al. 2020; Yoganandham et al. 2020; Sharma et al. 2021)

Process (Temperature range)	Operation condition	Products	Major product
Pyrolysis (300–1200 °C)	The inert atmosphere in the absence of oxygen	Biochar; syngas; pyrolytic oil	Biochar (in slow pyrolysis); bio-oil (in fast pyrolysis)
Hydrothermal carbonization (120–300 °C)	High pressure (14–22 MPa) in the presence of water	Hydrochar; nutrient-rich liquid; gaseous by-product predominantly CO_2	Hydrochar
Gasification (>500–700 °C)	Partially oxygenated condition to burn the biomass	Biochar (<20 wt.%); combustible gases (CO, H_2, CH_4)	Combustible gases (CO, H_2, CH_4)
Torrefaction (200–300 °C)	Low-temperature exposure for a short duration	Biochar; combustible gases (CO, H_2, CH_4); pyrolytic oil	Biochar

The *composition* of biochar significantly affects its carbon sequestration potential. Nan et al. (2019) states that the inherent minerals (e.g., Na, K, Ca, Mg, etc.) present in biochar interact with its organic component and affect its carbon sequestration potential. Demineralization of biochar removed the inherent minerals and oxygen-containing functional groups that resulted in 3.5–30.1% more carbon retention in biochar that was otherwise released as low molecular weight species like CO_2, CH_4, C_3H_6O, etc. Demineralization also enhanced the carbon stability in the biochar and made it less prone to decay (via Thermal oxidation or chemical decomposition). The alkaline nature of biochar and the presence of Ca and Mg aid the precipitation of carbonate on the surface of biochar (Lehmann et al. 2011). The biochar has a very porous structure with a high BET (Brunauer–Emmett–Teller) surface area that increases the uptake of nutrients and precipitates.

The decomposition of biochar in soil is *time-dependent*. The amount of biochar decomposition decreases exponentially, resulting in high CO_2 emissions during the early phase of biochar application to soil due to a burst of microbial activity. The decomposition rate of biochar decreases with the progress of time (El-Naggar et al. 2015; Wang et al. 2016).

Pyrolysis temperature affects the decomposition of biochar in soil. At higher pyrolysis temperatures, the aromatic carbon formation makes it recalcitrant (Wang et al. 2016). The carbon is concentrated during the thermal decomposition, which leads to the formation of aromatic-C bonds with the release of H and O (Kim et al. 2012), due to which the H:C, O:C ratios, and volatile matter decrease linearly with an increase in the pyrolysis temperature (Ferreira et al. 2021). Due to the loss of O-alkyl and di-O alkyl moieties present in wood, chars produced at temperatures >340 °C have low alkyl and high aromatic intensity (Czimczik et al. 2002).

The *age of biochar* also affects the release of CO_2 from the soil. Zhao et al. (2015) reported that soil treated with 10-year-old aged biochar did not release CO_2 during the 42-day incubation, while the soil treated with fresh and 4-month-old released CO_2. Fresh biochar increases the soil microbial biomass and soil pH and is responsible for carbon mineralization. A decreasing trend of carbon content in biochar is also observed with aging; 10-year-old biochar, 4-month-old biochar, and fresh biochar have 60.51%, 65.54%, and 66.66%, respectively.

A study by (Li et al. 2015) shows that the incorporation of biochar increases humic acids (HAs), fulvic acids (FAs), and humins (HMs) in the soil increased by 13.6%, 20.8%, and 22.6%, respectively, these contents vary for different soils. As the biochar addition increases the humus content of soil (Quan et al. 2020), i.e., it is more resistant to degradation; thus, the carbon sequestration potential of soil is improved.

A study specific to Helsinki, Finland, states that biochar in the soil of a tree plantation has the potential to sequester 330,000 t CO_2 for 50 years using i-Tree as an LCA (Life cycle assessment) tool. The biochar-enhanced growth of trees contributed 35% of the added carbon sequestration, while the biochar treatment itself carried a share of 65% (Ariluoma et al. 2021).

Fig. 12.2 Biochar preparation and field level experimental validation of biochar input in agricultural field replacing chemical fertilizers in north India. (Field photo credit: Pradeep Kumar Dubey)

12.4.3 Agro-Residue Utilization as Biochar Substrate

Different modes of pyrolysis and biochar preparation are followed for a variety of crop-based agricultural residues and are further used to enhance soil quality as well as plant health. For instance, rice husk-based biochar prepared through pyrolysis was used at the field level in eastern Uttar Pradesh and validated for its implication in nutrient uptake and balance over a period of 3 years duration, as shown in Fig. 12.2 (Dubey et al. 2022). An indicative list of examples of various crop residues reported to be sustainably utilized as biochar in agriscape is shown in Table 12.4.

12.5 Upscaling Biochar Production for Environmental Sustainability

Several studies reveal the benefits of using pyrolysis products (Biochar, syngas, and pyrolytic oil) as sustainable replacements for non-renewable fuels. Several industries and organizations are coming up with bulk manufacturing of quality pyrolytic products. The list of products manufactured by these industries are biochar-based manures, torrefied biomass as cleaner renewable fuel replacing non-renewable fuels, biochar-based carbon products, batteries, lubricants, etc. A few such organizations are listed in Table 12.5.

Table 12.4 Indicative examples of sustainable utilization of crop-based agro-residue as biochar for soil and plant health reclamation

Agro-residue substrate	Method of biochar preparation	Paybacks as carbon sequestered and improved plant nutrition
1. Rice straw and husk	Pyrolyzed at 450 °C to prepare biourea showing >90% sorption, significant desorption rate, and a predicted half-life in the range of 500–750 years.	The Biourea fabricates novel N fertilizer and acts in slow-release kinetics. Nitrogen use efficiency and crop yield are enhanced. The long half-life of biochar ensures remarkable carbon sequestration in soil (Singh et al. 2020).
2. White lead tree waste wood	Pyrolyzed at 700 °C	The soil pH (3.9–5.1), cation exchange capacity (7.41–10.8 cmol (+) kg^{-1})), base cation % (6.4–26%), microbial biomass carbon (835–1262 mg kg^{-1}), K_{sat} (1.8 times), mean weight diameter of soil aggregates (2.6–4.0 cm) is all enhanced. In contrast, bulk density (1.4–1.1 Mg m^{-3}) is reduced as well, as soil loss is reduced by 50% and 64% on biochar application at rates of 2.5% and 5%, respectively (Jien and Wang 2013).
3. Walnut shell; corn cob and straw; rice straw	Twelve varieties of biochar were prepared through pyrolysis for 4 h at 250 °C, 400 °C, and 600 °C.	At high pyrolysis temperatures, the available phosphorous in biochar prepared from corn and rice straw increases (from 2.14 to 28.35 mg kg^{-1}); available nitrogen reduces (from 21.76 to 14.96 mg kg^{-1}), and the ryegrass biomass showed maximum increase (i.e., from 0.07 for o 0.16 g per pot) at 400 °C pyrolysis temperature (Zhang et al. 2017b).
4. Sugarcane bagasse	Pyrolyzed at 600 °C for 4 h.	Biochar incubated in soil for 30 days at 80% water holding capacity. Biochar-amended soils had higher pH values (~2.5 pH units), electrical conductivity (~2.66 times), and cation exchange capacities (~42%) relative to the unamended acidic soil. Biochar application increased the affinity of soil to retain Zn ions (Haghighatjou 2020).
5. Wheat straw	Pyrolyzed at 500 °C @ 5 °C min^{-1} and held constant for 8 h.	The co-application of biofertilizer* and biochar suppresses the NH3 volatilization by 12.3%, and GHG intensity (GHGI) is mitigated by 15.2%. Biofertilizer and /or biochar increase rice grain yield by 16.5–38.3%. *The primary component of biofertilizer was straw biomass compost with a moisture content of 10.0% containing 12.3% total carbon and 0.74% total N, with a pH of 6.5 and incubated with *Bacillus Subtili*, *Bacillus licheniformi*, *Bacillus laterosporu*, and lactic acid bacteria (Sun et al. 2021).

(continued)

Table 12.4 (continued)

Agro-residue substrate	Method of biochar preparation	Paybacks as carbon sequestered and improved plant nutrition
6. Wheat straw	Pyrolyzed at 350–550 °C in a vertical kiln made of refractory bricks	With Biochar soil amendment @ 20 t ha^{-1} and 40 t ha^{-1}, seasonal total N_2O emission was significantly and greatly decreased by 45.1% and 39.5% of the rice growing season, and by 37.6% and 41.2% of the winter wheat growing season, respectively over the control (Zhang et al. 2013). Net GHGI reduced by 10–20%.
7. *Parthenium hysterophorus*	Pyrolyzed at temperatures varying between 370 and 417 °C	A 1:1 application of commercial fertilizer and biochar was performed at par with only commercial fertilizer treatment in terms of grain yield, 1000 grain weight, harvest index, etc., in wheat and rice crops. Biochar is a suitable soil conditioner that can reduce fertilizer requirements. Biochar is not the only preferred treatment (Qurat-ul-Aina, Firdaus-e-Bareena 2016).
8. Rice/wheat straw	Pyrolyzed at 500 °C @ 5 °C min^{-1} for 8 h	For biochar doses of 124 and 270 t ha^{-1}, the Zn concentrations were reduced to 36.6 and 37.5 mg kg-1, i.e., below the admissible concentration of 45 mg kg^{-1}. The enhanced SOC and pH of biochar-treated soil reduced the phyto-availability of Zinc by increased (Wu et al. 2018).
9. Organic waste substrates (mushroom production)	Slow pyrolysis at 400 °C for 4 h	Biochar amendment in soil aids carbon sequestration by: Reducing labile pool I (i.e., weak acid hydrolyzable) of native SOC by 11.2–47.7% (compared to the control treatment) Labile pool II (i.e., strong acid hydrolyzable) and recalcitrant pool (acid non-hydrolyzable) remained unaffected (Dong et al. 2018).
10. Wheat straw	Pyrolyzed at 500 °C	CH_4 and N_2O emissions were significantly reduced by 9.3% and 18.7%, respectively, and the yield for the rice-wheat rotation system was increased by 9.0% (Wu et al. 2019). The carbon stocks were increased by 10.3%, and GHG emissions were reduced by 10.4% for Chinese rice and wheat production over the duration of the 6-year study.
11. Sugarcane straw	Pyrolyzed at 450 °C @ 10 °C min^{-1} for 2 h	With biochar amendment, cumulative N_2O emissions were reduced by 71% compared to the fertilized, no-biochar soil, available P was increased by up to 30%, and there were eight-fold higher exchangeable K^+ concentrations. At a 1.9% biochar application rate, the grain yield and shoot biomass increased by 27 and 16%, respectively (Abbruzzini and Davies, CA 2019).

(continued)

Table 12.4 (continued)

Agro-residue substrate	Method of biochar preparation	Paybacks as carbon sequestered and improved plant nutrition
12. Wheat straw	Pyrolyzed at 500 °C	Biochar application reduced the methanogens to methanotrophs ratio by 11–31%, hence reducing methane emissions. Annual methane emissions are reduced by 20–51% (Wang et al. 2019).
13. Rice, maize, and wheat straw; rice husk; bamboo	Pyrolyzed at 550 °C for 3 h	Three crops (rice- rape- maize) at consecutive seasons treated with RB, MB, WB, HB, and BB showed the following physical transformations in the soil: Increased macroaggregates >5 mm diameter and aggregate stability. Increased soil water retention capacity at the metric potential of −0.033 and − 1.5 MPa, thus increasing the water holding capacity of the soil. Increase the plasticity index of soil, hence extending the soil cultivable water. Reduce tensile strength and cohesion of soil. The internal friction angle of the soil is unaffected. Yield increased for rice (first crop) by 6.3 ~ 13.3% by all biochars. Rape (second crop) yield was increased by only RH and HB. RB, WB, and HB increased the yield of maize (the third crop) (Yang and SG 2021).
14. Wheat straw	Pyrolyzed at 400 °C	Effect of biochar application on plants relative to nitrogen fertilizer: Roots of rice and wheat increased by 3–19% Straw increased by 10–19% Grain biomasses increased by 10–16% NUE increased by 20–53% PUE increased by 38–230% Effect of biochar application on soil properties relative to nitrogen fertilizer: Enhance SOC by 26–53% Increase Total N by 14–16% Increase Total P by 6–19% Increase 250–2000 mm aggregate class by 87–93% Increase aggregate stability by 43–48% (Zhang et al. 2020).
15. Poultry-litter	Pyrolyzed between 450 and 500 °C	Co-application of Azolla and biochar: Increased rice grain yield by 27.3–75.0%, Decreased both yield equivalent CH_4 emissions by 24.7–25.0% and N_2O emissions by 81.8–97.7% (Kimani et al. 2020).

Table 12.5 Indicative lists of global industries or organizations manufacturing biochar to ascertain environmental sustainability through specifically set targets

Industry/organization (website): specific target	Operation/strategy
Ecoera, Sweden (www.Ecoera.se): For soil carbon sequestration	Palletized biomass with a specific mineral composition is pyrolyzed. Gas generated is used for district heating. Biochar is added to the soil. Several field trials carried out between 2009 and 2013 show an increase in crop yield.
Bio green Woods, Ramada, Portugal (https://biogreenwoods.eu/en/): To ensure zero waste system	Forest surplus is processed by slow pyrolysis Prepare biochar, vegetable oil, and gases. Smoke-free coal is one of the products for clean burning.
Tanmiah Food Company, Riyadh (https://tanmiah.com/sustainability/): For soil conservation	Produce and supply meat and poultry products. Convert the poultry excreta into eco-char and add it to the soil to enhance soil fertility.
EnviraPAC, Monticello, Arkansas (https://envira-pac.com): Biomass-based carbon material	Prepare renewable carbon powder from the biomass and reduce carbon footprints.
British Columbia Biocarbon, Canada (https://www.bcbiocarbon.com): Produce alternative to fossil fuels	Pyrolysis and gasification products are used as low-emission alternatives to fossil fuels. Biochar, bio coal, wood vinegar, etc., are produced.
Applied Biomass Solutions, Houston, Texas (https://applied-biomass.com): Large-scale biochar supply	Produce and supply biochar on a large scale for mining, the agriculture industry, etc.
Carbofex Ltd., Tampere, Finland (https://www.carbofex.fi/?view=Home): Renewable energy	Produce biochar and clean energy.
Aries Clean Energy, Franklin, Tennessee (https://ariescleanenergy.com): Provide clean energy	Product from gasification of waste is used to generate electricity.
Circle Carbon S.L., Balears, España (https://circlecarbon.com): Soil regeneration, carbon capture and storage, and food security	Biomass waste residue is used to prepare biochar. Biochar is used as the soil additive.

12.6 Conclusions and Future Recommendations

The process of pyrolysis enhances the aromatic fraction of biomass. Therefore, when added to soil, the presence of a higher fraction of stable carbon makes it a long-lasting carbon sink despite increase in soil respiration due to the increased microbial activity in the soil. The carbon is also sequestered in the form of increased biomass gain due to conducive soil properties like reduced bulk density, nutrient retention, increased SOC, and microbial activity, among others. Further, a reduction in GHG emissions is also observed. The pyrolyzed agricultural biomass as soil additive holds immense potential to reintroduce minerals to the soil along with carbon sequestration, unlike the open-air burning that may reintroduce the minerals to the soil but is heavily taxing on the soil-friendly microbes, inherently available nutrients and the ambient air quality.

The incorporation of biochar can be a potential solution for achieving sustainable development goals. Owing to its potential to increase soil productivity, sequester carbon, and reduce GHG emissions, biochar technology can significantly help in meeting SDG goal number 2 (zero hunger) and 13 (Climate Action). Several indirect benefits of the use of biochar in soil remediation and water treatment can be one of the several stepping-stones towards sustainable development. Pyrolytic products other than biochar, i.e., syngas and pyrolytic oil, can be a source of renewable energy (affordable and clean energy). A win-win situation will be the pyrolysis of crop residue, which will increase the carbon content of the soil with biochar treatment, produce renewable fuels, and sustain the management of agricultural residue. The pyrolysis industry may also create more employment opportunities for people. With constant progress in biochar research since the inception of the twenty-first century and the ongoing climate crisis, quicker actions need to be taken to implement biochar-based soil amendments. More extensive Life Cycle Assessment studies are required to quantify the benefits of biochar in terms of carbon footprints, which can further navigate the avenues for voluntary carbon markets and agricultural sustainability.

Acknowledgments Authors gratefully acknowledge Mr. Robert Gilett, ex-Chief Editor (International Biochar Initiative), for his kind assistance during the compilation of this chapter. PKD is thankful to Prof. PK Mishra, Dept. of Chemical Engineering, IIT-BHU, Varanasi for providing rice-husk biochar for doing field experiments in selected sites in eastern Uttar Pradesh, India.

References

Abbruzzini T, Davies CA, Toledo F (2019) Dynamic biochar effects on nitrogen use efficiency, crop yield and soil nitrous oxide emissions during a tropical wheat-growing season. J Environ Manage 252:109638

Agegnehu G, Bass AM, Nelson PN, Bird MI (2016) Benefits of biochar, compost and biochar-compost for soil quality, maize yield and greenhouse gas emissions in a tropical agricultural soil. Sci Total Environ 543:295–306. https://doi.org/10.1016/j.scitotenv.2015.11.054

Amini S, Ghadiri H, Chen C, Marschner P (2016) Salt-affected soils, reclamation, carbon dynamics, and biochar: a review. Springer. https://doi.org/10.1007/s11368-015-1293-1

Anderson PS (2020) Climate intervention with biochar a white paper about biochar and energy (BC&E) for carbon dioxide removal (CDR) and emission reduction (ER) first edition dated 2020-12-07 (with minimal edits 2020-12-16)

Arai M, Minamiya Y, Tsuzura H et al (2014) Changes in water stable aggregate and soil carbon accumulation in a no-tillage with weed mulch management site after conversion from conventional management practices. Geoderma 221–222:50–60. https://doi.org/10.1016/j.geoderma.2014.01.022

Ariluoma M, Ottelin J, Hautamäki R et al (2021) Carbon sequestration and storage potential of urban green in residential yards: a case study from Helsinki. Urban For Urban Green 57:126939. https://doi.org/10.1016/j.ufug.2020.126939

Basso AS, Miguez FE, Laird DA et al (2013) Assessing potential of biochar for increasing water-holding capacity of sandy soils. GCB Bioenergy 5:132–143. https://doi.org/10.1111/gcbb.12026

Basu P (2010) Biomass gasification and pyrolysis: practical design and theory

Bationo A, Kihara J, Vanlauwe B et al (2007) Soil organic carbon dynamics, functions and management in West African agro-ecosystems. Agric Syst 94:13–25. https://doi.org/10.1016/j.agsy.2005.08.011

Besnard E, Chenu C, Balesdent J et al (1996) Fate of particulate organic matter in soil aggregates during cultivation. Eur J Soil Sci 47:495–503. https://doi.org/10.1111/j.1365-2389.1996.tb01849.x

Biederman LA, Stanley Harpole W (2013) Biochar and its effects on plant productivity and nutrient cycling: a meta-analysis. GCB Bioenergy 5:202–214. https://doi.org/10.1111/gcbb.12037

Bier H (2020) EBI Whitepaper Biochar-based carbon sinks to mitigate climate change

Biochar for healthy soils – LEISA INDIA. https://leisaindia.org/biochar-for-healthy-soils/. Accessed 24 Sept 2018

Brewer CE, Schmidt-Rohr K, Satrio JA, Brown RC (2009) Characterization of biochar from fast pyrolysis and gasification systems. Environ Prog Sustain Energy 28:386–396. https://doi.org/10.1002/ep.10378

Brown K (2021) 2020 tied for warmest year on record, NASA analysis shows

Burak Bir (2020) Wildfires, forest fires around world in 2020. https://www.aa.com.tr/en/environment/wildfires-forest-fires-around-world-in-2020/2088198. Accessed 30 Jan 2021

Chen HX, Du ZL, Guo W, Zhang QZ (2011) Effects of biochar amendment on cropland soil bulk density, cation exchange capacity, and particulate organic matter content in the North China Plain. Chinese J Appl Ecol 22:2930–2934

Chen D, Liu X, Bian R et al (2018) Effects of biochar on availability and plant uptake of heavy metals – a meta-analysis. J Environ Manag 222:76–85. https://doi.org/10.1016/j.jenvman.2018.05.004

Cheng CH, Lin TP, Lehmann J, Fang LJ, Yang YW, Menyailo OV, Chang KH, Lai JS (2014) Sorption properties for black carbon (wood char) after long term exposure in soils. Org Geochem 70:53–61

Crombie K, Mašek O (2015) Pyrolysis biochar systems, balance between bioenergy and carbon sequestration. GCB Bioenergy 7:349–361. https://doi.org/10.1111/gcbb.12137

Curtis Monger H, Kraimer RA, Khresat S et al (2015) Sequestration of inorganic carbon in soil and groundwater. Geology 43:375–378. https://doi.org/10.1130/G36449.1

Czimczik CI, Preston CM, Schmidt MWI et al (2002) Effects of charring on mass, organic carbon, and stable carbon isotope composition of wood. Org Geochem 33:1207–1223. https://doi.org/10.1016/S0146-6380(02)00137-7

Dai Y, Zheng H, Jiang Z, Xing B (2020) Combined effects of biochar properties and soil conditions on plant growth: a meta-analysis. Sci Total Environ 713:136635. https://doi.org/10.1016/j.scitotenv.2020.136635

Das SK, Avasthe RK, Singh R, Babu S (2014) Biochar as carbon negative in carbon credit under changing climate. academia.edu

Dong X, Singh B, Li G et al (2018) Biochar application constrained native soil organic carbon accumulation from wheat residue inputs in a long-term wheat-maize cropping system. Agric Ecosyst Environ:200–207

Dubey PK, Singh GS, Abhilash PC (2016) Agriculture in a changing climate. J Clean Prod 113:1046–1047

Dubey RK, Dubey PK, Abhilash PC (2019) Sustainable soil amendments for improving the soil quality, yield and nutrient content of Brassica juncea (L.) grown in different agroecological zones of eastern Uttar Pradesh, India. Soil Tillage Res 195:104418.

Dubey P, Singh G, Abhilash P (2020a) Adaptive agricultural practices: building resilience in a changing climate. Springer Nature Switzerland. Springer Brief in Environmental Science

Dubey RK, Dubey PK, Chaurasia R et al (2020b) Sustainable agronomic practices for enhancing the soil quality and yield of Cicer arietinum L. under diverse agroecosystems. J Environ Manag 262:110284. https://doi.org/10.1016/j.jenvman.2020.110284

Dubey PK, Singh A, Chaurasia R et al (2022) Animal manures and plant residue-based amendments for sustainable rice-wheat production and soil fertility improvement in eastern Uttar Pradesh, North India. Ecol Eng 177:106551. https://doi.org/10.1016/J.ECOLENG.2022.106551

Edrisi SA, Tripathi V, Chaturvedi RK, Dubey DK, Patel G, Abhilash PC (2021) Saline soil reclamation index as an efficient tool for assessing restoration progress of saline land. Land Degrad Dev 32(1):123–138

El-Naggar AH, Usman ARA, Al-Omran A et al (2015) Carbon mineralization and nutrient availability in calcareous sandy soils amended with woody waste biochar. Chemosphere 138:67–73. https://doi.org/10.1016/j.chemosphere.2015.05.052

El-Naggar A, Lee SS, Awad YM et al (2018) Influence of soil properties and feedstocks on biochar potential for carbon mineralization and improvement of infertile soils. Geoderma 332:100–108. https://doi.org/10.1016/j.geoderma.2018.06.017

EPA (2020) Global greenhouse gas emissions data | Greenhouse Gas (GHG) Emissions | US EPA. https://www.epa.gov/ghgemissions/global-greenhouse-gas-emissions-data. Accessed 14 Jan 2021

Ferreira A, Ribau J, Bioenergy MC-B (2021) A decision support method for biochars characterization from carbonization of grape pomace. Elsevier

Ford G, Trans DG-ISSS, 1968 U (1968) Dynamics of partly humified organic matter in some arable soils. agris.fao.org

Fu Q, Zhao H, Li TX et al (2019) Effects of biochar addition on soil hydraulic properties before and after freezing-thawing. Catena 176:112–124. https://doi.org/10.1016/j.catena.2019.01.008

Ghuman BS, Sur HS (2001) Tillage and residue management effects on soil properties and yields of rainfed maize and wheat in a subhumid subtropical climate. Soil Tillage Res 58:1–10. https://doi.org/10.1016/S0167-1987(00)00147-1

Glaser B, Parr M, Braun C, Kopolo G (2009) Biochar is carbon negative. Nat Geosci 2:2

Global Fires | Future Earth. https://futureearth.org/publications/issue-briefs-2/global-fires/. Accessed 30 Jan 2021

Godbout S, Verma M, Larouche JP et al (2010) Methane production potential (B_0) of swine and cattle manures – a Canadian perspective. Environ Technol 31:1371–1379. https://doi.org/10.1080/09593331003743096

Greenpeace (2020) Burning up: health impact of Indonesia's forest fires and implications for the Covid-19 pandemic – Greenpeace Southeast Asia. https://www.greenpeace.org/southeastasia/publication/43859/burning-up-health-impact-of-indonesias-forest-fires-and-implications-for-the-covid-19-pandemic/. Accessed 30 Jan 2021

Gulde S, Chung H, Amelung W et al (2008) Soil carbon saturation controls labile and stable carbon pool dynamics. Soil Sci Soc Am J 72:605–612. https://doi.org/10.2136/sssaj2007.0251

Haghighatjou MS (2020) Sugarcane bagasse biochar: preparation, characterization, and its effects on soil properties and zinc sorption-desorption. Commun Soil Sci Plant Anal

Hernandez-Soriano MC, Kerré B, Kopittke PM et al (2016) Biochar affects carbon composition and stability in soil: a combined spectroscopy-microscopy study. Sci Rep 6:1–13. https://doi.org/10.1038/srep25127

Herzog HJ (2001) What future for carbon capture and sequestration? Environ Sci, Technol, p 35

Hurtado C, Cañameras N, Domínguez C et al (2017) Effect of soil biochar concentration on the mitigation of emerging organic contaminant uptake in lettuce. J Hazard Mater 323:386–393. https://doi.org/10.1016/j.jhazmat.2016.04.046

IPCC (2005) IPCCSpecial report on carbon dioxide capture and storage
Jeffery S, Abalos D, Prodana M et al (2017) Biochar boosts tropical but not temperate crop yields. Environ Res Lett 12:053001
Ji B, Chen J, Li W et al (2021) Greenhouse gas emissions from constructed wetlands are mitigated by biochar substrates and distinctly affected by tidal flow and intermittent aeration modes. Elsevier
Jien SH, Wang CS (2013) Effects of biochar on soil properties and erosion potential in a highly weathered soil. Catena 110:225–233. https://doi.org/10.1016/j.catena.2013.06.021
Jones MW, Santín C, van der Werf GR, Doerr SH (2019) Global fire emissions buffered by the production of pyrogenic carbon. Nat Geosci 12:742–747
Kim KH, Kim JY, Cho TS, Choi JW (2012) Influence of pyrolysis temperature on physicochemical properties of biochar obtained from the fast pyrolysis of pitch pine (Pinus rigida). Bioresour Technol 118:158–162. https://doi.org/10.1016/j.biortech.2012.04.094
Kimani SM, Bimantara PO, Hattori S et al (2020) Co-application of poultry-litter biochar with Azolla has synergistic effects on CH4 and N2O emissions from rice paddy soils. Heliyon 6:e05042. https://doi.org/10.1016/j.heliyon.2020.e05042
Laird DA, Fleming P, Davis DD, Horton R, Wang B, Karlen DL (2010) Impact of biochar amendments on the quality of a typical Midwestern agricultural soil. Geoderma 158(3–4):443–449
Lal R (2008) Carbon sequestration. Philos Trans R Soc B Biol Sci 363:815–830. https://doi.org/10.1098/rstb.2007.2185
Lee JW, Kidder M, Evans BR et al (2010) Characterization of biochars produced from cornstovers for soil amendment. Environ Sci Technol 44:7970–7974. https://doi.org/10.1021/es101337x
Lehmann J, Gaunt J, Rondon M (2006) Bio-char sequestration in terrestrial ecosystems – a review. Mitig Adapt Strateg Glob Chang 11:403–427
Lehmann J, Rillig MC, Thies J et al (2011) Biochar effects on soil biota – a review. Soil Biol Biochem 43:1812–1836
Leng L, Huang H (2018) An overview of the effect of pyrolysis process parameters on biochar stability. Bioresour Technol 270:627–642
Li H, Yutong W, Tianpei W, Hongrui M (2015) Effect of biochar on organic matter conservation and metabolic quotient of soil. Environ Prog Sustain Energy 34:1467–1472. https://doi.org/10.1002/ep.12122
Li X, Wang T, Chang SX et al (2020) Biochar increases soil microbial biomass but has variable effects on microbial diversity: a meta-analysis. Sci Total Environ 749:141593. https://doi.org/10.1016/j.scitotenv.2020.141593
Liu S, Zhang Y, Zong Y et al (2016) Response of soil carbon dioxide fluxes, soil organic carbon and microbial biomass carbon to biochar amendment: a meta-analysis. GCB Bioenergy 8:392–406. https://doi.org/10.1111/gcbb.12265
Liu C, Sun B, Zhang X et al (2020) The water-soluble pool in biochar dominates maize plant growth promotion under biochar amendment. J Plant Growth Regul. https://doi.org/10.1007/s00344-020-10203-3
Lombardo U, Arroyo-Kalin M, Huisman H, Teixeira WG, Clement CR, Alho CFBV, Almeida F, Anjos LHC, Ramsey CB, Brown G, Costa M, Cunha L, Will SLW (2021) Evidence confirms an anthropic origin of Amazonian Dark Earths. https://eartharxiv.org/repository/view/2003/. Accessed 31 Jan 2021
Ma L, Min W, Guo H et al (2020) Response of soil organic C fractions and enzyme activity to integrating N fertilisation with cotton stalk or its biochar in a drip-irrigated cotton field. Acta Agric Scand Sect B Soil Plant Sci. https://doi.org/10.1080/09064710.2020.1853214
Marcucci A, Kypreos S, Panos E (2017) The road to achieving the long-term Paris targets: energy transition and the role of direct air capture. Clim Chang 144:181–193. https://doi.org/10.1007/s10584-017-2051-8
Mohan D, Abhishek K, Sarswat A et al (2018) Biochar production and applications in soil fertility and carbon sequestration–a sustainable solution to crop-residue burning in India. RSC Adv 8:508–520

Most forest fires in India on account of human activity. https://india.mongabay.com/2020/01/most-forest-fires-in-india-on-account-of-human-activity/. Accessed 23 Jan 2021

Nan H, Yang F, Zhao L et al (2019) Interaction of inherent minerals with carbon during biomass pyrolysis weakens biochar carbon sequestration potential. ACS Sustain Chem Eng 7:1591–1599. https://doi.org/10.1021/acssuschemeng.8b05364

Nath AJ, Lal R, Das AK (2015) Managing woody bamboos for carbon farming and carbon trading. Glob Ecol Conserv 3:654–663

Nie C, Yang X, Niazi NK et al (2018) Impact of sugarcane bagasse-derived biochar on heavy metal availability and microbial activity: a field study. Chemosphere 200:274–282. https://doi.org/10.1016/j.chemosphere.2018.02.134

NOAA (2019) Global climate report – annual 2019 | State of the Climate | National Centers for Environmental Information (NCEI). https://www.ncdc.noaa.gov/sotc/global/201913. Accessed 14 Jan 2021

Pan G, Smith P, Pan W (2009) The role of soil organic matter in maintaining the productivity and yield stability of cereals in China. Agric Ecosyst Environ 129:344–348. https://doi.org/10.1016/j.agee.2008.10.008

Payen F, Sykes A, Aitkenhead M et al (2020) Soil organic carbon sequestration rates in vineyard agroecosystems under different soil management practices: a meta-analysis. Elsevier

Prapagdee S, Tawinteung N (2017) Effects of biochar on enhanced nutrient use efficiency of green bean, Vigna radiata L. Environ Sci Pollut Res 24:9460–9467. https://doi.org/10.1007/s11356-017-8633-1

Quan G, Fan Q, Sun J et al (2020) Characteristics of organo-mineral complexes in contaminated soils with long-term biochar application. J Hazard Mater 384:121265. https://doi.org/10.1016/j.jhazmat.2019.121265

Qurat-ul-Aina, Firdaus-e-Bareena, Shafiq M (2016) Management of the Parthenium hysterophorus through biochar formation and its application to rice-wheat cultivation in Pakistan. Agric Ecosyst Environ. https://doi.org/10.1016/j.agee.2016.10.027

Santín C, Doerr SH, Kane ES et al (2016) Towards a global assessment of pyrogenic carbon from vegetation fires. Glob Chang Biol 22:76–91

Schulz H, Glaser B (2012) Effects of biochar compared to organic and inorganic fertilizers on soil quality and plant growth in a greenhouse experiment. J Plant Nutr Soil Sci 175:410–422. https://doi.org/10.1002/jpln.201100143

Sharma A, Jindal J, Mittal A et al (2021) Carbon materials as CO2 adsorbents: a review. Environ Chem Lett

Singh SV, Chaturvedi S, Dhyani VC, Kasivelu G (2020) Pyrolysis temperature influences the characteristics of rice straw and husk biochar and sorption/desorption behaviour of their biourea composite. Bioresour Technol 314:123674

Sun H, Zhang Y, Yang Y et al (2021) Effect of biofertilizer and wheat straw biochar application on nitrous oxide emission and ammonia volatilization from paddy soil. Elsevier

Sustainable Development Goals | UNDP. https://www.undp.org/content/undp/en/home/sustainable-development-goals.html. Accessed 23 Jan 2021

Van Zwieten L, Kimber S, Morris S, Chan KY, Downie A, Rust J, Joseph S, Cowie A (2010) Effects of biochar from slow pyrolysis of papermill waste on agronomic performance and soil fertility. Plant Soil 327(1):235–246

Vergragt PJ, Markusson N, Karlsson H (2011) Carbon capture and storage, bio-energy with carbon capture and storage, and the escape from the fossil-fuel lock-in. Glob Environ Chang 21:282–292. https://doi.org/10.1016/j.gloenvcha.2011.01.020

Walker XJ, Baltzer JL, Cumming SG et al (2019) Increasing wildfires threaten historic carbon sink of boreal forest soils. Nature 572:520–523. https://doi.org/10.1038/s41586-019-1474-y

Wang J, Xiong Z, Kuzyakov Y (2016) Biochar stability in soil: meta-analysis of decomposition and priming effects. GCB Bioenergy 8:512–523

Wang C, Shen J, Liu J et al (2019) Microbial mechanisms in the reduction of CH4 emission from double rice cropping system amended by biochar: a four-year study. Soil Biol Biochem:251–263

Wu P, Cui P, Fang G et al (2018) Biochar decreased the bioavailability of Zn to rice and wheat grains: insights from microscopic to macroscopic scales. Sci Total Environ 621:160–167

Wu Z, Zhang X, Dong Y et al (2019) Biochar amendment reduced greenhouse gas intensities in the rice-wheat rotation system: six-year field observation and meta-analysis. Agric For Meteorol

Wuddivira MN, Stone RJ, Ekwue EI (2009) Structural stability of humid tropical soils as influenced by manure incorporation and incubation duration. Soil Sci Soc Am J 73:1353–1360. https://doi.org/10.2136/sssaj2008.0080

Yang C, SG L (2021) Effects of five different biochars on aggregation, water retention and mechanical properties of paddy soil: a field experiment of three-season crops. Soil Tillage Res

Yang Q, Mašek O, Zhao L et al (2021) Country-level potential of carbon sequestration and environmental benefits by utilizing crop residues for biochar implementation. Appl Energy 282:116275. https://doi.org/10.1016/j.apenergy.2020.116275

Yoganandham ST, Sathyamoorthy G, Renuka RR (2020) Emerging extraction techniques: hydrothermal processing. In: Sustainable seaweed technologies. Elsevier, pp 191–205

You S, Li W, Zhang W et al (2020) Energy, economic, and environmental impacts of sustainable biochar systems in rural China. Crit Rev Environ Sci Technol. https://doi.org/10.1080/10643389.2020.1848170

Yu OY, Raichle B, Sink S (2013) Impact of biochar on the water holding capacity of loamy sand soil. Int J Energy Environ Eng 4:1–9. https://doi.org/10.1186/2251-6832-4-44

Yuan JH, Xu RK (2011) The amelioration effects of low temperature biochar generated from nine crop residues on an acidic Ultisol. Soil Use Manag 27:110–115. https://doi.org/10.1111/j.1475-2743.2010.00317.x

Zhang W, Xu M, Wang X et al (2012) Effects of organic amendments on soil carbon sequestration in paddy fields of subtropical China. J Soils Sediments 12:457–470

Zhang A, Bian R, Hussain Q et al (2013) Change in net global warming potential of a rice-wheat cropping system with biochar soil amendment in a rice paddy from China. Agric Ecosyst Environ 173:37–45. https://doi.org/10.1016/j.agee.2013.04.001

Zhang A, Zhou X, Li M, Wu H (2017a) Impacts of biochar addition on soil dissolved organic matter characteristics in a wheat-maize rotation system in Loess Plateau of China. Chemosphere 186:986–993. https://doi.org/10.1016/j.chemosphere.2017.08.074

Zhang G, Guo X, Zhu Y et al (2017b) Effect of biochar on the presence of nutrients and ryegrass growth in the soil from an abandoned indigenous coking site: the potential role of biochar in the. Elsevier

Zhang F, Liu M, Li Y, Che Y, Xiao Y (2019) Effects of arbuscular mycorrhizal fungi, biochar and cadmium on the yield and element uptake of Medicago sativa. Sci Total Environ 655:1150–1158

Zhang Q, Song Y, Wu Z et al (2020) Effects of six-year biochar amendment on soil aggregation, crop growth, and nitrogen and phosphorus use efficiencies in a rice-wheat rotation. J Clean Prod

Zhao R, Coles N, Wu J (2015) Carbon mineralization following additions of fresh and aged biochar to an infertile soil. Catena 125:183–189. https://doi.org/10.1016/j.catena.2014.10.026

Chapter 13
Strengthening Tree Nutrition Through the Application of Biofertilizers

Elizabeth T. Jojy and Abha Manohar K

Abstract The Nutritional requirements of trees have received less attention compared to agricultural crops. Tree species provide a long-term investment for many potential benefits for human well-being. Nutritional supplementation is a requisite for improvement in the growth and quality of planting stock. Biofertilizers are a boon for chemical fertilizers, which produce several environmental hurdles. Certain groups of bacteria and fungi are majorly used as biofertilizers in tree cultivation. It is a sustainable approach to support growth in these higher plants. Heavy reliance on the indiscriminate use of chemical fertilizers results in adverse effects on the soil ecosystem, environmental quality, and human and animal health risks. The practice of inorganic fertilizers can be greatly reduced by substituting them with eco-friendly microbial biofertilizers or bio-inoculants. Plant Growth-Promoting Rhizobacteria (PGPR) and mycorrhizal fungi are frequently involved in biofertilizer formulations, providing plants with nutrients required to enhance their growth, increase yield, manage abiotic and biotic stress, and prevent phytopathogens attacks. Being an alternate natural source of nutrients, biofertilizers help sustain soil health, fertility, and productivity while significantly improving the growth of plants. In addition, these microbial inoculants are organic, sustainable, and environmentally friendly, thereby reducing the ill effects on human health and the soil biome. Thus, bio-inoculants form an attractive and pecuniary approach towards the development of sustainable forestry. The availability of pertinent microbial strains can increase the survival as well as the quality of seedlings and other propagation units. Hence, the chapter aims to encompass the importance, roles, and application of biofertilizers in improving tree nutrition.

E. T. Jojy
College of Agriculture, Kerala Agricultural University, Thrissur, Kerala, India

A. Manohar K (✉)
MS Swaminathan School of Agriculture, Centurion University of Technology and Management, R.Sitapur, Odisha, India
e-mail: abha.manohar@cutm.ac.in

S. A. Edrisi et al. (eds.), *Sustainable Plant Nutrition and Soil Carbon Sequestration*, Sustainable Plant Nutrition in a Changing World,
https://doi.org/10.1007/978-3-031-53590-1_13

Keywords Fertilizer alternatives · Tree nutrition · Yield · Mycorrhiza · Sustainability

13.1 Introduction

Living beings like animals and plants need sufficient nutrition for proper growth and survival in terms of carbohydrates, proteins, fats, minerals, etc. Plants have the capability to produce food because of photosynthesis - autotrophic nutrition. Some plants also exhibit heterotrophic nutrition (parasitic plants, insectivorous plants, symbiotic plants, and saprophytic plants). Most of the animals tend to depend on plants for their food needs directly or indirectly. Nutritional requirements vary from plant species to species, and not all times are they sufficiently obtained from soil. The soil supplements to the plants are not enough many times, and thus, an external supply of nutrition is necessary. In order to double food production, especially as the human population is increasing day by day, plant health and nutrition are very important (Vance 2001; FAO 2017). The optimum amount of macronutrients (N, P, K, Ca, Mg, S) and micronutrients (B, Cu, Cl, Fe, Mn, Mb, Z) are the main requisite for plants proper growth. When the soil health is poor, then alternative sources need to be found to reduce the nutritional crisis. Fertilizer application is an easy alternative for supplementing nutritional requirements. Chemical alternatives to fertilizer are harmful to the soil plant as well as to the environment in the long run compared to organic/biofertilizers (Srivastava et al. 2002). Biological fertilizers provide essential nutrients in the long run without harming nature (Mahdi et al. 2010).

Trees have received less attention when compared to agricultural crops. These higher plants are multifunctional in nature by acting as carbon sinks and shelters for biodiversity, and they provide food, timber, fuel, etc. (Rai et al. 2021). Thus, trees are gaining more importance presently. Tree planting activities are happening throughout the world, owing to trees gaining importance in the current scenario. Biofertilizer utilization in forest trees is less compared to the Agri-Horti sector, except for the usage of mycorrhizae (Liu and Poobathy 2021). Usage of chemical fertilizers in forest plantations and forest trees is also less compared to agricultural ecosystems (Smethurst 2010). In tree plantation practice, fertilizers are usually applied only once or, if it has 20–30 years of rotation, twice or a few times fertilizer is applied (Davis et al. 2015; Smaill and Clinton 2016). Chemical fertilizers can harm forest microbiomes adversely in the long run (Han et al. 2016). Hence, it is a responsible act that will be helpful in ecosystem restoration as well as incorporate the application of organic biofertilizers for the growth and development of trees. This chapter hereby throws light on the potential applications of various biofertilizers to utilize them as a promising tool for upgrading nutrition for tree populations.

13.2 Necessity of Tree Nutrition

Like all other living beings, trees also need an adequate amount of nutrition support to survive. Tree nutrition got less attention as compared to agricultural crops. However, from a silvicultural aspect, each tree has its unique nutritional needs for development, which vary in different environmental contexts. A nutrient or an element is essential when it is required to carry out the metabolism or completion of the life cycle in the plant (Epstein and Bloom 2005; Havlin et al. 2005). By diligently managing the soil nutrition to the higher plants or trees, we can achieve higher yields. The usage of NPK fertilizers has successively caused higher yields in tropical countries (Stewart et al. 2005). Artificial nitrogen fertilizers reportedly account for the world's 40–60% protein supply (Erisman et al. 2008). Many multipurpose trees are rich sources of protein, considering that tree cultivation also holds great potential (Foroughbakhch et al. 1988). It has been established that though leguminous trees fix nitrogen, the amount is less than nitrogen deposition (Barron et al. 2011). This shows a need to provide nutrients to the trees for their growth and development.

Trees utilize three categories of nutrients, namely macronutrients, secondary nutrients, and micronutrients. Macronutrients comprising nitrogen, phosphorus, and potassium are required in substantial amounts. Trees do not need as much of the secondary nutrients such as calcium, magnesium, and sulfur. Finally, micronutrients, including iron, zinc, copper, manganese, and boron, are required in very small quantities. In agroforestry systems and in systems where fruit trees are utilized, trees have great potential to cure food insecurity and add income to people experiencing poverty (Akinnifesi et al. 2004; Leakey et al. 2005). Fertilizer recommendation for the trees is not an easy task as it requires huge quantities of resources, and research is also less in this particular area. In addition, it is highly dependent on the area, species, soil regimes, etc. As we are moving towards sustainability and the usage of inorganic fertilizers is not recommended, organic fertilizers are preferred; however, the fact that not many organic fertilizers have complex nutrient status has revealed it, creating difficulty in developing good practices of optimized organic fertilizer dosage application. Specific research is yet to be done to advance in the arena of higher plant nutrition quality.

Good alternatives include soil management and the application of indigenous mycorrhizae and plant growth-promoting rhizobacteria (PGPR) in tree production areas (Ortas and Coskan 2016). Mycorrhizae are beneficial associations in the tree roots, which help in the uptake of nutrients and are most widely utilized in tree populations (Alipour et al. 2016). Mycorrhizal inoculation of micro-propagated plantlets in the earlier stages had a beneficial effect on overall plant survival (Ortas et al. 2017). Arbuscular mycorrhizal fungi (AMF) form mutualistic associations with most terrestrial plants, mainly trees. They are beneficial mutual partners that cause certain changes in the host tree physiology. A tree with mycorrhizal association helps better absorption of water and nutrients from soil. These trees have higher survival rates. The roots can absorb 80% of the available phosphorus from the soil.

Nitrogen is also supplied from the decaying organic matter by the mycorrhizal fungus. In areas of heavy metal toxicity, the roots are protected from damage by the fungus. Mycorrhiza also produces certain hormones, which stimulate the growth of fine root hairs.

PGPRs also play a major role in improving tree growth and development. These are soil-dwelling bacteria inhabiting the rhizosphere of trees. They are found associated with the roots or leaves or within the host. The potential benefits offered by these bacteria include nitrogen fixation, solubilization of inorganic phosphates, auxins, cytokinins, gibberellins, and siderophore production. These properties of PGPR help the host trees grow and develop properly. To reach sustainable development goals, we shall opt for the usage of biological fertilizers, which will have a less negative impact on nature with less carbon emission. Therefore, fertilizer alternatives for higher plant nutrition are given in detail in the following paragraphs.

13.3 Biofertilizers

The term biofertilizer defines various biologically active microorganisms that increase the fertility of soil. Biofertilizers are the materials or products that contain microorganisms benefiting plants that are applied to soil, seed, or plant tissues to supply nutrients for plant development and growth. They are actually formulations comprising living or latent cells of competent microbe strains helping in efficient plant nutrient uptake (Vessey 2003; Mahanty et al. 2017). These microorganisms facilitate nutrient attainment in plants via the modulation of plant hormones. These microbial inoculants enhance plant growth by increasing the supply of nutrients, followed by fixation of nitrogen, availability of unavailable nutrients, and release of antimicrobial substances.

Biofertilizers are either applied through seed or soil. Plant or rhizosphere colonization through inoculation of seeds, plants, or soil results in enhanced nutrient availability to the host plants. Application of these organisms to seeds or directly to the soil or dipping seedlings in the slurry is carried out to help in population expansion. As a result, there is an accelerated breakdown of organic wastes, which releases nutrients into the soil. The fixed and non-usable nutrients will be converted to their usable forms by the beneficial microbes. This further improves the fertility and productivity of soil (Guo et al. 2020). These efficient strains of micro-organisms function as atmospheric nitrogen fixers, nutrient mobilizers, phosphorus solubilizers, and decomposers.

In crop nurseries, there is a surging demand for microbial inoculants for the production of quality seedlings, hereby increasing forest productivity. Being sustainable, cheap, environment-friendly, cost-effective, and renewable nutrient resources, the use of these bioinoculants at the nursery stage of tree species has become prevalent. In trees, the application of biofertilizers is limited to the nursery stage, mainly owing to the need for larger quantities and following high cost, while most of the studies have focused on mycorrhiza (Karličić et al. 2016; Mohan and Rajendran

2019). Native strains of biofertilizers play a major role in nutrient mobilization and decomposition in the forest ecosystem and aid in the improved growth, development, and survival of tree groups. Hence, microbial inoculants aid in the production of good quality planting stocks that can be successfully used in the promotion of various developmental programs related to afforestation, social forestry, and wasteland reclamation.

13.3.1 Classification of Biofertilizers

Biofertilizers useful in tree cultivation are broadly classified into bacterial, fungal, and algal inoculants. Each of these microbial inoculants is further divided into symbiotic and asymbiotic bioinoculants. Biofertilizers are also classified on the basis of their function in plants as nitrogen fixers, phosphorus mobilizers and solubilizers, micronutrient solubilizers, and plant growth promoters (Table 13.1).

Table 13.1 Different types of biofertilizers (Singh et al. 2015; Wong and Yeh 2021)

Types of biofertilizer	Category	Examples of microbial used in biofertilizers
Nitrogen-fixing	Free-living (Non-symbiotic)	*Azotobacter, Clostridium, Cyanobacteria (Anabaena), Klebsiella, Rhodopseudomonas, Rhodospirillum, Bacillus*
	Free-living (Symbiotic)	*Rhizobia* (*Allorhizobium, Azorhizobium, Bradyrhizobium, Mesorhizobium, Rhizobium, Sinorhizobium*), *Actinomycete* (*Frankia*)
	Associative	*Azospirillum, Herbaspirillum, Alcaligenes, Bacillus, Pseudomonas, Enterobacter, Klebsiella, Acetobacter*
Phosphorus	Phosphate solubilizer	*Bacillus, Pseudomonas, Penicillum, Aspergillus, Trichoderma, Rhizobium, Burkholderia, Flabobacterium, Pantoea*
	Phosphate mobilizer	*Arbuscular mycorrhiza* (*Acaulospora, Glomus, Gigaspora, Sclerocystis, Scutellaspora*), ectomycorrhizal (*Amanita, Boletus, Laccaria, Pisolithus*)
Micronutrients	Potassium solubilizers	*Bacillus, Paenibacillus*
	Silicate and zinc solubilizers	*Bacillus, Thiobacillus, Saccharomyces*
Promotion of Growth	Plant Growth Promoters	*Alcaligenes, Azotobacter, Actinomycetes, Rhizobium, Bradyrhizobium, Bacillus, Pseudomonas, Enterobacter, Pantoea, Xanthomonas, Flavobacterium*

13.3.1.1 Arbuscular Mycorrhizal (AM) Fungi

AM fungi are involved in the mobilization of various nutrients in the various tree and plant species. These fungi produce vesicles and arbuscles, which are used for food storage and supply of nutrients to the plant cells, respectively. The important genera of AM fungi are *Glomus, Gigaspora, Acaulospora, Sclerocysts*, and Endogone, of which Glomus is the most common. AM fungi extend their mycelia into the soil to access the insoluble nutrients for the plants (Bhat et al. 2010; Singh and Giri 2017). They provide better soil aeration and enhance soil quality, reduce the susceptibility of the plants to phytopathogens, and promote tolerance of host plants to environmental stresses (Thakur and Singh 2018). The different mycorrhizal fungi colonizing the specific host plants and the benefits are illustrated in Table 13.2.

13.3.1.2 Plant Growth Promoting Rhizobacteria (PGPR)

PGPR is defined as a group of beneficial bacteria in soil that colonizes the rhizosphere roots of many plant and tree species. They benefit crop plants in many ways-enhance growth and development, disease resistance, nutrient fixation and mobilization, and production of phytohormones and antimicrobial substances such as antibiotics. Some of the well-known PGPR are listed in Table 13.3, viz., *Azotobacter, Arthrobacter, Bacillus, Alcaligenes, Azospirillum, Pseudomonas, Burkholderia, Serratia, Klebsiella,* and *Enterobacter*. PGPR helps in soil enrichment by breaking down complex matter into simple and usable forms. These PGPR are further classified as bio-stimulants, biofertilizers, and bio-protectants (Saravanakumar et al. 2007).

Rhizobium

Rhizobium is a host-specific, symbiotic nitrogen fixer found to inhabit the soil. This soil bacterium colonizes and symbiotically fixes atmospheric nitrogen in the roots of leguminous plants and trees (Afzal 2006). It is considered the most competent biofertilizer, fixing the highest quantity of nitrogen. *Rhizobium* forms nodules in the rhizospheric roots. These bacteria are highly specific to legumes in which they form nodules. *Rhizobium* at the rate of 5 g per tree is usually essential for the growth and development of trees and the fixation of nitrogen.

Azotobacter

Numerous species of *Azotobacter* are usually found in cultivable soil. The main dominant among them is *A. chroococcum*. Slime production by the bacterium leads to the aggregation of soil particles, helping to reduce soil erosion to an extent. The population of *A. chroococcum* is influenced by the presence of organic matter and antagonistic microorganisms (Gupta 2005).

Table 13.2 Impact of mycorrhizal fungi on various plants

Mycorrhizal fungi	Host plant	Benefits to plant	References
Glomus spp.	Sunflower (*Helianthus annuus*)	Enhances root growth and yield	Adewole et al. (2010)
Hymenoscyphus ericae	Highbush blueberry (*Vaccinium corymbosum*), Calluna (*Calluna vulgaris*)	Improves the growth and improves amino acid uptake	Eccher et al. (2009) and Sokolovski et al. (2002)
Acaulospora scrobiculata	Indian Beech (*Pongamia pinnata*)	Improves seedling growth	Jha et al. (2014)
Scutellospora heterogama	Passion fruit (*Passiflora alata*)	Stimulates plant growth and inhibits nematode infection	Anjos et al. (2010)
Scutellospora fulgida	Common milkweed (*Asclepias syriaca*)	Improves biomass, foliar P, and trichome density	Vannette and Hunter (2013)
Glomus mosseae	Tomato (*Solanum lycopersicum*)	Improves growth and resistance against bacterial wilt	Tahat et al. (2012)
Glomus clarum	Cucumber (*Cucumis sativus*)	Improves yield and P and Zn uptake	Ortas (2010)
Glomus etunicatum	Lime (*Citrus aurantifolia*)	Improves growth via the increase in chlorophyll contents, photosynthesis rate	Shahsavar et al. (2016)
Glomus fasciculatum	Bhringraj (*Wedilia chinensis*)	Improves total biomass and nutrition uptake	Nisha and Rajeshkumar (2010)
Scutellospora spp.	Maize (*Zea mays*), Soybean (*Glycine max*)	Improves the root and shoot growth	Jeong et al. (2006)
Archaeospora trappei	Ribgrass (*Plantago lanceolata*)	Increases in growth rate resulting in endurance to actions of herbivores.	Bennett and Bever (2007)
Rhizophagus fasciculatus	Tomato (*Solanum lycopersicum*), Chilli (*Capsicum annuum*)	Improves fruit yield, biomass, and phosphorus accumulation in shoots	Padmavathi et al. (2015)
Rhizophagus clarus	Chilli (*Capsicum annuum*)	Improves nitrogen and phosphorus uptake	Lee and Eom (2015)
Funneliformis mosseae	Mulberry (*Morus alba*)	Improves growth via the increase in chlorophyll contents, photosynthesis rate, and stomatal conductance	Shi et al. (2016)

Azospirillum

Azospirillum sp. is a gram-negative bacteria establishing an associative symbiotic relation with the roots of graminaceous plants. *A. lipoferum* and *A. brasilense* are the commonly found bacteria colonizing the intercellular spaces of the rhizosphere root cortex of monocots and soil. Abiotic and biotic resistance in inoculated plants

Table 13.3 Impact of PGPR on various plants

PGPR	Host plant	Benefits to plant	References
Pseudomonas putida	Beans (*Phaseolus vulgaris*)	Alteration of plant defense potential by increased lignin	Anderson and Guerra (1985)
Pseudomonas fluorescens	Pea (*Pisum sativum*)	Polymorphic wall appositions along epidermal cells against *Pythium ultimum*	Benhamou et al. (1996)
Azospirillum brasiliense	Beans (*Phaseolus vulgaris*)	Nod gene induction	Burdman et al. (1996)
Azospirillum brasiliense	Wheat (*Triticum aestivum*)	Production of Indole-3 acetic acid (IAA) to alter root morphology	Dobbelaere et al. (1999)
Achromobacter sp.	Canola (*Brassica napus*)	Ionic transport stimulation	Bertrand et al. (2001)
Bacillus pumilis and *B. licheniformis*	Alder (*Alnus glutinosa*)	Promotion of stem elongation by bacterial gibberellins	Gutierrez-Manero et al. (2001)
Azospirillum brasiliense	Tomato (*Lycopersicum esculentum*)	Enhance lateral root and root hair development	Creus et al. (2005)
Burkholderia phytofirmans	Grape (*Vitis vinifera*)	Increase root growth and plantlet biomass	Ait Barka et al. (2006)
Azospirillum lipoferum	Maize (*Zea mays*)	Enhance root growth and development	El Zemrany et al. (2007)
Pseudomonas fluorescens	Catharanthus roseus	Biomass production	Jaleel et al. (2007)
Chryseobacterium balustinum	Soybean (*Glycine max*)	Production of flavonoids	Dardanelli et al. (2010)
Herbaspirillum seropedicae	Rice (*Oryza sativa*)	Modulation of plant defense responses	Brusamarello-Santos et al. (2012)
Phyllobacterium brassicacearum	*Arabidopsis thalliana*	Stimulate root hair elongation	Galland et al. (2012)
Exiguobacterium oxidotolerans	Brahmi (*Bacopa monnieri*)	Increase yield and content of secondary metabolites	Bharti et al. (2013)
Pseudomonas spp., *Bacillus* spp. and *Acinetobacter* sp.	Mung bean (*Vigna radiata*)	IAA production, phosphate solubilization, siderophore production, and antagonism against pathogens	Kumari et al. (2018)
Acinetobacter calcoaceticus	*Sambucus williamsii*	Mitigate reduction of chlorophyll and delay restriction caused by non-stomatal factors.	Liu et al. (2019)

is favored by *Azospirillum* sp. in addition to nitrogen fixation and mobilization with growth promotion (Cassan et al. 2014).

Phosphate Solubilizing Microorganisms (PSM)
PSMs increase the availability of phosphorus to plants by solubilization and mineralization of complex phosphorus-containing compounds (Wani et al. 2007). The PSMs secrete various organic acids, which further enhances the phosphorus solubilization process (Marra et al. 2012). The solubilization efficacy is dependent on the type, concentration, and quality of organic acids released into the soil (Scervino et al. 2010). Phosphate solubilizing bacteria function in solubilizing insoluble phosphorus to plant usable form. This group includes several soil bacteria and fungi such as *Pseudomonas* sp., *Bacillus* sp., *Penicillium* sp., and *Aspergillus* sp. PSM produces certain organic acids that reduce the soil pH, causing the degradation of fixed phosphates. The phosphate-solubilizing bacteria include *Bacillus licheniformis*, *Serratia marcescens,* and *Staphylococcus warneri*.

13.3.2 Mechanisms of Biofertilizers

The different mechanisms (direct and indirect) associated with biofertilizer application are highlighted in Fig. 13.1. The direct mechanisms stimulates the growth of plants directly that involves N_2 fixation, phytohormone production, solubilzation of PO_4^{2-}, and others. Moreover, the indirect processes do not influence the plant growth directly; however, it shields the plant from the adverse impacts of plant pathogens (Mahmud et al. 2021).

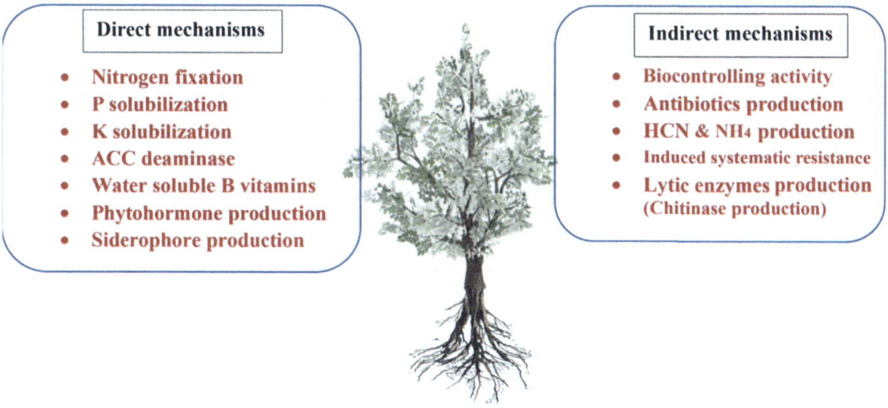

Fig. 13.1 Direct and indirect mechanisms in the application of Biofertilizers. (Adopted from Mahmud et al. 2021)

Fig. 13.2 Sequential steps in the development of Biofertilizers

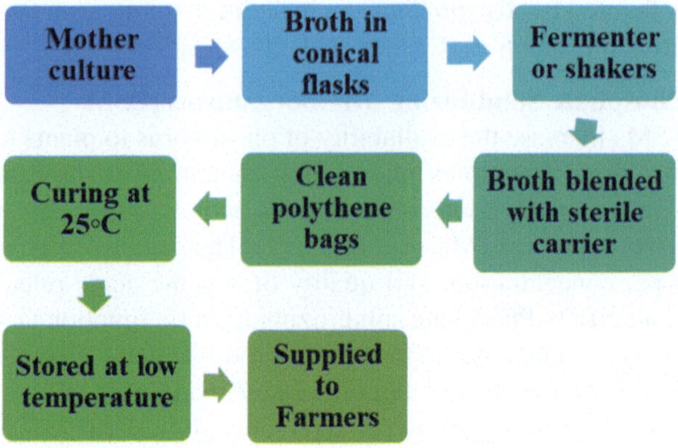

13.3.3 Commercial Production of Biofertilizers

Initially, microbes are isolated from the roots or soil. These microbes are screened in the laboratory for stimulation in plant growth. Greenhouse screening is carried out in potted soil for any plant growth promotion. Later, field screening is done in the cropped soil with a crop variety and in different soil types. Then, the microbes are refined and tested for any environmental impacts and substantiation of microbes. The selected microbes are carried on for mass production. The isolated microbial culture is sub-cultured onto a suitable broth under shaking conditions. The culture is incubated until it reaches a maximum cell population of 10^{10} to 10^{11}. The culture thus obtained in the flask is known as starter culture.

For large-scale production, inoculum from the starter culture is transferred to large flasks or fermenters, where it is grown until a certain cell count is reached. The cells are harvested with a population load of about 10^9 per ml. The microbial cells are mixed with neutralized and sterilized carrier material manually or using a mechanical mixer. The inoculant mixed carrier material is then filled in polythene bags sealed with an electric sealer. These polythene packets should be properly labeled with the name of the manufacturer, name of the product, strain number, crops to which it is recommended, method of inoculation, date of manufacture, batch number, date of expiry, price, full address, and storage instructions (Suthar et al. 2017). Figure 13.2 represents the different steps involved in the commercial production of biofertilizers.

13.3.4 Methods of Application of Biofertilizer

Biofertilizers are inoculated onto the plant surfaces (seeds and roots) and soil by different methods.

13.3.4.1 Seed Treatment

The most common effective method for application of bioinoculants is seed treatment. The biofertilizers are mixed with a 10% solution of jaggery. The slurry is poured over the seeds and mixed such that the seeds are uniformly coated with a thin layer of biofertilizers. The treated seeds are dried in the shade before planting (Thomas and Singh 2019). Generally, the legume seeds are treated with 750 g of the biofertilizer per hectare. The seed treatment allows the delivery of bacterial combinations on the seed surface to achieve the necessary results (Chen 2006).

13.3.4.2 Seedling Root Dip

The roots of the seedlings are dipped in a slurry made by mixing bioinoculants in water before transplanting them. The duration for which the roots are dipped varies from crop to crop. This method is popular in the forest nurseries. Usually, the seedling roots are treated for half an hour in the biofertilizer solution before they are transplanted. This method uses 2–2.5 kg of biofertilizers to inoculate seeds required for one acre of area. The microbial inoculants are mixed with 5–10 l of water to prepare the water suspension for root dipping.

13.3.4.3 Soil Application

Here, biofertilizers are directly or applied to the soil with other biofertilizers. Soil application enhances the delivery rates, avoids seed mixing, reduces contact with biofertilizer-treated seeds, and improves survival in low moisture conditions compared to powder form (Zahran 1999). About 20 g of the biofertilizers mixed with compost are added to the ring of fruit saplings when planted. The bio-inoculants can be mixed with farmyard manure (FYM) in a ratio of 1:25 for 24 h and applied to nursery beds and pots. This FYM acts as both a nutrient medium and a carrier for biofertilizers (Kar et al. 2020).

13.3.5 Benefits of Biofertilizers

Biofertilizers are natural, cheap, and eco-friendly bio-additives applied to plants. Application methods are easy and require less time and energy. They increase the soil quality and fertility while making the available nutrients. As a result, it improved crop yields by 3–39%. They are required in very small quantities, economical, and highly efficient. They act as nitrogen fixers, help mobilize essential nutrients, and further decompose the complex organic materials into simpler forms (Fig. 13.3). These bioinoculants leave considerable residual effects, benefitting crop growth. In addition to the above advantages, they enhance resistance to the abiotic and biotic

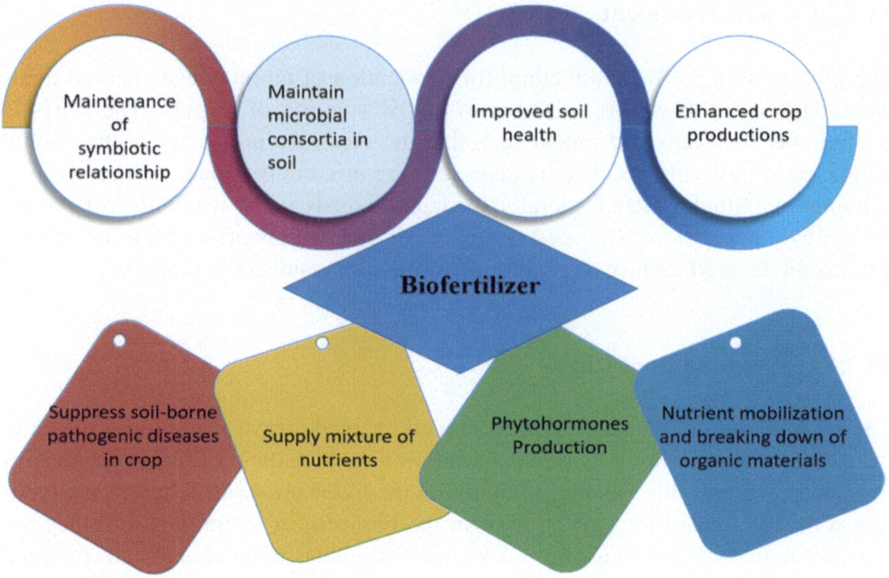

Fig. 13.3 Broad areas for Biofertilizer utilization

factors affecting the crop. Phytohormones or growth regulators are secreted, further improving plant growth and development. Production of antifungal and antibacterial compounds helps in the suppression of pathogenic microorganisms. It reduces environmental pollution by playing a role in the detoxification of heavy metals. Integrating biofertilizers with organic substrates, manures, and chemical fertilizers enhances soil and crop productivity with improved nutrient use efficiency.

13.4 Limitations of Biofertilizers

The functioning of biofertilizers is a slow process, taking more time to act on the problem when compared with chemical fertilizers. These microbial inoculants can only supplement the chemical fertilizers but can never replace them. Long-term storage of biofertilizers can pose a problem, as they are sensitive to changes in temperature and humidity (Yadav and Sarkar 2019). Non-availability of suitable production facilities, lack of technically qualified staff, and inadequate and improper transport and storage facilities can result in the large-scale production of biofertilizers. The challenges and recommendations for improvising biofertilizers' potentialities are listed below (Table 13.4).

Table 13.4 Major challenges and recommendations to improve biofertilizer efficiency

Challenges	Recommendations	Reference
Microbial inoculant interaction with biotic and abiotic stress	Future studies on the key genes involved in the interaction process between inoculants and native microbes to understand the metabolic potential of soil microbial communities in response to inoculation, which allow the design of biofertilizers for specific soil/crops Future study to study the root exudates of the crop of various genotypes, physiological stages and health conditions, which could directly affect microbial colonization	Micallef et al. (2009a, b), Win et al. (2019), and Olanrewaju et al. (2019)
Knowledge transfer to industrial level for optimum formulation	Engaging researchers-industry collaboration to produce quality biofertilizers	Malusa and Vasilev (2014)
Awareness of farmers	Engaging stakeholders in educating farmers in biofertilization Monitoring farmers' and stakeholder's requirements and feedback on the biofertilization efficiency on crop production for continuous improvement	Burbi and Rose (2016) and Chaudhary and Chauhan (2016)
Fertilization scheme	Optimization study of suitable dosages in combined treatments of fertilizers and inoculants to achieve positive outcomes Choosing suitable organic fertilizers except for sewage sludge, which reduces soil microbial community	Adesemoye et al. (2019), Viti et al. (2010), and Tojlander et al. (2008)
Cultural practices in farm	Conservation tillage enhances the stability of rhizospheric microbial community which directly improves crop production	Wang et al. (2017) and Piazza et al. (2019)
Application methods	Automated application of biofertilizer to ensure equal distribution Repeat application of biofertilizer as boosters to improve microbial colonization Further optimization work needed for biofertilizer application in different agriculture field production setting to meet farmers' requirement	Keswani et al. (2019), Malusa et al. (2012), and Malusa and Vassilev (2014)
Plant genotype, physiological factors, and plant health	Proper optimization study needed to study the effect of crop genotypes and physiological stages of the colonization behavior of inoculants Future study to study the root exudates of the crop of various genotypes, physiological stages and health conditions, which could directly affect microbial colonization	Micallef et al. (2009a) and Win et al. (2019)

13.5 Conclusions

Higher plant nutrition is a relatively less explored field. Tree nutrition is very important based on their varied environmental and ecological benefits. Trees have the capacity to cure hunger in the present world by creating multiple incomes for the cultivaters if they are effectively managed. To achieve sustainable development goals, we need to minimize the utilization of inorganic fertilizers and need to focus on organics. In addition, a recent hike in the price of inorganic fertilizers has resulted in a drastic decline in their use, underlining the requirement for integrated nutrient management in forestry and agriculture. The present world scenario has a higher demand for organic foods, minimizing the use of chemicals in every way possible. Biofertilizers form a vital part of sustainable crop production in this situation, which can help deal with climate change, food scarcity, and the deprivation of available land for food production. Biofertilizers play an important role in the supply and making nutrients available to various crops, improving the yield and nutrient dynamics in trees. Thus, they can be a promising tool for nutrient management. However, inadequate biofertilizer research, lack of technological development, and an ineffective regulatory framework have contributed to challenges in biofertilizer development. Capacity building needs to be done collaboratively by tree scientists, silviculturists, agronomists, and pathologists to develop effective biological fertilizers to benefit humanity in support of tree growth and development.

References

Adesemoye AO, Torbert HA, Kloepper JW (2019) Plant growth-promoting rhizobacteria allow reduced application rates of chemical fertilizers. Microbial Ecol 58:921–929

Adewole MB, Awotoye OO, Ohiembor MO, Salami AO (2010) Influence of mycorrhizal fungi on phytoremediating potential and yield of sunflower in Cd and Pb polluted soils. J Agric Sci 55:17–28

Afzal M (2006). Effect of Rhizobium, PSB with different fertility levels on green gram under temperate conditions of Kashmir. P.G. Thesis, pp 103–104

Ait Barka E, Nowak J, Clement C (2006) Enhancement of chilling resistance of inoculated grapevine plantlets with a plant growth-promoting rhizobacterium, *Burkholderia phytofirmans* strain PsJN. Appl Environ Microbiol 72:7246–7252

Akinnifesi F, Kwesiga F, Mhango J, Mkonda A, Chilanga T, Swai R (2004) Domesticating priority miombo indigenous fruit trees as a promising livelihood option for small-holder farmers in Southern Africa. Acta Hortic 632:15–30

Alipour H, Nikbakht A, Etemadi N, Nourbakhsh F, Rejali F (2016) The efficiency of mycorrhizal fungi on growth characteristics and some nutrients uptake of plane tree seedling (*Platanus orientalis* L.). J Hort Sci 29(4):537–546.

Anderson AJ, Guerra D (1985) Responses of bean to root colonization with *Pseudomonas putida* in a hydroponic system. Phytopathology 75:992–995

Anjos ECT, Cavalcante UMT, Goncalves DMC, Pedrosa EMR, Santos VF, Maia LC (2010) Interactions between an arbuscular mycorrhizal fungus (*Scutellospora heterogama*) and the root-knot nematode (*Meloidogyne incognita*) on sweet passion fruit (*Passiflora alata*). Brazilian Archives Biol Technol 53:801–809

Barron AR, Purves DW, Hedin LO (2011) Facultative nitrogen fixation by canopy legumes in a lowland tropical forest. Oecologia 165(2):511–520

Benhamou N, Belanger RR, Paulitz TC (1996) Pre-inoculation of Ri T-DNA-transformed pea roots with *Pseudomonas fluorescens* inhibits colonization by *Pythium ultimum* Trow: an ultrastructural and cytochemical study. Planta 199:105–117

Bennett E, Bever JD (2007) Mycorrhizal species differentially alter plant growth and response to herbivory Alison. Ecology 88:210–218

Bertrand H, Nalin R, Bally R, Cleyet-Marel JC (2001) Isolation and identification of the most efficient plant growth-promoting bacteria associated with canola (*Brassica napus*). Biol Fertil Soils 33:152–156

Bharti N, Yadav D, Barnawal D, Maji D, Kalra A (2013) *Exiguobacterium oxidotolerans*, a halotolerant plant growth promoting rhizobacteria, improves yield and content of secondary metabolites in *Bacopa monnieri* (L.) Pennell under primary and secondary salt stress. World J Microbiol Biotechnol 29:379–387

Bhat MI, Rashid A, Rasool F, Mahdi SS, Haq SA, Bhat RA (2010) Effect of Rhizobium and V A mycorrhizae on green gram under temperate conditions. Res J Agricul Sci 1(2):113–116

Brusamarello-Santos LCC, Pacheco F, Aljanabi SMM, Monteiro RA, Cruz LM, Baura VA (2012) Differential gene expression of rice roots inoculated with the diazotroph *Herbaspirillum seropedicae*. Plant Soil 356:113–125

Burbi S, Rose KH (2016) The role of internet and social media in the diffusion of knowledge and innovation among farmers. [Paper presentation] 12th European IFSA symposium: social and technological transformation of farming systems: diverging and converging pathways, Harper Adams University, Newport, UK

Burdman S, Volpin H, Kigel J, Kapulnik Y, Okon Y (1996) Promotion of nod gene inducers and nodulation in common bean (*Phaseolus vulgaris*) roots inoculated with *Azospirillum brasilense* Cd. Appl Environ Microbiol 62:3030–3033

Cassán F, Vanderleyden J, Spaepen S (2014) Physiological and agronomical aspects of phytohormone production by model plant-growth-promoting rhizobacteria (PGPR) belonging to the genus *Azospirillum*. J Plant Growth Regul 33(2):440–459

Chaudhary D, Chauhan NM (2016) Extent of knowledge and adoption of biofertilizer use by biofertilizer users of Navsari district in Gujarat state. Int J Manag Soc Sci 4:151–155

Chen JH (2006) The combined use of chemical and organic fertilizers and/or biofertilizer for crop growth and soil fertility. International workshop on sustained management of the soil-rhizosphere system for efficient crop production and fertilizer use 16, pp 1–11

Creus CM, Graziano M, Casanovas EM, Pereyra MA, Simontacchi M, Puntarulo S (2005) Nitric oxide is involved in the *Azospirillum brasilense*-induced lateral root formation in tomato. Planta 221:297–303

Dardanelli MS, Manyani H, Gonzalez-Barroso S, Rodriguez-Carvajal MA, Gil-Serrano AM, Espuny MR (2010) Effect of the presence of the plant growth promoting rhizobacterium (PGPR) *Chryseobacterium balustinum* Aur9 and salt stress in the pattern of flavonoids exuded by soybean roots. Plant Soil 328:483–493

Davis M, Xue J, Clinton P (2015) Planted-forest nutrition. Science Publication (Publicly available)

Dobbelaere S, Croonenborghs A, Thys A, Vande Broek A, Vanderleyden J (1999) Phytostimulatory effect of *Azospirillum brasilense* wild type and mutant strains altered in IAA production on wheat. Plant Soil 212:153–162

Eccher T, Bacchetta M, Granelli G (2009) Long term effects of ericoid endomycorrhizae on the growth of micropropagated plants of *Vaccinium corymbosum* L. in the field. Acta Horticulture 810:657–664

El Zemrany H, Czarnes S, Hallett PD, Alamercery S, Bally R, Jocteur-Monrozier L (2007) Early changes in root characteristics of maize (*Zea mays*) following seed inoculation with the PGPR *Azospirillum lipoferum* CRT1. Plant Soil 291:109–118

Epstein E, Bloom AJ (2005) Mineral nutrition of plants: principles and perspectives, 2nd edn. Sinauer Associates, Inc., Sunderland

Erisman JW, Sutton MA, Galloway J, Klimont Z, Winiwarter W (2008) How a century of ammonia synthesis changed the world. Nat Geosci 1:636–639

FAO (2017). STAT. http://www.fao.org/faostat/en

Foroughbakhch R, Hauad LA, Maiti RK, Hernández RB, Ramirez RG (1988) Traditional uses and potential of *Helietta parvifolia* in northeastern Mexico. Forest Farm Commun Tree Res Rep 3:38–42

Galland M, Gamet L, Varoquaux F, Touraine B, Desbrosses G (2012) The ethylene pathway contributes to root hair elongation induced by the beneficial bacteria *Phyllobacterium brassicacearum* STM196. Plant Sci 190:74–81

Guo S, Wang P, Wang X, Zou M, Liu C, Hao J (2020) Microalgae as biofertilizer in modern agriculture. In: Microalgae biotechnology for food, health and high value products. Springer, pp 397–411

Gupta PK (2005) Vermicomposting for sustainable agriculture. Bharat Printing Press, Jodhpur, pp 11–14

Gutiérrez-Mañero FJ, Ramos-Solano B, Probanza A, Mehouachi J, Tadeo FR, Talon M (2001) The plant-growth promoting rhizobacteria *Bacillus pumilus* and *Bacillus licheniformis* produce high amounts of physiologically active gibberellins. Physiol Plant 111:206–211

Han SH, An JY, Hwang J, Bin KS, Park BB (2016) The effects of organic manure and chemical fertilizer on the growth and nutrient concentrations of yellow poplar (*Liriodendron tulipifera* Lin.) in a nursery system. For Sci Technol 12(3):137–143

Havlin JL, Beaton JD, Tisdale SL (2005) Soil fertility and fertilizers: an introduction to nutrient management. Pearson Prentice Hall, Upper Saddle River

Jaleel CA, Manivannan P, Sankar B, Kishorekumar A, Gopi R, Somasundaram R (2007) *Pseudomonas fluorescens* enhances biomass yield and ajmalicine production in *Catharanthus roseus* under water deficit stress. Colloids Surf B Biointerfaces 60:7–11

Jeong HS, Lee J, Eom AH (2006) Effects of interspecific interactions of arbuscular mycorrhizal fungi on growth of soybean and corn. Mycobiology 34:34–37

Jha A, Kamalvanshi M, Kumar A, Chakravarty N, Shukla A, Dhyani SK (2014) The effects of arbuscular mycorrhizal inoculations and cotyledon removal on early seedling growth of *Pongamia pinnata*. Turkish J Bot 38:526–535

Kar S, Rout S, Sahu ML, Sharma Y, Patra SS (2020) Biofertilizer in forest nursery – a review. Int J Industrial Biotechnol Biomaterials 6(2):1–15

Karličić V, Golubović-Ćurguz V, Raičević V (2016) The alleviation of reforestation challenges by beneficial soil microorganisms. Reforesta 1(1):238–260

Keswani C, Prakash O, Bharti N, Vílchez JI, Sansinenea E, Lally RD, Borriss R, Singh SP, Gupta VK, Fraceto LF, de Lima R, Sing HB (2019) Re-addressing the biosafety issues of plant growth promoting rhizobacteria. Sci Total Environ 690:841–852

Kumari P, Meena M, Gupta P, Dubey MK (2018) Plant growth promoting rhizobacteria and their biopriming for growth promotion in mung bean (*Vigna radiata* (L.) R. Wilczek). Biocatalysis Agric Biotechnol 16:163–171

Leakey RR, Tchoundjeu Z, Schreckenberg K, Shackleton SE, Shackleton CM (2005) Agroforestry tree products (AFTPs): targeting poverty reduction and enhanced livelihoods. Int J Agric Sustain 3(1):1–23

Lee EH, Eom AH (2015) Growth characteristics of *Rhizophagus clarus* strains and their effects on the growth of host plants. Mycobiology 43:444–449

Liu WYY, Poobathy R (2021) Biofertilizer utilization in forestry. In: Inamuddin, Ahamed MI, Boddula R, Rezakazemi M (eds) Biofertilizers

Liu F, Ma H, Peng L, Du Z, Ma B, Liu X (2019) Effect of the inoculation of plant growth promoting rhizobacteria on the photosynthetic characteristics of *Sambucus williamsii* Hance container seedings under drought stress. AMB Express 9(1):1–9

Mahanty T, Bhattacharjee S, Goswami M, Bhattacharyya P, Das B, Ghosh A, Tribedi P (2017) Biofertilizers: a potential approach for sustainable agriculture development. Environ Sci Pollut Res 24(4):3315–3335

Mahdi SS, Hassan GI, Samoon SA, Rather HA, Dar SA, Zehra B (2010) Biofertilizers in organic agriculture. J Phytol 2(10):42–54
Mahmud AA, Upadhyay SK, Srivastava AK, Bhojiya AA (2021) Biofertilizers: a nexus between soil fertility and crop productivity under abiotic stress. Curr Res Environ Sustain 3:100063
Malusà E, Vassilev N (2014) A contribution to set a legal framework for biofertilizers. Appl Microbiol Biotechnol 98:6599–6607
Malusà E, Sas-Paszt L, Ciesielska J (2012) Technologies for beneficial microorganisms inocula used as biofertilizers. Sci World J:491206
Marra LM, Soares CRFS, de Oliveira SM, Ferreira PAA, Soares BL, de Fráguas CR, de Lima JM, de Souza MFM (2012) Biological nitrogen fixation and phosphate solubilization by bacteria isolated from tropical soils. Plant Soil 357(1–2):289–307
Micallef SA, Channer S, Shiaris MP, Colón-Carmona A (2009a) Plant age and genotype impact the progression of bacterial community succession in the Arabidopsis rhizosphere. Plant Signal Behav 4:777–780
Micallef SA, Shiaris MP, Colón-Carmona A (2009b) Influence of Arabidopsis thaliana accessions on rhizobacterial communities and natural variation in root exudates. J Exp Bot 60:1729–1742
Mohan E, Rajendran K (2019) Sustainable development of horticulture and forestry through bio-inoculants. In: Hasanuzzaman M, Fujita M, Filho MCMT, Nogueira TAR (eds) Sustainable crop production. IntechOpen
Nisha MC, Rajeshkumar S (2010) Effect of arbuscular mycorrhizal fungi on growth and nutrition of *Wedilia chinensis* (Osbeck) Merril. Indian J Sci Technol 6:676–678
Olanrewaju OS, Ayangbenro AS, Glick BR, Babalola OO (2019) Plant health: feedback effect of root exudates-rhizobiome interactions. Appl Microbiol Biotechnol 103:1155–1166
Ortas I (2010) Effect of mycorrhiza application on plant growth and nutrient uptake in cucumber production under field conditions. Spanish J Agric Res 8:116–122
Ortas I, Coskan A (2016) Precipitation as the most affecting factor on soil-plant environment conditions affects the mycorrhizal spore numbers in the three different ecological zones in Turkey. Acta Agriculturae Scandinavica, Section B Soil and Plant Sci 66(4):369–378
Ortaş I, Lal R, Kapur S (2017) Carbon sequestration and mycorrhizae in Turkish soils. In: Erşahin S, Kapur S, Erhan A, Namli A, Erdogan HE (eds) Carbon management, technologies, and trends in Mediterranean ecosystems. Springer International, pp 139–151
Padmavathi T, Dikshit R, Seshagiri S (2015) Effect of *Rhizophagus* spp. and plant growth promoting *Acinetobacter junii* on *Solanum lycopersicum* and *Capsicum annuum*. Braz J Bot 38:273–280
Piazza G, Ercoli L, Nuti M, Pellegrino E (2019) Interaction between conservation tillage and nitrogen fertilization shapes prokaryotic and fungal diversity at different soil depths: evidence from a 23-year field experiment in the Mediterranean area. Front Microbiol 10:2047
Rai P, Shukla VG, Manohar KA, Bhat JA, Kumar A, Kumar M, Cabral-Pinto M, Chakravarty S (2021) Carbon storage of single tree and mixed tree dominant species stands in a reserve forest—case study of the eastern sub-Himalayan Region of India. Land 10:435
Saravanakumar D, Vijayakumar C, Kumar N, Samiyappan R (2007) PGPR-induced defense responses in the tea plant against blister blight disease. Crop Prot 26:556–565
Scervino JM, Mesa MP, Della Mónica I, Recchi M, Moreno NS, Godeas A (2010) Soil fungal isolates produce different organic acid patterns involved in phosphate salts solubilization. Biol Fertil Soils 46(7):755–763
Shahsavar AR, Refahi A, Zarei M, Aslmoshtagh E (2016) Analysis of the effects of *Glomus etunicatum* fungi and *Pseudomonas fluorescence* bacteria symbiosis on some morphological and physiological characteristics of Mexican lime (*Citrus aurantifolia* L.) under drought stress conditions. Adv Horticult Sci 30:39–45
Shi SM, Chen K, Gao Y, Liu B, Yang XH, Huang XZ, Liu GX, Zhu LQ, He XH (2016) Arbuscular mycorrhizal fungus species dependency governs better plant physiological characteristics and leaf quality of mulberry (*Morus alba* L.) seedlings. Front Microbiol 7:1–11

Singh I, Giri B (2017) In: Uptake M-N, Biocontrol E, Varma A, Prasad R, Tuteja N (eds) Arbuscular mycorrhiza mediated control of plant pathogens. Springer, pp 131–160

Singh B, Bauddh K, Bux F (2015) Algae and environmental sustainability, vol 7. Springer

Smaill SJ, Clinton PW (2016) Overview of the issues affecting fertiliser use in New Zealand's radiata pine forests. New Zeal J For 61(2):11–15

Smethurst PJ (2010) Forest fertilization: trends in knowledge and practice compared to agriculture. Plant Soil 335(1–2):83–100

Sokolovski SG, Meharg AA, Maathuis FJM (2002) *Calluna vulgaris* root cells show increased capacity for amino acid uptake when colonized with the mycorrhizal fungus *Hymenoscyphus ericae*. New Phytol 155:525–530

Srivastava AK, Singh S, Marathe RA (2002) Organic citrus: soil fertility and plant nutrition. J Sustain Agric 19(3):5–29

Stewart WM, Dibb DW, Johnston AE, Smyth TJ (2005) The contribution of commercial fertilizer nutrients to food production. Agron J 97:1–6

Suthar KP, Patel RM, Singh D, Khunt MD (2017) Efficacy of *Bacillus subtilis* isolate K18 against chickpea wilt *Fusarium oxysporum* f. sp. *ciceri*. Int J Pure App Biosci 5(5):838–843.

Tahat MM, Sijam R, Othman R (2012) The potential of endomycorrhizal fungi in controlling tomato bacterial wilt, *Ralstonia solanacearum* under glasshouse conditions. Afr J Biotechnol 11:13085–13094

Thakur P, Singh I (2018) Biocontrol of soilborne root pathogens: an overview. In: Giri B, Prasad R, Varma A (eds) Root biology. Springer, pp 181–220

Thomas L, Singh I (2019) Microbial biofertilizers: types and applications. In: Biofertilizers for sustainable agriculture and environment, pp 1–19

Tojlander JF, Santos-Gonzalez JC, Tehler A, Finlay RD (2008) Community analysis of arbuscular mycorrhizal fungi and bacteria in the maize mycorrhizosphere in a long-term fertilization trial. FEMS Microbiol Ecol 65:323–328

Vance CP (2001) Symbiotic nitrogen fixation and phosphorus acquisition. Plant nutrition in a world of declining renewable resources. Plant Physiol 127(2):390–397

Vannette RL, Hunter MD (2013) Mycorrhizal abundance affects the expression of plant resistance traits and herbivore performance. J Ecol 101:1019–1029

Vessey JKJP (2003) Plant growth promoting rhizobacteria as biofertilizers. Plant Soil 255(2):571–586

Viti C, Tatti E, Decorosi F, Lista E, Rea E, Tullio M, Sparvoli E, Giovannetti L (2010) Compost effect on plant growth-promoting rhizobacteria and mycorrhizal fungi population in maize cultivations. Comp Sci Utiliz 18:273–281

Wang Z, Liu L, Chen Q, Wen X, Liu Y, Han J, Liao Y (2017) Conservation tillage enhances the stability of the rhizosphere bacterial community responding to plant growth. Agron Sustain Dev 37:44

Wani P, Khan M, Zaidi A (2007) Co-inoculation of nitrogen-fixing and phosphate-solubilizing bacteria to promote growth, yield and nutrient uptake in chickpea. Acta Agron Hungarica 55(3):315–323

Win KT, Okazaki K, Ookawa T, Yokoyama T, Ohwaki Y (2019) Influence of rice-husk biochar and *Bacillus pumilus* strain TUAT-1 on yield, biomass production, and nutrient uptake in two forage rice genotypes. PLoS One 14

Wong CKF, Yeh CY (2021) Impact of biofertilizers on horticultural crops. In: Inamuddin, Ahamed MI, Boddula R, Rezakazemi M (eds) Biofertilizers

Yadav KK, Sarkar S (2019) Biofertilizers, impact on soil fertility and crop productivity under sustainable agriculture. Environ Ecol 37(1):89–93

Zahran HH (1999) Rhizobium-legume symbiosis and nitrogen fixation under severe conditions and in an arid climate. Microbiol Mol Biol Rev 63:968–989

Chapter 14
Role of Novel Microorganisms for Maintaining Soil Carbon Pool and Plant Health

Sonia Petersen

Abstract Microorganisms in soils are ubiquitous, yet it is considered that the majority of soil microbes are awaiting discovery. This chapter provides a synthesis of the methods of detecting new soil microbes, their role in the carbon cycle, mainly as they contribute to total SOC stores, and their influence on the health of plants. Most recently, the detection and isolation of novel microbes have become the domain of metagenomic methods, while in vitro growth of these microbes depends on new ways of culturing based on phylogenic ties that indicate physical, chemical, and metabolic requirements for the specificity of culturing medium. Further, novel microbes can be grouped in commensal or symbiotic guilds to formulate inocula to amend soils to produce physicochemical responses in plants for biocontrol, such as bio-fertilizers and bioremediation. Transitory survival, functional redundancy, and competitiveness of microbial inoculum on the indigenous plant holobiont, which determines the rate and quantity of microbial necromass, can affect stable carbon stores in soil. This chapter describes the formation of stable organic matter through microbial reproduction and decline, introduces the effect of novel microbes in indigenous soil, strategies to augment microbiota in soil, and details developments in the detection of soil microorganisms. These aims will emphasize the diverse functions of microorganisms within the below-ground ecology of land systems and the significance of the role of novel microorganisms in maintaining soil carbon pools and plant health.

Keywords Novel microorganisms · Soil carbon · Necromass · Holobiont · Inocula

S. Petersen (✉)
Institute for Future Farming Systems, Central Queensland University, Rockhampton, Australia
e-mail: sonia@vinetime.com.au

S. A. Edrisi et al. (eds.), *Sustainable Plant Nutrition and Soil Carbon Sequestration*, Sustainable Plant Nutrition in a Changing World,
https://doi.org/10.1007/978-3-031-53590-1_14

14.1 Introduction

Since Girolamo Fracastoro's assertion in the 1540s that microscopic organisms cause disease and the burgeoning concept of scientific germ theory (Haas 1991), each new microorganism discovered has been considered novel. With an estimated 1% of microbes detected thus far, the 99% yet to be recognized presents microbiologists with a wealth of possibilities in the field of biotechnology (Amann et al. 1995; Riesenfeld et al. 2004). Although the ubiquity of microscopic organisms is well known, the diverse range of functions and roles within the soil profile is poorly understood. The gap in knowledge about microbial functionality and contribution to soil carbon pool and plant health drives research to continue to characterize the complex relationships between soil microbes, plants, and their environment. To determine their function, microbiologists must identify novel microbes and culture them in quantities large enough to extract usable compounds.

Technology has advanced so species recognition through genomic sequencing tools and genome databanks can help identify novel microbes along phylogenic lines within their taxonomy without the need to culture them in the laboratory. The traits that microbes display can be classified as functional groups and partly predicted by their evolutionary history. Rather than exploring individual microbes, an ecological approach may be taken whereby a community of microbes is identified within a microbiome, representing functions that multiple species perform together. Functional redundancy occurs when different microbes within a community perform similar and sometimes identical functions. These may include fixing nitrogen, lignin degradation, humus formation, and plant growth-promoting qualities. It is called phylogenics to organize groups of microbial species into their functional similarity rather than their taxonomy.

Novel organisms are not only those that occur naturally but also those derived from modifying their genes or those that are invasive, non-indigenous, and range-expanding species (Jeschke et al. 2013). Genetically engineered microorganisms (GEM) may also impact SOC maintenance and plant health as microbial functionality becomes more evident. Therefore, there is the potential for novel microorganisms to be engineered to include selected functional traits. Soil microorganisms store carbon in their cells, and when they die, they form necromass. This process alters the composition of biomolecules. The interaction of intra/extracellular biomolecules with mineral and inorganic compounds in the soil contributes to forming and maintaining stable carbon stores by forming half-lives of more than the separate biomolecules. Soil microorganisms also contribute to 'soil health,' which implies well-functioning nutrient cycling and soil physical properties, improving plant health. This interrelationship between plants and their microbiomes (the holobiont) impacts carbon sequestration (Liang et al. 2020).

The global soil carbon pool is larger below the ground than in aboveground biomass. Soil management practices have impacted the decline of SOC. For example, in Australia's semi-arid bioregion, commonplace for vertosol soils referred to as brigalow country, the clearing, cultivating, and grazing has resulted in a decline in

SOC by 25% within approximately 30 years (Pringle et al. 2016). Soil health may be improved by using soil conservation strategies to increase soil microorganism biomass, which provides a higher turnover of microorganisms, necromass formation, and stable SOC accumulation. Strategies such as erosion control, pesticide and fertilizer reduction, soil organic amendments, and alteration of physicochemical properties of soil are used to provide optimal habitat for microorganisms with the use of methods such as low/no-till, integrated pest management (IPM), compost applications and green manures. Although the aim is to increase microbial abundance and diversity, the success of these methods is specific to each soil and climate conditions for the bioregion. Applying inocula with specific microorganism consortiums may assist in increasing the augmentation of specific soil microbiota to increase plant health and maintain soil carbon accumulation.

14.2 Functions of Soil Microorganisms

Microbes' functions in microbiomes may relate to the production of enzymes, proteins, or other compounds. Other functions may assist ecosystem services by managing nutrient cycles or eliciting a plant response to alleviate abiotic and biotic stress. In particular, extracting aromas, surfactants, lipids, and polysaccharides from novel microorganisms has generated much interest in their discovery. These functions of each new microbe detected present scientists with the prospect of discovering compounds for use in pharmacology, the food industry, or agriculture as biocontrol, biofertilizers, and bioremediation.

A study modeling diverse microbiomes using the bioinformatic database KEGG (Starke et al. 2020) found that when microbial diversity was increased, so too were the functions of the microbiome. Although the functions included the decomposition of organic and recalcitrant carbon, other rare functions were poorly understood. The researchers estimated that only 0.02% and 0.14% of functions performed from the 35.5 million functions of bacteria and 3.2 million functions of fungi had been discovered so far. In a recent literature review by (Singh et al. 2019), the researchers provide a comprehensive list of compounds and their corresponding functions derived from novel microorganisms. Functions relating to plant health, such as local and systemic plant protection, resistance to diseases, enhanced plant growth, and promotion of nutrient and water uptake, can be directly attributed to microbially produced compounds such as indole acetic acid, auxins, and cytokinins (phytohormones), and N-acyl-L homoserine lactones.

The relevance of microbially produced compounds to healthy plants in agriculture has led to the study of Plant Growth Promoting Rhizobacterium (PGPR) such as *Bacillus*, *Klebsiella*, *Pseudomonas*, *Azobacter*, *Enterobacter*, *Serratia*, *Variovorax*, *Azospirillum*. These PGPRs have been made commercially available for use in agriculture as bioinoculants and are viewed as a sustainable alternative to replace the use of synthetic fertilizers and pesticides that are often used in conventional farming systems (Rai et al. 2020).

14.2.1 Necromass—A Leading Role in Maintaining Carbon Stores

The complex relationship between soil microbial communities and the environment has eluded soil scientists. To gain more understanding, many methods have been developed to characterize microbial communities, identifying functional groups within and outside the plant rhizosphere. Gene databases can help determine the phylogenic relationship between microbes, which can enhance knowledge of biotic and abiotic properties that control the formation of SOC.

SOC is the carbon fraction of SOM. The SOM was once referred to as humus and humic-like substances. Still, soil microorganisms' role in humus formation has been re-evaluated, and the language changed accordingly. Originally, it was thought that microbes contributed 1–2% (w/w) of SOM, with their prime function being the decomposition of plant residues (Schaeffer et al. 2015). Fresh organic matter, such as plant roots and micro and macroorganisms, comprise approximately 10% of SOM. Over 60% of SOM in soils comprises necromass, dead microorganisms, and their biomolecular residues (cell envelopes and lysates). The mechanics of fragmentation and adsorption to mineral surfaces within the soil aggregates form stable carbon stores that provide retention of SOM in soils (Liang et al. 2019). Figure 14.1 illustrates the process of stable SOM formation from necromass fragments (cytoplasm components), with stability increasing as residence time increases. A portion of these fragments is recycled as a food source by rhizobial microbial metabolism.

The long-held theory behind plant-based carbon formation in the soil as the primary source of SOM overlooks the integral role of microorganisms and simplifies complex natural processes. The molecular analysis of stable SOM has revealed links to the cells, cytoplasm, and excretions of microorganisms rather than plant matter. Contrary to plant chemical recalcitrance, microbially produced SOM may be stimulated by abiotic stresses such as a rise in temperature and increased soil disturbances, the extent to which requires further study (Kallenbach et al. 2016). Further, a recent publication with the interesting title "Sticky dead microbes: Rapid abiotic retention of microbial necromass in soil" by (Buckeridge et al. 2020) proposed that biotic immobilization was the process of biotic immobilization less necessary than abiotic adsorption in the short-term retention of microbial necromass. The method of necromass cell chemistry in forming stable aggregates was more prevalent within microbially diverse soils. The research describes how dead soil microbes fragment and adsorb to mineral surfaces and chemically to other necromass fragments to form stable soil carbon aggregates retained within the soil structure.

The quantity of SOC can be estimated across a larger landscape by measuring the changes in biotic and abiotic properties. Wiesmeier et al. (2019) quantify SOC storage at different spatial scales across bioregional, climatic, and topographical parameters. The study suggests that changes in biotic and abiotic properties that impact microbial biomass indicate SOC storage. Land management changes SOC stores, which can also be used as indicators of the SOC pool. In farming systems that rely

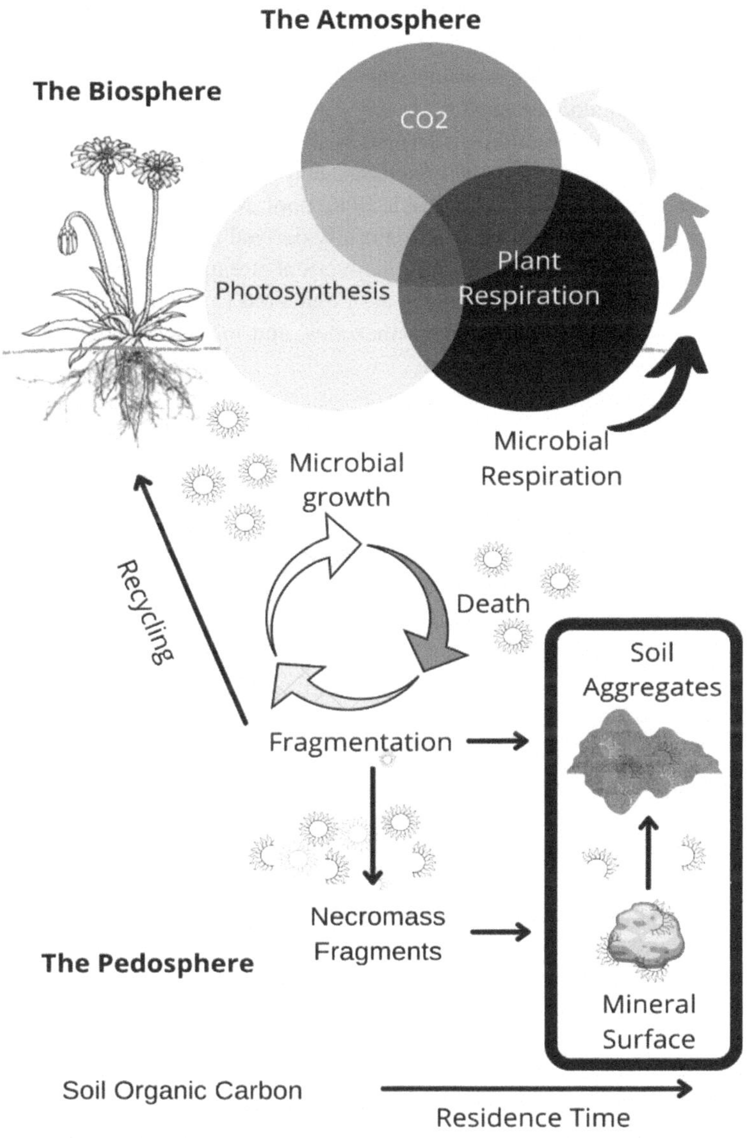

Fig. 14.1 The formation of organic carbon soil medited by the microorganisms. (Adapted from Liang et al. 2019)

on external inputs of carbon or systems that use methods such as reduced tilling to maintain SOC, the importance of increasing microbial biomass to stabilize SOC is often overlooked. The transformation of carbon through the plant system into microbial biomass was an alternate mechanism for the accumulation of SOC in a study by (Kallenbach et al. 2015). In particular, the paradox was observed whereby an increase in necromass was found to occur in systems that have reduced external carbon inputs and increased tillage. It was hypothesized that the physiological processes that increased microbial biomass turnover could be the most significant contribution to the maintenance of SOC.

Further to this concept, adding a diverse microbial population through inocula or soil amendments may increase the abundance and turnover of microbial necromass and, therefore, increase the overall stable SOC pool. Microorganisms can be selectively harvested to produce inoculum, naturally derived through biomimicry to create soil amendments, or added through inadvertent streams such as off-farm inputs or irrigation. These complex processes are connected to microbial functions, with scope for novel microbes, indigenous microbes, and inocula implicated in abiotic adsorption and biotic immobilization.

14.2.2 Microbial Life—Indigenous vs. Inoculated, a Critical Comparison

As the importance of microbial function to the necromass pool becomes more apparent, so does the knowledge of what occurs when novel microorganisms are applied to the soil and how the indigenous microbes interact with the inoculated species. As novel microbial inocula can react with the resident community of microbes in unpredictable ways, it continues to be a complex field of research. Figure 14.2 illustrates some of the reactions that can take place within the plant holobiont. Novel microbes vie for survival against indigenous microbes. Still, they can be overcome based on the competitiveness of microbes within the microbiome. Indigenous microbes may be more adapted to the physicochemical properties of the soil, or the microbiome may already possess a high level of microbial diversity that produces a high degree of functional redundancy.

Due to this mechanism, altering the microbial community may prove difficult in microbially diverse soils. Additional and sustained applications of inoculants may be required to counter this phenomenon. In depauperate soils, resilience may be lower, responding well to the introduced microbial inocula. Taxonomically diverse organisms often form symbiotic or commensal relationships with one another. Having evolved together, they can function simultaneously within a microbiome to perform specific tasks. The diversity of microbes can factor redundancy into the system. This biodiversity protects the microbiome as it becomes resilient to functional losses if one group of microbes that share the same taxon do not survive. A high level of taxonomical diversity will result in a higher functional system. A microbiome containing many unrelated species produces a healthier ecosystem.

Inoculum plays a role in increasing microbial diversity. The efficacy of minute amounts of biodiverse inoculum on soil and plant health has been an active research topic. (Golueke 1977) surmised that commercial inocula could be rivaled for effectiveness by small amounts of rich forest soil or horse manure. This concept is long-held by followers of Rudolph Steiner's biodynamic practices, whereby inoculants,

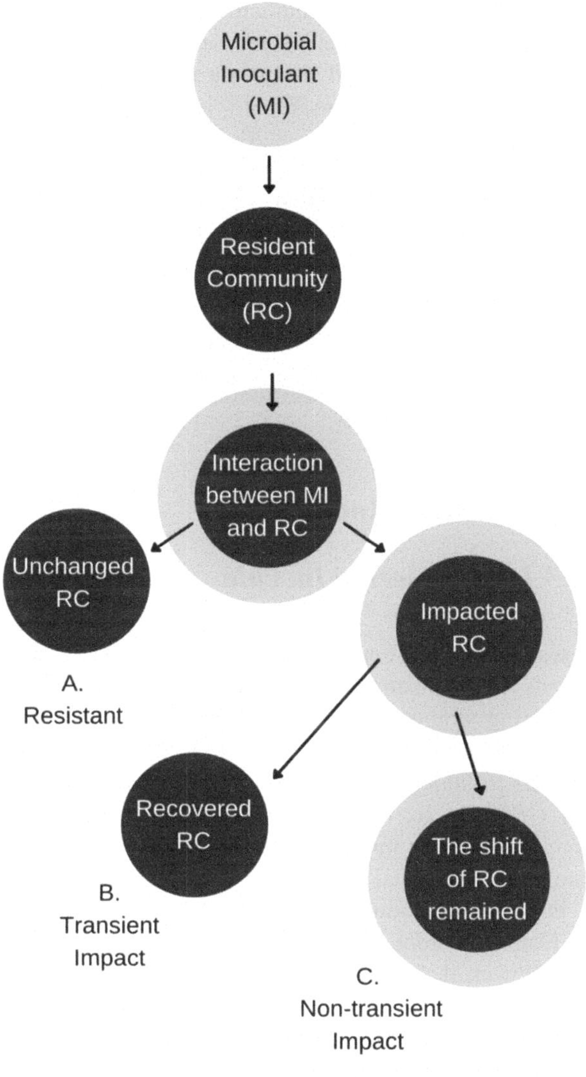

Fig. 14.2 Transitory survival, functional redundancy, and competitiveness of microbial inoculum on the indigenous plant holobiont as adapted from Mawarda et al. (2020)

including the fermented BD 500 (cow horn compost) and 501 (cow horn silica) preparations, are diluted with water and applied to the plant holobiont. The followers of biodynamic farming believe that small amounts of these extracts can increase soil quality and stimulate plant health, forming the basis of a study (Carpenter-Boggs et al. 2000). On the open market, inocula are available to promote plant health, with many making claims of the specific microbes present, such as *Bacillus sp.*, *Streptomyces sp.*, *Lactobacillus sp.*, and arbuscular mycorrhizal fungi (AMF). AMF is a well-known constituent of microbial inoculants. Its ability to form symbiotic relationships with plant roots within the rhizosphere has been studied. Its function is to increase nutrient uptake of N, P, S, K, Ca, Cu, and Zn.

Mainly, inocula are manufactured based on the target microbe. However, the metagenome of inoculants is rarely studied. In research of a commercial AMF inoculant by (Agnolucci et al. 2019), through phylogeny, they identified microbes that performed PGPR functions, identifying 36 species with 14 novel strains. The inoculum was calculated to have a 20–30% mycorrhizal potential. Still, the impact of novel strains on this process is not known. The study suggests that microbial inoculants may be more culturally diverse than thought of and contribute to various plant health-inducing functions. Microbial inoculants can be applied as foliar sprays, encapsulated to release slowly, or applied directly to the root zone. (Qiu et al. 2019) proffer encapsulation of microbial inoculants to extend their long-term viability. The most common form of bioencapsulation of inoculants is using a gelling agent, sodium alginate, for ionic or inverse gelation. This technique is most suitable for bacteria but not all microbes (Schoebitz et al. 2013).

14.3 Forms of Inocula

14.3.1 Engineered Inocula

Researchers have engineered inocula to exploit soil microbes' bioactive compounds' metabolism, enzyme production, and production. Recent studies show that combining strains increases biocontrol and plant health. One such study by (Liu et al. 2017) combined the previously identified individual novel PGPR strains, *Bacillus altitudinis* strain *AP69* and *B. velezensis* strains *AP197, AP199, AP298, and GB03,* to form a foliar spray inoculum that suppressed pathogenic fungi. Combining two or more PGPRs has significantly affected disease suppression more than using individual PGPRs.

PGPRs also produce acetoin and 2,3-butanediol, directly and indirectly affecting the plant holobiont. These compounds can ease biotic and abiotic stress on plants, and (Wu et al. 2018) found that the novel microbe strain *B. amyloliquefaciens* FZB42 stimulated the plant's response to drought by closing the stomata in *Arabidopsis thaliana* and *Nicotiana benthamiana.*

A strategy to create inocula based on beneficial indigenous (and often novel) microbes is described in Fig. 14.3. The *in-situ* method to isolate and select plant health-inducing microbes begins with selecting the plant that performs best traits when biotic or abiotic stress is applied. Samples are taken from the rhizosphere, and microbes are selected based on their beneficial qualities toward plant health. These indigenous microbes are applied to the rhizosphere, and the process is repeated. An indigenous beneficial microbial inoculum can be developed from this method.

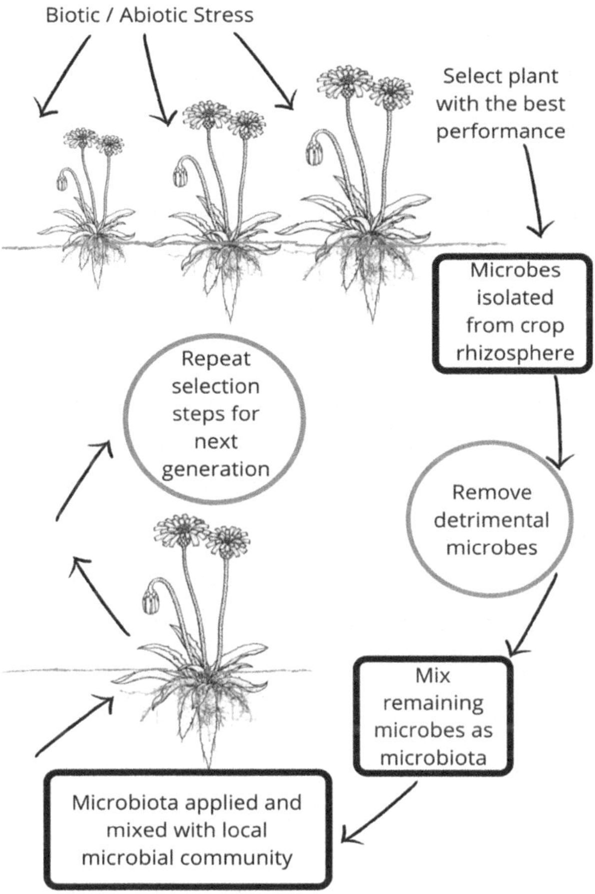

Fig. 14.3 Strategy to improve microbial inoculants. (Modified from Qiu et al. 2019)

14.3.2 Mimicking Natural Processes to Form Inocula

Another form of inocula production uses known natural processes to degrade organic matter to generate microbial communities. Some examples are compost, compost tea, biodynamic preparations, and worm juice/castings. Within these methods, microbial species selection becomes less accurate based on wide variations in processing and feedstock; microbes are often not selected for their functions, and therefore, incidental or inadvertent microbiomes are created. The research into the efficacy of the inocula created this way has thus far focused on nutrient qualities rather than microbial diversity so that novel microbes generally go undiscovered. Due to the variability of the microbial communities within naturally forming inocula, the way the microbes interact with those in the plant holobiont is relatively unknown.

14.4 Methods for Augmenting Microbiota in Soils

The origins of farming are said to have begun when our hunter-gatherer ancestor groups grew in numbers to the point that semi-nomadic or stationary communities evolved. Feeding the community resulted in the exploitation of wild food sources initially with minimal environmental impact, and these methods were considered supportive of ecosystem services (Edrisi et al. 2021). Varieties of plants were selected for the size of their fruit, vegetable, or grain, the ease with which they grew, and their ability to be stored. Over time, these varieties became dominant, and plant food biodiversity decreased.

Land management methods rapidly changed with the advent of synthetic fertilizers and pesticides. Mechanization for applying these new compounds and harvesting the produce meant vast expanses of the landscape could grow the one type of plant that shared the growth cycle with each other, enabling the coordination of spray and harvest timing. The clearing of land, the compacting effect of heavy machinery, and the compounds used have minimized plant diversity, resulted in deficiencies of macrofauna and microbial diversity in the soil, and reduced SOC.

Globally, best practices for land management attempt to address this deficit by advocating techniques to increase microbial diversity and/or organic matter in soils, yet the abundance and diversity of microbial populations and SOC changes using these methods are rarely quantified. Strategies for augmenting microbiota in soil can be found in Table 14.1 below.

14.5 Soil Conservation Methods

Methods such as no or low tillage, stubble retention, and 'trash blankets' have become popular ways to conserve topsoil from erosion. This usually occurs in crops with annual or biannual harvest, such as sugarcane, corn, sunflowers, and wheat. The theory explores the use of harvest stubble and trash as a permeable barrier to raindrops or wind, or in the instance of no/low tillage, maintaining the soil profile (aggregates) so that the topsoil is not washed away in rain events.

Research shows that these methods can assist in sustaining SOM levels. However, it can be argued that the SOC is derived from plant organic matter mechanisms rather than necromass. Without microbial detection, there is no way to determine which novel microorganisms may be present and their role in sequestering carbon in the soil.

Table 14.1 Strategies for augmenting microbiota in soil

Strategies	Methods	Targets
Soil conservation	No/low till, stubble retention	To maintain a physical barrier to prevent runoff from rainfall events
	Green manure crops/crop rotation	To maintain roots in the soil structure to prevent fallow and runoff from rainwater events
	Contour ploughing	To slow the flow of water across the landscape to prevent erosion from rainfall events
	Rotational grazing	To prevent runoff from rainfall events due to overgrazing of pastures
Pesticide and fertilizer reduction	Rotational grazing	To minimize pest pressure and increase soil nutrients from animal manures and urine
	Integrated pest management (IPM)	To utilize beneficial insect species to minimize pest pressure to reduce use of pesticides
	Green manure crops	To be re-incorporated in soils to provide macro/micronutrients or soil fumigant
Soil organic amendments	Compost/mulch applications	To increase microbial diversity and organic matter for microbial reproduction To maintain soil moisture, encouraging microbial growth.
	Inocula	To increase microbial diversity by selecting specific microorganisms
	Organic amendments, including animal manures and urine, compost, worm castings, kelp, biochar, molasses, and food residues and wastes	To increase micro/macronutrients in soil, decrease erosion, and increase microbial communities
Altering the physicochemical properties of soil	Biofertilizers	To provide optimal soil nutrients for microbial and plant growth
	Green manure crops	To be re-incorporated in soils to provide soil structure
	Chemical and mineral additions such as Ca or Mg-based limes, potash, rock dust	To alter the physical or chemical composition of soil to provide the optimum environment for microbial growth
	Keyline plowing	To open compacted soils without exposing microbes

14.5.1 Pesticide and Synthetic Fertilizer Reduction

As microbial diversity increases resilience in plants and the environment, so does the diversity of macrofauna, insects, and animals. Integrated pest management (IPM) is a method to minimize pesticide use by using natural processes to increase the diversity of bats and birds, predatory insects, and macrofauna such as mites, wasps, spiders, ladybirds, lacewings and praying mantis, to keep the pest insect and microbial species populations at bay. This can be accomplished by providing habitat for the predators, including biodiversity, plantings of cover and interrow crops, insectaries, or bat/bird boxes. Other methods incorporated into IPM introduce beneficial insects (predators and parasites) into the landscape at the egg stage in the form of commercial release of insect eggs or live larvae so that the different growth stages can predate pests throughout their growth cycles (Dubey et al. 2022).

IPM can include rotational grazing, which performs a dual purpose by increasing nutrient densities in soils, resulting in the need for lower synthetic fertilizers. A succession of species such as cattle, sheep, pigs, and chickens follow each other to new paddocks in a grazing schedule; they can break insect pest cycles and reduce the need for insecticides for livestock and plants, along with depositing manures of differing constitutions on the pasture. Cell grazing is often associated with a mixed species pasture, increasing paddock biodiversity, with livestock moving to the next pasture. In contrast, the first paddock (cell) is rested, often allowing it to go to seed. The succession of animals through the paddocks depends on the pasture, with the focus being maintaining 100% ground cover. Refuge crops, physical barriers, traps, and mechanical removal are other methods of IPM. However, they do not necessarily promote microbiota in soils.

14.5.2 Soil Organic Amendments

Organically derived amendments have been added to farming land throughout the centuries. As well as supporting microbial populations, animal manures and urine, compost, worm castings, kelp, biochar, molasses, and food residues and wastes have added nutrients to the soil, increased moisture-holding capabilities, and decreased erosion.

Although compost applications can be used as an inadvertent inoculant for the plant holobiont, the complexity of the microbiome of composts due to variations of feedstock and processing in composts assures that the microbes within one compost may not be the same as microbes in another. Therefore, compost advocates encourage applying a wide range of composts and use terms such as 'microbially rich' without specifying which microbes may be present.

14.5.3 Amendments to Alter the Physicochemical Properties of Soil

Altering the physicochemical properties of soil can influence a wider range of microbial diversity in the soil. Soils that are sodic, saline, or have limited pore space do not have the physical and chemical parameters that plants prefer to grow in and usually do not contain the microbial diversity that promotes plant health. Soil amendments that alter pHs, such as gypsum (calcium sulfate), limestone (calcium carbonate), and aluminum sulfate, are used, which positively affect soil microorganisms. While soil microbes may prefer the concentration of hydrogen ions in the soil, a functioning microbiome will have the right types of diversity of microbe species to survive changes in pH.

Primarily, mechanical methods are used to increase aeration and reduce the compatibility of soils, such as keyline tilling, a method with which soils are not turned over as they are in traditional plowing. Instead, a furrow is cut into the soil, usually on a contour with the land, breaks open the compaction and does not destroy the soil's microbial communities by exposing them.

14.6 Holobiont—Microbial Effect on Plant Health

Novel microorganisms are conducive to plant health. The plant holobiont encompasses the whole range of plant microbiomes containing the atmospheric, phyllospheric, seed, and soil microbiomes. Figure 14.4 illustrates the plant holobiont mechanism in which atmospheric, seed, and soil microbiomes form part of the phyllospheric microbiome over time. As a seed is introduced to the soil, it is coated with microbes from the soil rhizosphere. These form relationships with microbes in the atmosphere, and as the seedling grows, it continues to acquire a larger diversity of microbiology, eventually forming a unique microbial community within its phyllosphere microbiome. The principles of transitory survival, functional redundancy, and competitiveness play out in the phyllosphere whenever a novel microbe is introduced, and the plant holobiont may change over time due to biotic and abiotic stress.

The plant holobiont, therefore, determines the type, abundance, and turnover of microbial necromass, which directly affects stable carbon stores in soil. Plants can interact with their microbiomes by forming defense compounds that attract or repel microbes or pests or alter plant physiology. To do so, plants can produce antimicrobial components or enzymes such as salicylic acid, jasmonic acid, and ethylene or ACC deaminase. The development of plant physiology, such as germination, stem, leaf, flower, fruit, and root growth, bud dormancy, and senescence timing, can be influenced by the production of gibberellins (GA) by microorganisms in the plant holobiont (Bottini et al. 2004).

Microorganisms in the holobiont can also perform quorum sensing, whereby they perceive when other microorganisms have reached a critical density or as a

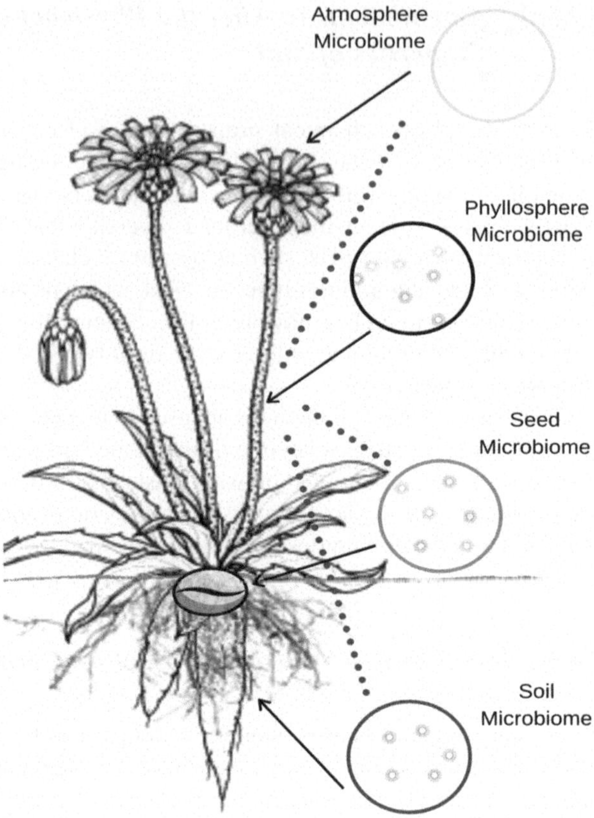

Fig. 14.4 The holobiont mechanism. (Adapted from Sánchez-Cañizares et al. 2017)

response to other changes in the surrounding environment. For example, the microbial community can make coordinated reactions as defenses, which often occur in the presence of pathogenic microbes that have reached critical densities. Quorum sensing describes how individual microbes signal to each other by releasing molecules that generate a response from other microbes. Quorum sensing is a form of social behavior by microbes that includes antibiotic resistance, swarming motility, and biofilm formation.

14.7 The Role of Plants in Selecting Microorganisms Within the Rhizosphere

Plants can recruit and interact with microbes within their rhizosphere to help them evade attacks by pathogens and insects and perform other functions. Studies have revealed that different species of plants establish and manipulate microbial populations in their root zone, producing very different rhizobial microbiomes. For example, in a review of literature, (Berendsen et al. 2012) drew similarities between the rhizosphere microbiome and the human gut microbiome in that they shared important characteristics such as nutrient uptake, pathogen prevention, immunity responses, recognition of symbiont and commensal microorganisms, and microbial abundance and diversity. As with the human microbiomes, the rhizobial microbiome can vary greatly according to the individual plant. Microbes often appear plant-dependent and can perform different tasks depending on the surrounding environment and the plant's needs.

14.8 Plant Reactions to Microbial Changes Outside the Rhizosphere

The endosphere is the soil portion situated outside the rhizosphere. The phyllosphere, biosphere, and atmosphere are the above-ground regions. These terms are used in microbiology to refer to the habitat for microbes. The phyllosphere comprises the plant's stem, leaves, flowers, and fruit. It is called the caulosphere, phylloplane, anthosphere, and carposphere.

An alteration in the microbiomes of the areas outside the rhizosphere can also impact plant health. For example, spatially distant plants can signal to one another by producing volatile organic compounds (VOCs) such as β-caryophyllene. A recent study of this phenomenon (Kong et al. 2021) found that after treating a tomato seedling with a common PGPR *Bacillus amyloliquefaciens* GB03, the neighboring plant released salicylic acid (SA) in its root zone, which is a microbe-induced plant volatile (MIPV). The use of airborne signaling between plants by producing VOCs to stimulate MIPV highlights the significance of changes in microbial diversity throughout the whole plant holobiont.

14.9 Microbial Formulations for Biocontrol, Biofertilizer, Bioremediation

Knowledge of symbiotic, commensal, and parasitic microbial guilds can assist in formulating inocula to apply to the plant holobiont to produce physicochemical responses that enable biocontrol, biofertilization, and bioremediation. Bioinformatic tools can compare genomic sequencing with gene databases to predict microbial functions. A common method of grouping species based on their DNA profile is using operational taxonomic units (OTUs), whereby species are considered directly related when their sequence distance is less than 3% from each other.

Using the data from functional analysis can lead researchers to discover novel microbes that can be assessed *in vivo* for antagonistic properties to pests and pathogens such as root-knot nematodes and pathogenic fungi. Often, based on the

Table 14.2 Bioremediation—novel bacteria for degradation of pollutants

Novel Bacteria	Pollutant	References
Bacillus (JN897279) KC3, pseudomonas (KJ541832) 5	Polycyclic aromatic hydrocarbons (PAH)	Kuppusamy et al. (2016)
Escherichia fergusonii KLU01	Hydrocarbon	Sriram et al. (2011)
Lysinibacillus boronitolerans MH2	Halogenated compounds	Heidarrezaei et al. (2020)
Methylosinus trichosporium OB3b, Geobacter bemidjiensis	Methylmercury	Zhou et al. (2020)
Proteus vulgaris strain CPY1, P. Aeruginosa strain LPY1	Pyrene	Behera and Ray (2021)
Pseudomonas putida MHF 7109	Benzene	Behera and Ray (2021)
Raoultella sp. strain X13	Cadmium	Xu et al. (2019)
Rhodanobacter sp. PHE-2	Phenanthrene	Li et al. (2017)

sampling region of the holobiont, OTUs can be grouped into 'micro-habitats.' The propensity of disease in plants is directly related to diminished microbial diversity. In the past, microbial biofertilizers have been selected based on their nutrient-producing functions. Microbes that can fix nitrogen from the air, solubilize phosphorous, oxidize sulfur, and mobilize metals such as zinc, iron, and molybdenum for plant nutrition have been the focus. When each function is combined with another, the efficacy of the biofertilizer on plant health is increased (Ahmed et al. 2015). Bioinformatics can also identify novel strains of microbes that possess antimicrobial qualities or enzymatic compounds that break down heavy metals and other pollutants in soil to inhibit their undesirable impact on plants. Table 14.2 lists novel microbes and the pollutants they degrade.

Further to heavy metal and pollutant bioremediation, it was recently discovered that bacteria and archaea that can be isolated simultaneously are implicated in ammonia-oxidization in soils, and each performs this function depending on abiotic/biotic stresses. In soils that have a low pH, *Candidatus Nitrosotalea sp. GC1* may be advantageous over the bacteria genus *Nitrospira* in denitrification due to its preference for acidic soils. With over 30% of the world's soils classed as acidic, the novel microbe has the potential for remediation (Pu 2019).

14.10 Advanced Technologies for Detecting Novel Microorganisms in Soil

14.10.1 Metagenomics

Advances in technology have enabled the extraction and cloning of DNA from microbiomes. Metagenomic tools and databases have been shared throughout scientific communities to build knowledge of newly discovered microbes, often in a

Table 14.3 Bioinformatic databases and tools for metagenomics

Name	Web reference	Function
BLAST	https://blast.ncbi.nlm.nih.gov/Blast.cgi	Statistical significance calculator for protein and nucleotide sequences. Primer design tool and database
ELIXIR	https://elixir-europe.org/	A collaborative network of databases and tools
KEGG	https://www.genome.jp/kegg/	Kyoto Encyclopedia of Genes and Genomes (KEGG) database for functional classes
PIT Bioinformatics Group	https://pitgroup.org/amphoranet/	Hosts AMPHORA2 database and metagenomic analysis tools
QIIME	https://qiime2.org/	Microbiome bioinformatics platform

regulated open-source format. Novel genomic sequences can be submitted to the database host, who adds them. Bioinformatic software tools assist users of the databases in quickly and accurately comparing genomic sequences they generate from the polymerase chain reaction (PCR) processes. Users can access these tools to generate graphical images of complex data to display their findings.

The discovery rate of novel microbes and new database listings is increasing rapidly. For example, in 1988, the National Center for Biotechnology Information (NCBI) was founded, which hosts GenBank, one of the many collaborative open-source databases of DNA nucleotide sequences, adenine (A), thymine (T), guanine (G), and cytosine (C). It is a collaboration between the European Nucleotide Archive (ENA) and the DNA Data Bank of Japan (DDBJ). In 1996, GenBank formally described >15,500 species of microorganisms, and in 2018, this list increased enormously to over 400,000 species (Sayers et al. 2018; Benson et al. 1996).

The scientific community has reorganized database hosting over the past few years to accommodate the growing magnitude of genomic information collected. In 2009 (Singh et al. 2009) produced a table listing 13 Bioinformatics tools and databases in metagenomics, with only seven on the list still in use today, being ARB, GenBank, GOLD, Megx.net, RefSeq, SILVA, SINA, and UniFrac. Aside from those mentioned, Table 14.3 lists current bioinformatic tools and databases commonly used by micro/molecular biologists.

14.10.2 Advances in Bioinformatics

The genetic databases have further developed as the focus has led to phylogenics based on enzyme and protein production and functions of microbes, and as such, database search facilities and bioinformatic tools will continue to evolve to include these traits. Bioinformatic tools rely on computer algorithms to detect, compare, and organize biological gene information. One approach scientists can use is applying a protein comparison whereby a reference microorganism that fills a particular niche habitat is selected. The output from the bioinformatic tools includes a phylogenic tree that lists microbes that distantly compare with the signature protein domain. The microbe may not share physical or metabolic traits but produces the protein being compared. The diversity of microbes in this example may contribute to functional redundancy and increase the resilience to external abiotic and biotic stresses the plant may experience.

For example, when studying the phylum *Verrucomicrobia*, a group of bacterium that defies culturing yet is prevalent in water, soil, and gut microbiota, (Bünger et al. 2020) found the culture-independent method to analyze signature proteins using bioinformatics revealed significant findings. In addition to the production of the proteins, four novel strains were isolated, and two were found to be endophytic, that is, a symbiont of plant roots. The bacteria share their metabolic exudates with the plant to promote health, similar to the well-known group of AMF. Bioinformatic tools can provide a more comprehensive understanding of micro-habitats' diversity by comparing the functions of distantly related phyla.

14.11 Developing Technologies to Culture the 'Unculturable' Novel Microorganisms

The culturability of novel microorganisms depends on the knowledge of the habitats in which the microbes thrive, their interactions with abiotic and biotic stress, and their functional role within soil microbiomes. Bioinformatic tools can assist in identifying these parameters, and recently, (Chaudhary et al. 2019) expanded on this knowledge and developed a diffusion bioreactor technique to mimic natural environments to elicit novel microorganisms. The samples were taken across the seasons. As well as successfully cultivating 35 novel strains, they were also able to determine the seasonal variability of the microbiota from the sampling site. Other variables, such as incubation period and growth substrate, were applied to the bioreactor, and insights were made on the effect of microbe growth. This study expands the scope of the computational algorithms used in bioinformatics. It advances the discovery of novel microbes and their reproduction needs when applied to new technologies.

14.12 Way Forward

Going beyond the parameters of physicochemical principles, innovation in metagenomics, particularly bioinformatic algorithms, has brought gains in understanding the complex metabolic processes between plants, microbes, and their environment. Continuous progress in this field can assist in developing the environmental conditions suitable to culture novel microorganisms based on predictions of their functions. Growing microbes *in vitro* in these environments within the communities they

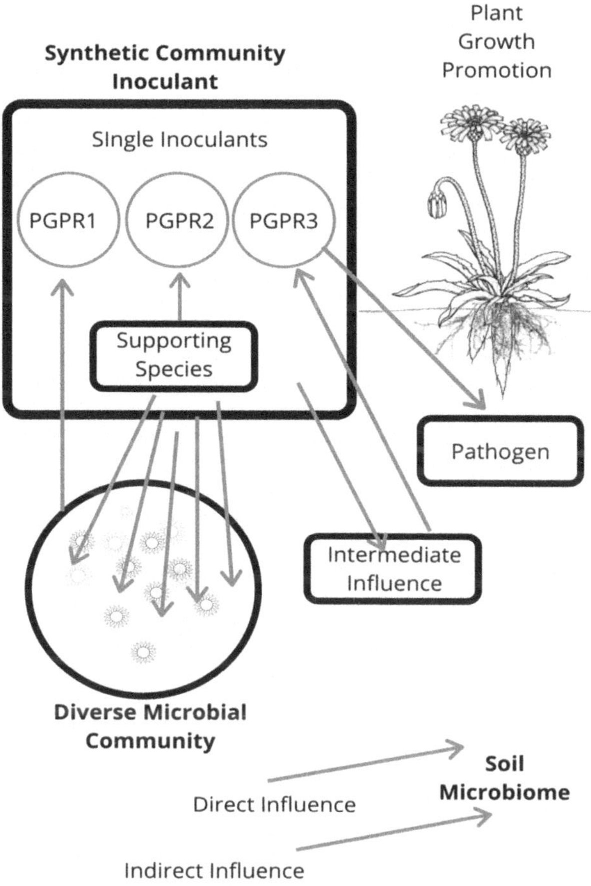

Fig. 14.5 The development of synthetic community inoculants. (Adapted from Sánchez-Cañizares et al. 2017)

have evolved with can help our understanding of the biological and chemical processes of symbiosis and commensal relationships, transiency, functional redundancy, and competitiveness within microbiomes (Natarajan and Bhatt 2020).

The introduction of novel microorganisms into a holobiont has traditionally occurred by manufacturing inoculum from the root rhizosphere. Traditional inoculant formation can be adapted by combining diverse microbes from different microecologies to form 'synthetic' inoculants. The microbial consortium within the stylized inoculum does not naturally form independently. It is integrated to form a new consortium. In Fig. 14.5, (Sánchez-Cañizares et al. 2017) propose combining keystone species with a diversity-based approach to broaden single or mixed inoculant production techniques.

Researchers have taken a step further and proposed *in-situ* microbial engineering. Using methods (Qiu et al. 2019), genetically modified RNA can be synthesized from beneficial indigenous microbes based on their effect on plant health. Figure 14.6 illustrates combining synthetic community inoculant methods with genetically

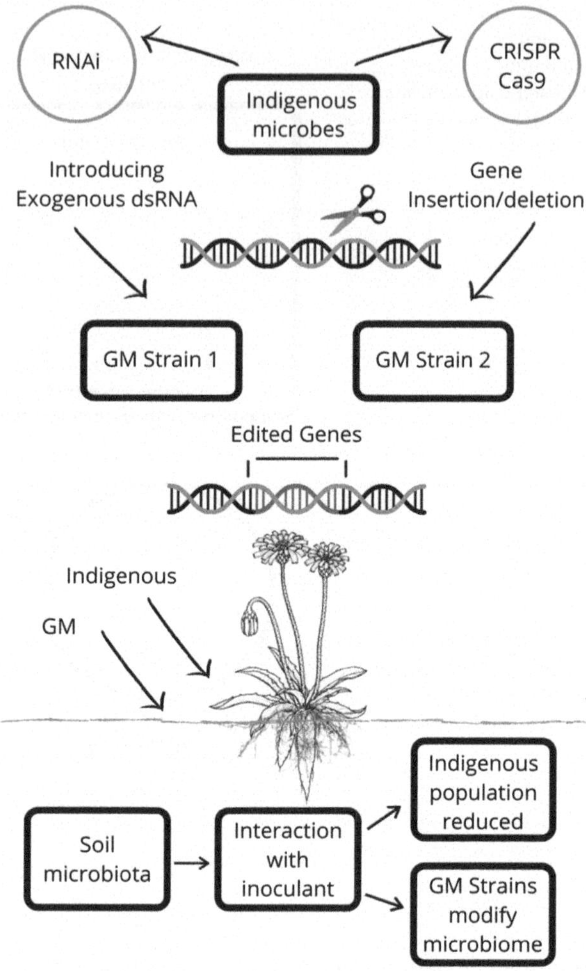

Fig. 14.6 *In situ* microbial engineering. (Adapted from Qiu et al. 2019)

modified methods to produce genetically engineered microorganisms (GEM). It indicates the microbial engineering process of modifying indigenous microbes with genomic tools such as RNAi, CRISPR, and Cas9 to edit genes through gene splicing or exogenous dsRNA to form genetically modified strains of soil microbes. Once applied to the soil, transitory survival, functional redundancy, and competitiveness are thought to be less than what occurs through natural mechanisms, and (Qiu et al. 2019) propose the GEM strains will dominate. Further considerations are required on how these GM microbes react to the plant holobiont. The ethical and environmental implications of releasing GEM strains of soil microbes into the landscape are yet to be deliberated by the scientific community, regulatory bodies, and ethical processes. The effect of GEM inoculants on the rhizosphere is not well understood. A recent research by (Azizoglu et al. 2020) suggests further in situ study to clarify the effect on the rhizosphere of the addition of engineered inoculants with recombinant DNA, particularly horizontal and vertical gene transfer. The risk of DNA transfer and the potential for indigenous biodiversity loss must be considered when introducing novel microorganisms into the ecosystem.

14.13 Conclusions

Novel microorganisms play an integral role in maintaining SOC stores and plant health. Not only can the addition of new microbes increase nutrient uptake by the plants, it can enhance the plant's resilience to biotic and abiotic stress. Plants display an increased ability to repel pests and diseases, mitigate moisture loss, or exhibit other beneficial chemical or physiological effects.

Improving plant health leads to an increase in the turnover of microorganisms within the rhizosphere and the abundance of necromass that contributes to stable organic compounds and the maintenance of SOC stores. Traditional methods to augment microbiota in soils, such as soil conservation, pesticide and synthetic fertilizer reduction, soil organic amendments, and amendments to alter the physicochemical properties of soil, assist in maintaining microorganisms by improving the soil environment for microbial habitation, but abundance and diversity of microorganisms through these strategies are not readily quantified. Ultimately, the biodiversity of microbe species in soils depleted of the abundance and diversity of microbes requires inoculation from sources outside of the plant holobiont.

Inoculum can be engineered through selective isolation, extraction, and culturing or cloning, or through GM techniques and applied to the plant to interact with the phyllosphere, the rhizosphere, and/or the endosphere to increase microbial colonies. Applying a species-diverse microbial inoculum improves functional redundancy by increasing the taxa diversity within the holobiont of plants. Microbiological technologies have advanced so that strains of soil microbes can be identified rapidly, and advancement in bioinformatic databases and tools that are readily available supports the metagenomic characterization of novel microbes and their functions within inoculum. The recent developments in the field of soil microbiological science and new understandings of the functional role of microbes within microbiomes and the evolutionary and biological significance of microorganisms bring a new perspective to the applications of inoculum for further exploration of opportunities to enhance soil carbon and soil health through the use of soil inocula.

Acknowledgments The author wishes to thank Dr. S. Bhattarai (Institute for Future Farming Systems, Central Queensland University) for comments on the manuscript and technical editing and Dr. K. Walsh (Institute for Future Farming Systems, Central Queensland University) for suggestions, useful discussions, and support. The author acknowledges Emeritus Professor Stuart B. Hill (Foundation Chair of Social Ecology, School of Education Western Sydney University) for the indirect assistance through a generous supply of cross-disciplinary reading materials.

References

Agnolucci M, Avio L, Pepe A, Turrini A, Cristani C, Bonini P, Cirino V, Colosimo F, Ruzzi M, Giovannetti M (2019) Bacteria associated with a commercial Mycorrhizal inoculum: community composition and multifunctional activity as assessed by Illumina sequencing and culture-dependent tools. Front Plant Sci 9:1956

Ahmed D-N, Babita, Thakur M (2015) Organic farming: a holistic approach towards sustainable fruit production. Eur J Pharm Med Res 2:108–115

Amann RI, Ludwig W, Schleifer KH (1995) Phylogenetic identification and in situ detection of individual microbial cells without cultivation. Microbiol Rev 59:143–169

Azizoglu U, Jouzani GS, Yilmaz N, Baz E, Ozkok D (2020) Genetically modified entomopathogenic bacteria, recent developments, benefits and impacts: a review. Sci Total Environ 734:139169

Behera SS, Ray RC (2021) Bioprospecting of cowdung microflora for sustainable agricultural, biotechnological and environmental applications. Curr Res Microb Sci 2:100018

Benson DA, Boguski M, Lipman DJ, Ostell J (1996) GenBank. Nucleic Acids Res 24:1–5

Berendsen RL, Pieterse CMJ, Bakker PAHM (2012) The rhizosphere microbiome and plant health. Trends Plant Sci 17:478–486

Bottini R, Cassan F, Piccoli P (2004) Gibberellin production by bacteria and its involvement in plant growth promotion and yield increase. Appl Microbiol Biotechnol 65:497–503

Buckeridge KM, La Rosa AF, Mason KE, Whitaker J, Mcnamara NP, Grant HK, Ostle NJ (2020) Sticky dead microbes: rapid abiotic retention of microbial necromass in soil. Soil Biol Biochem 149:107929

Bünger W, Jiang X, Müller J, Hurek T, Reinhold-Hurek B (2020) Novel cultivated endophytic Verrucomicrobia reveal deep-rooting traits of bacteria to associate with plants. Sci Rep 10:8692

Carpenter-Boggs L, Reganold JP, Kennedy AC (2000) Effects of biodynamic preparations on compost development. Biol Agric Horticult 17:313–328

Chaudhary DK, Khulan A, Kim J (2019) Development of a novel cultivation technique for uncultured soil bacteria. Sci Rep 9:6666

Dubey PK, Singh A, Merah O, Abhilash PC (2022) Managing agroecosystems for food and nutrition security. Curr Res Environ Sustain 4:100127. https://doi.org/10.1016/j.crsust.2022.100127

Edrisi SA, Rakshit A, Dubey PK, Abhilash PC, Singh SK, Patra AK, Pathak H (2021) Managing soil resources for human health and environmental sustainability. In: Soil science: fundamentals to recent advances, pp 3–11

Golueke CG (1977) Biological reclamation of solid wastes. Rodale Press, Emmaus

Haas LF (1991) Girolamo Fracastoro 1484-1553. J Neurol Neurosurg Psychiatry 54:855–855

Heidarrezaei M, Shokravi H, Huyop F, Rahimian Koloor SS, Petrů M (2020) Isolation and characterization of a novel bacterium from the marine environment for Trichloroacetic acid bioremediation. Appl Sci 10:4593

Jeschke JM, Keesing F, Ostfeld RS (2013) Novel organisms: comparing invasive species, GMOs, and emerging pathogens. Ambio 42:541–548

Kallenbach CM, Grandy AS, Frey SD, Diefendorf AF (2015) Microbial physiology and necromass regulate agricultural soil carbon accumulation. Soil Biol Biochem 91:279–290

Kallenbach CM, Frey SD, Grandy AS (2016) Direct evidence for microbial-derived soil organic matter formation and its ecophysiological controls. Nat Commun 7:13630

Kong HG, Song GC, Sim HJ, Ryu CM (2021) Achieving similar root microbiota composition in neighbouring plants through airborne signalling. ISME J 15:397–408

Kuppusamy S, Thavamani P, Singh S, Naidu R, Megharaj M (2016) Polycyclic aromatic hydrocarbons (PAHs) degradation potential, surfactant production, metal resistance and enzymatic activity of two novel cellulose-degrading bacteria isolated from koala faeces. Environ Earth Sci 76:14

Li J, Zhang D, Song M, Jiang L, Wang Y, Luo C, Zhang G (2017) Novel bacteria capable of degrading phenanthrene in activated sludge revealed by stable-isotope probing coupled with high-throughput sequencing. Biodegradation 28:423–436
Liang C, Amelung W, Lehmann J, Kästner M (2019) Quantitative assessment of microbial necromass contribution to soil organic matter. Glob Chang Biol 25:3578–3590
Liang C, Kästner M, Joergensen RG (2020) Microbial necromass on the rise: the growing focus on its role in soil organic matter development. Soil Biol Biochem 150:108000
Liu K, Mcinroy J, Hu C-H, Kloepper J (2017) Mixtures of plant growth-promoting Rhizobacteria enhance biological control of multiple plant diseases and plant growth promotion in the presence of pathogens. Plant Dis 102:67–72
Mawarda PC, Le Roux X, Dirk Van Elsas J, Salles JF (2020) Deliberate introduction of invisible invaders: a critical appraisal of the impact of microbial inoculants on soil microbial communities. Soil Biol Biochem 148:107874
Natarajan A, Bhatt AS (2020) Microbes and microbiomes in 2020 and beyond. Nat Commun 11:4988
Pringle MJ, Allen DE, Orton TG, Bishop TFA, Butler DW, Henry BK, Dalal RC (2016) Effects of land-use change and management on soil carbon and nitrogen in the Brigalow Belt, Australia: II. Statistical models to unravel the climate-soil-management interaction. Rangeland J 38:453–466
Pu J (2019) Enrichment and characterisation of novel microorganisms involved in the nitrogen cycle. Doctor of Philosophy, University of Queensland
Qiu Z, Egidi E, Liu H, Kaur S, Singh BK (2019) New frontiers in agriculture productivity: optimised microbial inoculants and in situ microbiome engineering. Biotechnol Adv 37:107371
Rai PK, Singh M, Anand K, Saurabh S, Kaur T, Kour D, Yadav AN, Kumar M (2020) Chapter 4: Role and potential applications of plant growth-promoting rhizobacteria for sustainable agriculture. In: Rastegari AA, Yadav AN, Yadav N (eds) New and future developments in microbial biotechnology and bioengineering. Elsevier
Riesenfeld CS, Schloss PD, Handelsman J (2004) Metagenomics: genomic analysis of microbial communities. Annu Rev Genet 38:525–552
Sánchez-Cañizares C, Jorrín B, Poole PS, Tkacz A (2017) Understanding the holobiont: the interdependence of plants and their microbiome. Curr Opin Microbiol 38:188–196
Sayers EW, Cavanaugh M, Clark K, Ostell J, Pruitt KD, Karsch-Mizrachi I (2018) GenBank. Nucleic Acids Res 47:D94–D99
Schaeffer A, Nannipieri P, Kaestner M, Schmidt B, Botterweck J (2015) From humic substances to soil organic matter–microbial contributions. In honour of Konrad Haider and James P Martin for their outstanding research contribution to soil science. J Soils Sediments 15. https://doi.org/10.1007/s11368-015-1177-4
Schoebitz M, Lopez M, Roldán A (2013) Bioencapsulation of microbial inoculants for better soil-plant fertilization. A review. Agron Sustain Dev 33:751–765
Singh J, Behal A, Singla N, Joshi A, Birbian N, Singh S, Bali V, Batra N (2009) Metagenomics: concept, methodology, ecological inference and recent advances. Biotechnol J 4:480–494
Singh D, Raina TK, Kumar A, Singh J, Prasad R (2019) Plant microbiome: a reservoir of novel genes and metabolites. Plant Gene 18:100177
Sriram MI, Gayathiri S, Gnanaselvi U, Jenifer PS, Mohan Raj S, Gurunathan S (2011) Novel lipopeptide biosurfactant produced by hydrocarbon degrading and heavy metal tolerant bacterium Escherichia fergusonii KLU01 as a potential tool for bioremediation. Bioresour Technol 102:9291–9295
Starke R, Capek P, Morais D, Callister SJ, Jehmlich N (2020) The total microbiome functions in bacteria and fungi. J Proteome 213:103623
Wiesmeier M, Urbanski L, Hobley E, Lang B, Von Lützow M, Marin-Spiotta E, Van Wesemael B, Rabot E, Ließ M, Garcia-Franco N, Wollschläger U, Vogel H-J, Kögel-Knabner I (2019) Soil organic carbon storage as a key function of soils – a review of drivers and indicators at various scales. Geoderma 333:149–162

Wu L, Li X, Ma L, Borriss R, Wu Z, Gao X (2018) Acetoin and 2,3-butanediol from bacillus amyloliquefaciens induce stomatal closure in Arabidopsis thaliana and Nicotiana benthamiana. J Exp Bot 69:5625–5635
Xu S, Xing Y, Liu S, Huang Q, Chen W (2019) Role of novel bacterial Raoultella sp. strain X13 in plant growth promotion and cadmium bioremediation in soil. Appl Microbiol Biotechnol 103:3887–3897
Zhou X-Q, Hao Y-Y, Gu B, Feng J, Liu Y-R, Huang Q (2020) Microbial communities associated with methylmercury degradation in Paddy soils. Environ Sci Technol 54:7952–7960

Part IV
Attaining Environmental Sustainability Under the Climate Change Scenario

Chapter 15
Identification of Priority Areas for Cropland Restoration in India Through Agroforestry and Soil Carbon Enrichment

Aritra Bandopadhyay, A. Arya, and Rajiv Kumar Chaturvedi

Abstract SOC is a crucial driver of plant growth and crop yield in agricultural fields. Croplands worldwide are estimated to potentially sequester the carbon from 0.90 to 1.85 Pg C year^{-1}. Due to excessive and unsustainable usage, Indian soil has depleted nutrient content, including SOC, which in turn is adversely affecting the agricultural yield and lowering carbon stocks. This study focuses on assessing the current state of SOC density distribution in Indian croplands by analyzing SOC stock data from the Food and Agriculture Organization of the United Nations and Land use and land cover (LULC) data from ESRI, segmenting according to agro-climatic zones of India. We identified priority cropland areas for SOC restoration and segmented them into ten distinct classes (Deciles). The first six deciles with relatively low SOC levels are priority areas for SOC restoration. The first decile representing the worst SOC density is considered for the agro-forestry option, as it will have the least implications regarding food security trade-offs. We identified about 14 million ha of area for agroforestry options. Hence, we project that these areas will be able to sequester about ~400–500 million tons of additional CO_2 in soils by 2050. Next, deciles from second to sixth are considered for SOC enrichment and restoration within cropland-remaining-cropland (CL-CL) land use, projected to yield 800–900 million tons of additional CO_2 in soils by 2050. Thus, our identified priority areas for agroforestry and soil carbon enrichment in croplands are projected to yield more than 1300 million tons of additional CO_2 in soils by 2050.

Keywords SOC · Soil mapping · Agroclimatic zones · Agroforestry · Cropland restoration · Carbon sequestration

A. Bandopadhyay · A. Arya · R. K. Chaturvedi (✉)
Birla Institute of Technology and Science Pilani, Pilani, Goa, India
e-mail: rajivc@goa.bits-pilani.ac.in

S. A. Edrisi et al. (eds.), *Sustainable Plant Nutrition and Soil Carbon Sequestration*, Sustainable Plant Nutrition in a Changing World,
https://doi.org/10.1007/978-3-031-53590-1_15

15.1 Introduction

Globally, soil health degrades due to over-utilization, climate change, land cover change, biodiversity loss, etc. (IBPES 2018). SOC is a major component of the terrestrial carbon pool as it helps regulate various processes such as the carbon cycle, water supplies, and biodiversity, thereby providing quality ecosystem services for the well-being of humans (Scharlemann et al. 2014). The SOC plays a significant role in the formation of soil. It is the only carbon component of organic soil compounds. SOC helps maintain the physical, chemical, and biological properties of soil in equilibrium and cropland productivity and food security in a sustainable manner (Ghosh et al. 2010).

Moreover, it plays a significant role in soil structure stabilization and improves aggregation to reduce soil loss through erosion (Zhou et al. 2020). SOC is significant in the carbon cycle as the top 1-meter layer contains about 1500 Gt C and is the top contributor to the terrestrial carbon pool (Jobbágy and Jackson 2000). The past decade has witnessed a declining trend in soil quality, resulting in increased atmospheric temperature, changing weather patterns, soil depletion, accelerated soil erosion, and mineralization (Lal 2012). Besides, anthropogenic activities like environmental pollution, deforestation, biomass burning, and land use changes alter the soil's organic carbon content. Hence, to increase carbon sequestration in soil, an initiative called 'the 4th per thousand' was initiated at 21st COP 21 in December 2015, aiming to mitigate climate challenges by restoration of soil carbon in the agricultural lands at the rate of 0.4% year^{-1} (Lal 2004).

15.1.1 Soil Organic Carbon Dynamics

The combined process of photosynthesis, respiration, and decomposition, as shown in Fig. 15.1 regulates the levels of SOC. The process of soil formation converts the parent material or geological strata into soil catalyzed by the combined influence of physical, chemical, and biological processes.

This process causes the decomposition of mineral particles into tiny fragments, which creates pore spaces for the microfaunas and organisms, even in the root systems of plants. This process eventually leads to the colonization of microscopic organisms beneath the ground of the parent plant. As half of the biomass of these microorganisms consists of carbon-bound organic compounds, when these organisms die, the carbon will remain in the soil. Moreover, the carbon in the soil is subjected to various chemical processes, such as oxidation and fermentation, which lead to the formation of CO_2. Some part of CO_2 thus formed will be released into the atmosphere through respiration, and the organisms will use the remaining to increase biomass. Part of the remaining carbon in the soil will be converted as SOC, defined as the excess of carbon in the soil due to the inflow and outflow of carbon by various micro-organisms (Gougoulias et al. 2014). Global temperature has steeped by about

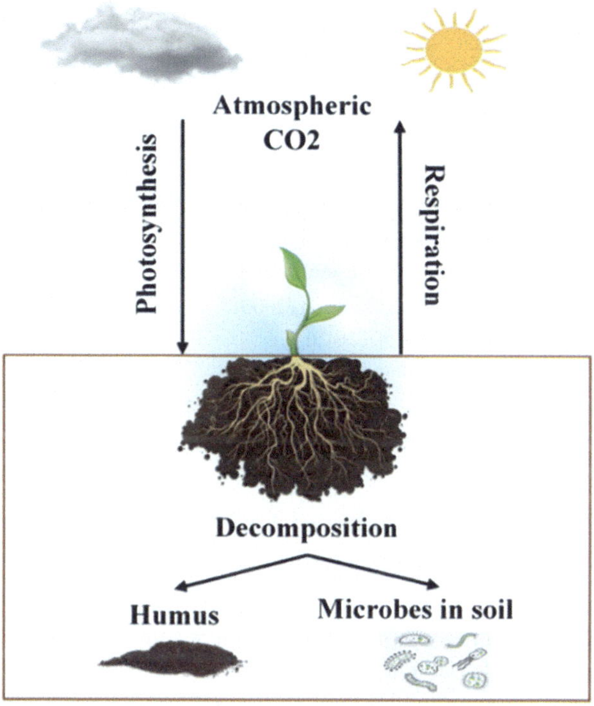

Fig. 15.1 The interaction among the natural photosynthesis, respiration, and decomposition processes. (Source: Gougoulias et al. 2014)

1.1 °C in the last century due to the uncontrolled emission of GHGs in the atmosphere (IPCC 2022). The rise in soil temperature levels of atmospheric CO_2 increases the decomposition rate and alters the Net Primary Productivity (NPP).

15.1.2 Global Distribution of Soil Organic Carbon in Croplands

The total concentration of SOC in the top 1-meter layer was estimated to be 1500–1600 Pg, which was 56% more than the top 3 meters of soil, which had 2344 Pg C (Jobbágy and Jackson 2000). Stockmann et al. 2015, based on spatio-temporal data, provide a global SOC assessment. The assessment results projected the impact of landcover changes as a primary factor and climatic factors such as temperature and precipitation as secondary agents influencing the SOC change. The estimate of global SOC concentration calculated by various studies is shown in Table 15.1. Different modeling approaches have been incorporated to assess the impact of climate change and global warming on SOC sequestration. They are effective tools for assessing soil SOC stocks (Falloon et al. 1998; Jones et al. 2005). A list of previous studies in this regard is shown in Table 15.1, and the SOC stock is estimated in each of these studies.

Table 15.1 List of studies previously conducted to estimate global SOC stocks

Global estimate of SOC			
Sl No:	Reference	SOC Stock (PgC)	Remarks
1	Jobbágy and Jackson (2000) and Batjes (2002)	1502	Upto 1 m depth
2	Harmonized World Soil Database (HWSD; FAO/IIASA/ISRIC/ISS-CAS/JRC (2009))	1417	For the top 1 m of soil
		716	Top 30 cm
3	Jobbágy and Jackson (2000)	2344	Depth of 3 m
4	Jansson et al. (2010)	3000	Depth of 3 m
5	Scharlemann et al. (2014)	504–3000	Intercomparison of 27 studies on global SOC stock between 1951 and 2011

15.1.3 Distribution of Soil Organic Carbon in Indian Croplands

With 20 agroecological regions, India's agricultural land area is the second largest in the world. Hence, agriculture is crucial in rural households as they depend on the agrarian economy for their livelihood (Agricultural Statistics 2021). Green revolution and increased production from agriculture by introducing hybrid crops altered the SOC status of Indian soils (Eliazer Nelson et al. 2019). Initial attempts to estimate the SOC concentration in Indian soil were carried out by (Bhattacharyya et al. 2000), and they revealed that 9.5 Pg SOC was confined below 0–3 meter depth. Chhabra et al. (2003) estimated the SOC concentration in the Indian forest to be 70 t ha^{-1} for tropical dry deciduous forests and 162 t/ha in the montane temperate forest up to 1 m soil depth. Bhattacharyya et al. (2007) predicted the SOC stocks and changes in India, having largely rice and wheat cropping systems using the GEFSOC Modeling System, and compared the output with soil survey data. They predicted an estimated stock of 1.27 PgC for 1990, 1.32 PgC for 2000, and 1.27 PgC for 2030. Singh et al. (2007) estimated that 2.13 PgC stock was present in the 0–100 cm soil depth of arid and semi-arid regions of Rajasthan. Dharumarajan et al. (2021) mapped the SOC stock over the Western Ghats region of south India. They considered variables like Landsat terrain attributes and bioclimatic and used covariates QRF algorithm to estimate the SOC stock.

The AFOLU sector accounts for 23 Gt of CO_2 emissions in India (MoEFF 2019). In India, out of the total 328.7 million hectares (M ha) of geographical extent, 49% of the land is treated as arable land, 17% as irrigated land, 20% is covered by forest and woodland, 3% permanent pasture and 2% by permanent cropland (Lal 2021). The rising food production and extensive cultivation of wheat and rice across the country have prompted unsustainable natural resources management, leading to degradation in soil health and quality. It is estimated that in cultivated soils, SOC concentration is 5 g kg^{-1} or less than in uncultivated soils, which can be attributed to continuous plowing, tillage, and frequent use of fertilizers (Lal 2004). Such intensive cultivation of crops, along with the profound use of synthetic fertilizers, has resulted in a reduction of 0.3–0.4% SOC content in the soil. Hence, it is essential to monitor the SOC content in soil to improve the soil quality and crop production, thus ensuring a sustainable future. It is estimated that worldwide, atmospheric carbon can be sequestered between 0.90 and 1.85 Pg C $year^{-1}$ in the croplands, which is 26–53% of the target of the "4p1000 Initiative (Zomer et al. 2017). In this study, we intend to explore the status of SOC in Indian cropland and assess its sequestration potential in the future through agroforestry and crop management practices.

15.2 Materials and Methods

15.2.1 Study Area

The geopolitical boundary of India is considered for the current study as shown in Fig. 15.2. India has diverse agricultural practices and cropping systems, and climate change is affecting the soil health, crop production, and the environmental and economic status of the nation. Due to its extensive dependency on agriculture, almost 54.6% of the Indian population (annual report, Ministry of Agriculture, 2020–21) has knowledge of soil carbon stock in Indian agricultural land. A study area map is shown in Fig. 15.2, including the Cropland coverage of India in 2020 and the range of spatial distribution of SOC within the same.

15.2.2 Data Used

The data for estimating the status of SOC in Indian Croplands was adopted from the Global SOC Sequestration Potential Map (GSOCseq) at 0–30 cm depth. The GSP secretariat developed it to provide basic knowledge and technical steps for simulating and mapping SOC sequestration potential. GSOCseq is a global product that depicts the simulated SOC stocks in tons of Carbon per hectare (tC/h) 20 years into the future, at a 0–30 depth and 1 × 1 km spatial resolution in current agricultural lands under different soil management scenarios. The 20-year period was adopted because it is assumed that after 20 years, the SOC stocks will appear in a new steady state. In this study, the total SOC stocks in Indian agricultural lands to 30 cm depth were obtained from the GSOCmap (30 arc seconds; ~1 × 1 km resolution grid) (FAO and ITPS 2018). A recent LULC map from the ESA (European Space Agency, http://www.esa-landcovercci.org/) land cover Global dataset (2019–2020) generated by ESRI was overlaid on the SOC stock map. The Data sets are mentioned in Table 15.2 below.

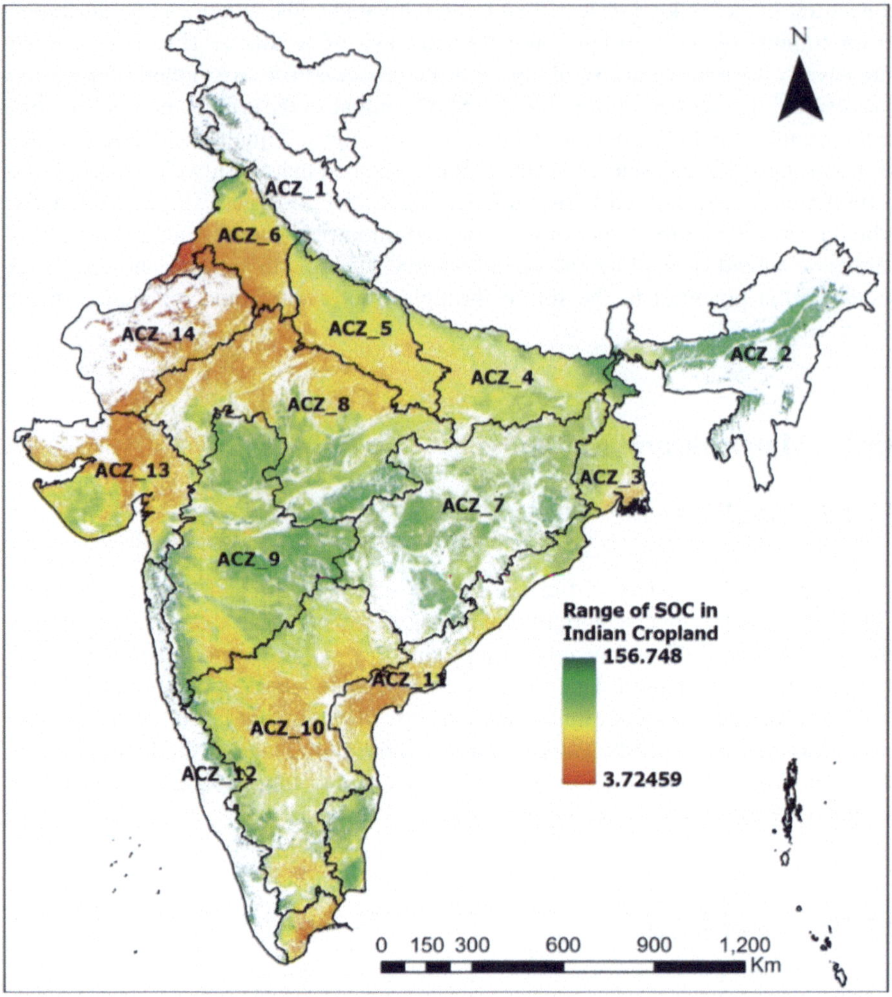

Fig. 15.2 Study area map, representing the cropland of India, the range of SOC and Agro Climatic zones

Table 15.2 Data sets under the study

Sl. No	Data set	Spatial resolution	Agency	Date
1	Soil organic carbon (tons/hectare)	1Km	FAO	Up to 2015
2	LU/LC product of sentinel II	10 m	ESRI	2020
3	Agro-climatic zones	–	ICAR	–

15.2.3 Methodology Employed

As the land use and land cover vary considerably between data sources and estimates of past and current LULC, a recent LULC map from the ESA (http://www.esa-landcovercci.org/) land cover Global dataset (2019–2020) generated by ESRI was overlaid on the SOC stock map. The present study only focuses on the SOC stocks in the croplands of India; thus, ArcGIS Pro was used to mask out SOC stock using the Cropland area as per the ESRI LULC. The country is divided into 15 agro-climatic zones (ACZ), classified into unit lands in major climates suitable for various crops and cultivars. Due to sparse cropland, we have considered only 14 out of 15 ACZ excluding Zone 15, i.e., the Andaman and Nicobar Islands region. Hence, to analyze the status of SOC in Indian croplands, the SOC stock extracted from LULC was divided into 14 ACZ zones. The maximum and minimum values of SOC stock were obtained from the Zonal Statistics tool in Arc GIS Pro. Each agro-climatic zone (ACZ) was further reclassified into ten classes or deciles according to the probabilistic distribution and reclassified using the spatial reclassification tool available in ArcGIS Pro to obtain the area statistics. The area under each decile of SOC stock in croplands was calculated concerning the ACZ, and a spatial map was created for the Indian mainland. Carbon sequestration rate of 9 t C ha^{-1} $year^{-1}$ for agro-forestry. Whereas for the 2nd, 3rd, 4th, 5th, and 6th decile classes, a rate of 5, 4, 3, 2, and 1 t C ha^{-1} $year^{-1}$ sequestration rate was considered for maintaining a no depletion scenario for SOC based on the recommendation by Minasny (2017). Finally, we estimated the total sequestration potential in Indian cropland between 2020 and 2050, considering the growth rates to be fixed. A schematic diagram of the methodology followed in this study is shown in Fig. 15.2. The potential pathways for restoring SOC were analyzed from the results obtained from the methodology described above. A schematic diagram of the methodology followed in this study is shown in Fig. 15.3. The potential pathways for restoring SOC were analyzed from the results obtained from the previously mentioned methodology.

15.3 Results

The total area under cropland in India is calculated to be around 149.26 million Hectare as of 2019–2020, according to the LULC data set obtained from Esri. The Pie chart (Fig. 15.4) represents the area coverage of cropland in each ACZ. Zone 9 has a maximum area under cropland with 25.3 mha, accounting for 17% of the total cropland, followed by zones 10, 8, and 7. The zonal statistics graph (Fig. 15.5) and Table 15.3 of minimum, mean, and maximum show that in all the agro-climatic zones, the mean value of SOC is closer to the minimum, compelling us to anticipate a further decline in the SOC stocks. However, it is hard to conclude the same due to unavailability of spatial time series data on SOC. The total cropland present in each agro-climatic zone is as follows: the Western Himalayan division (zone 1) covers

Fig. 15.3 Flow diagram of the methodology adopted

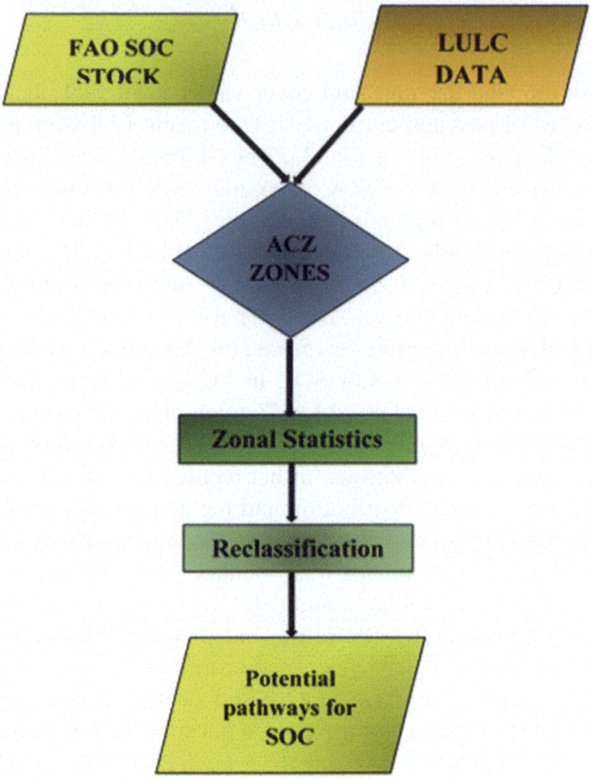

1.22 mha; the Eastern Himalayan division (zone 2) cover 3.45 mha; the Lower Gangetic plain region (Zone 3) covers 2.36 mha; the Middle Gangetic plain region (zone 4) 11.58; the Upper Gangetic plain region (zone 5) 11.72; the Trans-Gangetic plain region (zone 6) covers 9.45 mha of area; the Eastern plateau and hill region (zone 7) has 12.72 mha of area; the Central plateau and hill region (zone 8) has 19.48 mha of area; the Western plateau and hill region (zone 9) has 25.39 mha of area; the Southern plateau and hill region (zone 10) has 21.61 mha of area; the East coast plain and hill region (zone 11) has 7.81 mha of area; the West coast plain and hill region (zone 12) has 4.92 mha of area; the Gujarat plain and hill region (zone 13) has an area of 12.05 mha; the Western plain and hill region (zone 14) has 5.50 mha of area as shown in Fig. 15.3. From the same pie chart, we see that zone 9 has 17% of the total cropland, the Gangetic plain regions together account for 24% of the croplands and rest of the area is distributed mostly over the peninsular India. The SOC stock of the Indian cropland, as shown in Fig. 15.6, we see that most of the Gangetic plain region has low to moderate SOC stock due to extensive farming, except for the upper stretch along the Himalayan region, the central Indian region shows mostly moderate to very high values of SOC stock. However, the agricultural practices are not studied well in the present study; some sustainable agricultural practices may have been adopted in this region.

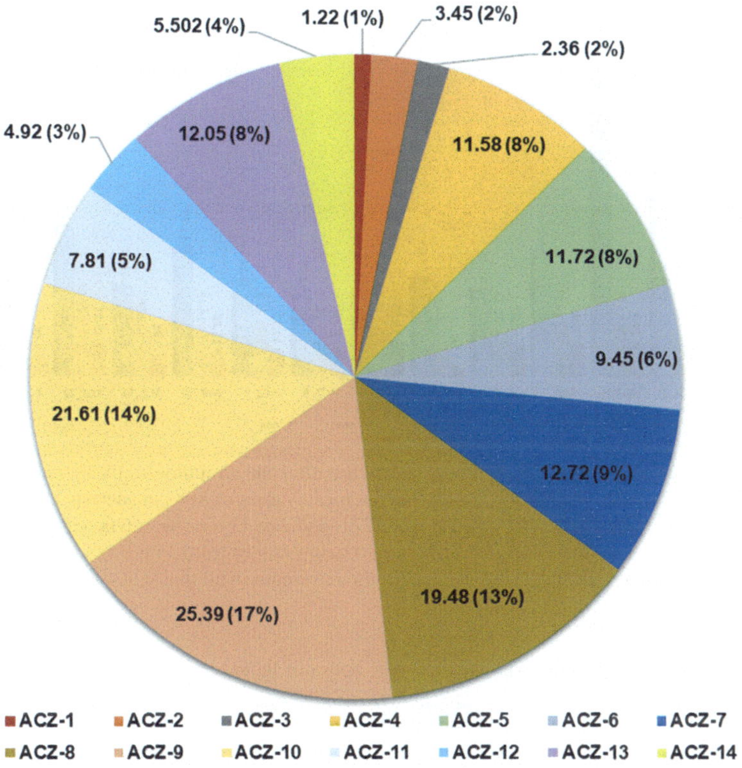

Fig. 15.4 Cropland area coverage under each agroclimatic zone (ACZ). Zone 9 has a maximum area under cropland with 25.3 mha, accounting for 17 percent of the total cropland

Zone 9, or the Maharashtra region, is seen to have been heavily depleted in SOC stock due to heavy stress on the croplands. The coastal regions show a mixed pattern of SOC stock, depending on the local agricultural practices. The map (Fig. 15.6) alarmingly shows a very low SOC level in the Punjab region, one of the most crucial regions for ensuring food security, as reduced SOC will adversely affect the yield. The region of Sundarbans, which is ecologically sensitive, also has a depleting level of SOC in the croplands and might need attention before it hampers the normal functioning of this ecosystem. In the Gangetic plains (zones 3, 4, 5) and zone 7, the mean has dropped very closely to the minimum, implying that long-time extensive farming has depleted the SOC stock in most of the croplands in this region and needs attention.

The deciles are grouped from low to high values of SOC stock in all agro-climatic zones combined (Fig. 15.7) as 10–30 percentile as very low; 30–50 percentile as low; 50–60 percentile as moderate; 60–80 percentile as high and 80–100 percentile as very high, reveals that 14.5 mha land is under ten percentile of SOC level. Only 12.4 mha of land comprises of the most fertile crop fields. 48 mha of the total 149 mha of cropland is in the 40–60 percentile of SOC. If proper planning and cropping pattern is not adopted, this area will slip into the low SOC category soon. A total distribution of agricultural land concerning SOC level is given in Table 15.4.

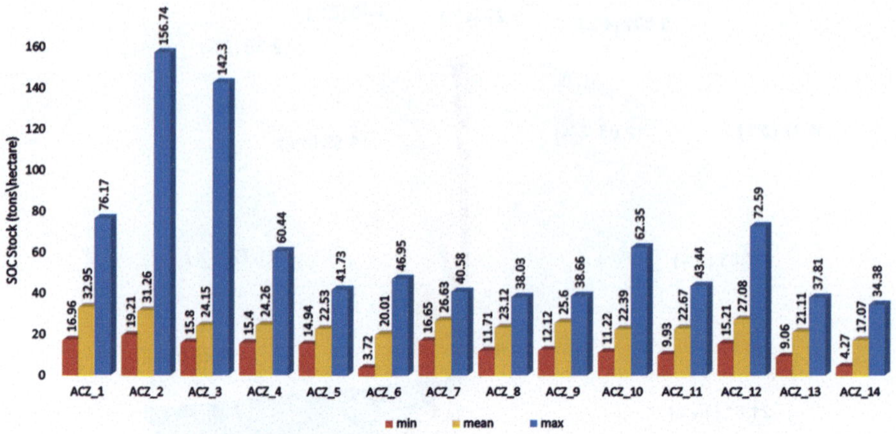

Fig. 15.5 Statistical representation of each agro-climatic zone's minimum, mean, and maximum values. The bar graph representation shows that the mean values of SOC in each ACZ are closer to the minimum value. The mean value varies from 17 tons/ha to 33 tons ha^{-1}, whereas the minimum ranges from 3.7 tons ha^{-1} to 19 tons/ha, and the maximum ranges from 34 tons/ha to 156 tons ha^{-1}. The noticeable thing is that the mean is closer to the minimum in all the ACZs, indicating a depletion of SOC in all the ACZs

Table 15.3 Zonal statistics SOC in agro-climatic zones of India

Agro climatic zones	Minimum	Mean	Maximum
ACZ-1	16.96	32.95	76.17
ACZ-2	19.21	31.26	156.74
ACZ-3	15.8	24.15	142.3
ACZ-4	15.4	24.26	60.44
ACZ-5	14.94	22.53	41.73
ACZ-6	3.72	20.01	46.95
ACZ-7	16.65	26.63	40.58
ACZ-8	11.71	23.12	38.03
ACZ-9	12.12	25.6	38.66
ACZ-10	11.22	22.39	62.35
ACZ-11	9.93	22.67	43.44
ACZ-12	15.21	27.08	72.59
ACZ-13	9.06	21.11	37.81
ACZ-14	4.27	17.07	34.38

The map showing the spatial distribution of cropland with only ten percentiles of SOC stock (Fig. 15.8). It gives an overview of regions that are most deficit in SOC and almost unfit for agriculture. However, this also mostly includes regions of Gangetic Plains, croplands of Maharashtra and Punjab, the regions where most extensive farming is done, and much livelihood depends on the yields of the croplands. Thus, we can suggest agroforestry practices, which will sequester organic carbon into the soil and become a source of income for the families dependent on SOC deficit croplands.

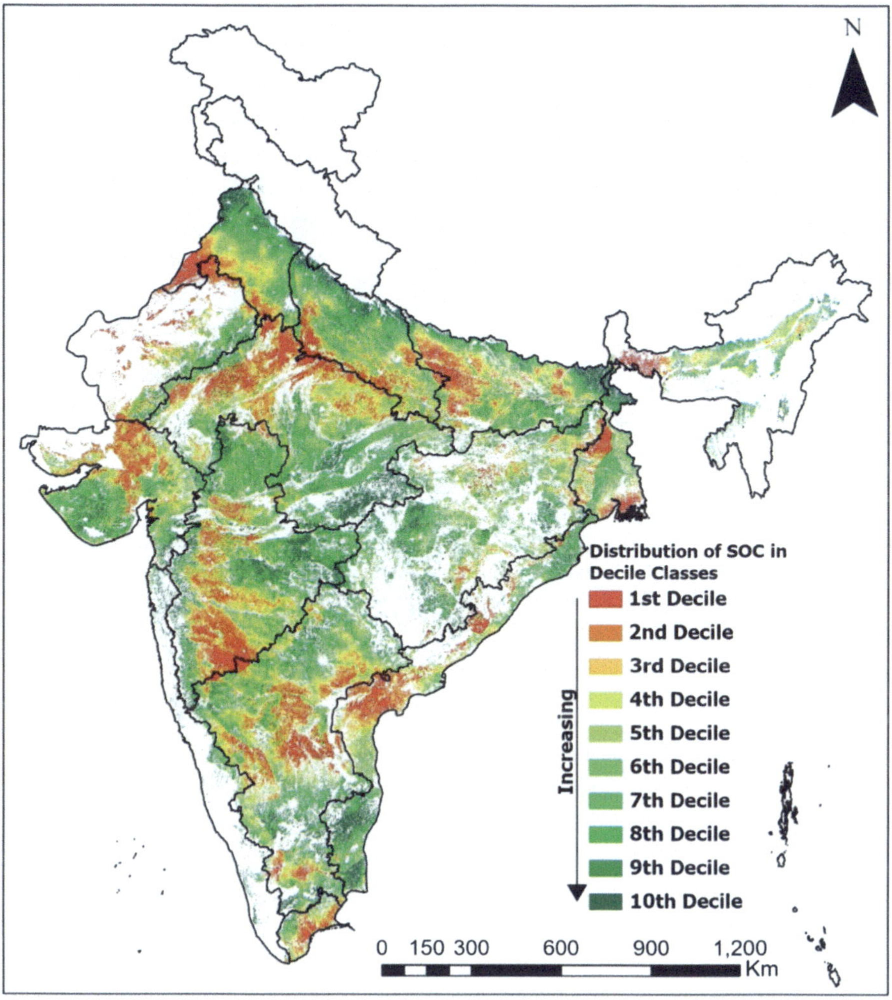

Fig. 15.6 The above image shows the spatial distribution of soil organic carbon in India's cropland. The Indo-Gangetic Plain, parts of Punjab, Rajasthan, and Gujrat, parts of Maharashtra, and some parts of Eastern Ghats identify a huge area deficit in SOC stock. The central Indian region has mostly moderate to very high stock of SOC. The uppermost parts of the Indo-Gangetic Plains also show a high to very high stock

The map (Fig. 15.8) suggests such potential zones for agro-forestry. The first five deciles with relatively low SOC levels are priority areas for SOC restoration. The first decile representing the worst SOC density is considered the agro-forestry option, as it will have the least implications regarding food security trade-offs. Thus, agroforestry will help enrich the soil by sequestering CO_2 and provide income for the dependent families in the long run. We identified about 14 million ha of area for agro-forestry options. We project that these areas will be able to sequester about ~400–500 million tons of additional CO_2 in soils by 2050. Next, deciles from 2nd to

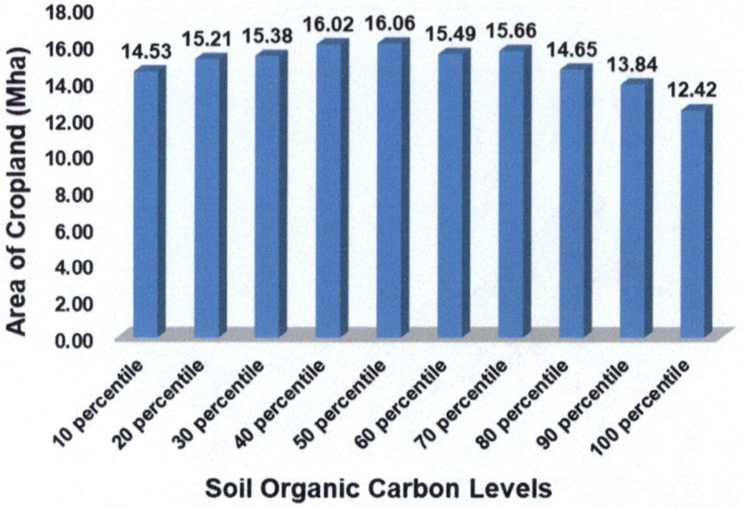

Fig. 15.7 Cropland area under each SOC decile combined from all the agro-climatic zones. It gives a clear picture of India's total cropland distribution in each decile. A 14.5 mha of land has less than ten percentiles of SOC, while the maximum area has moderate SOC stock. The most fertile land with the highest SOC accounts for the least area coverage

Table 15.4 Distribution of area under each decile in every Agro-climatic zone

	Area in million hectare										
Agro climatic zones	Decile classes										
	1st	2nd	3rd	4th	5th	6th	7th	8th	9th	10th	Total area
ACZ-1	0.11	0.14	0.13	0.13	0.13	0.12	0.11	0.12	0.12	0.11	**1.22**
ACZ-2	0.34	0.25	0.21	0.38	0.47	0.44	0.56	0.32	0.26	0.22	**3.45**
ACZ-3	0.23	0.22	0.10	0.18	0.28	0.35	0.32	0.33	0.17	0.17	**2.36**
ACZ-4	1.15	1.36	1.24	1.44	1.02	1.02	1.39	1.03	1.08	0.85	**11.58**
ACZ-5	1.16	1.18	1.19	1.22	1.34	1.31	1.15	1.17	1.05	0.97	**11.72**
ACZ-6	0.93	0.93	1.04	0.94	0.94	1.14	0.90	0.90	0.90	0.82	**9.45**
ACZ-7	1.22	1.24	1.25	1.26	1.35	1.31	1.40	1.35	1.26	1.08	**12.72**
ACZ-8	1.93	2.02	2.03	2.04	1.94	1.99	1.92	1.92	1.83	1.85	**19.48**
ACZ-9	2.48	2.62	2.77	2.77	2.55	2.72	2.57	2.55	2.36	1.99	**25.39**
ACZ-10	2.05	2.04	2.17	2.56	2.72	2.01	2.38	1.98	1.97	1.72	**21.61**
ACZ-11	0.77	0.80	0.85	0.77	0.86	0.77	0.75	0.79	0.75	0.70	**7.81**
ACZ-12	0.42	0.58	0.55	0.55	0.56	0.47	0.45	0.49	0.44	0.41	**4.92**
ACZ-13	1.18	1.25	1.27	1.19	1.32	1.26	1.21	1.18	1.14	1.06	**12.05**
ACZ-14	NI [a]	0.58	0.60	0.59	0.58	0.58	0.55	0.51	0.51	0.46	**5.502**
Total	**14.52**	**15.21**	**15.4**	**16.02**	**16.06**	**15.49**	**15.66**	**14.64**	**13.84**	**12.41**	**149.262**

[a]*NI* Not Included

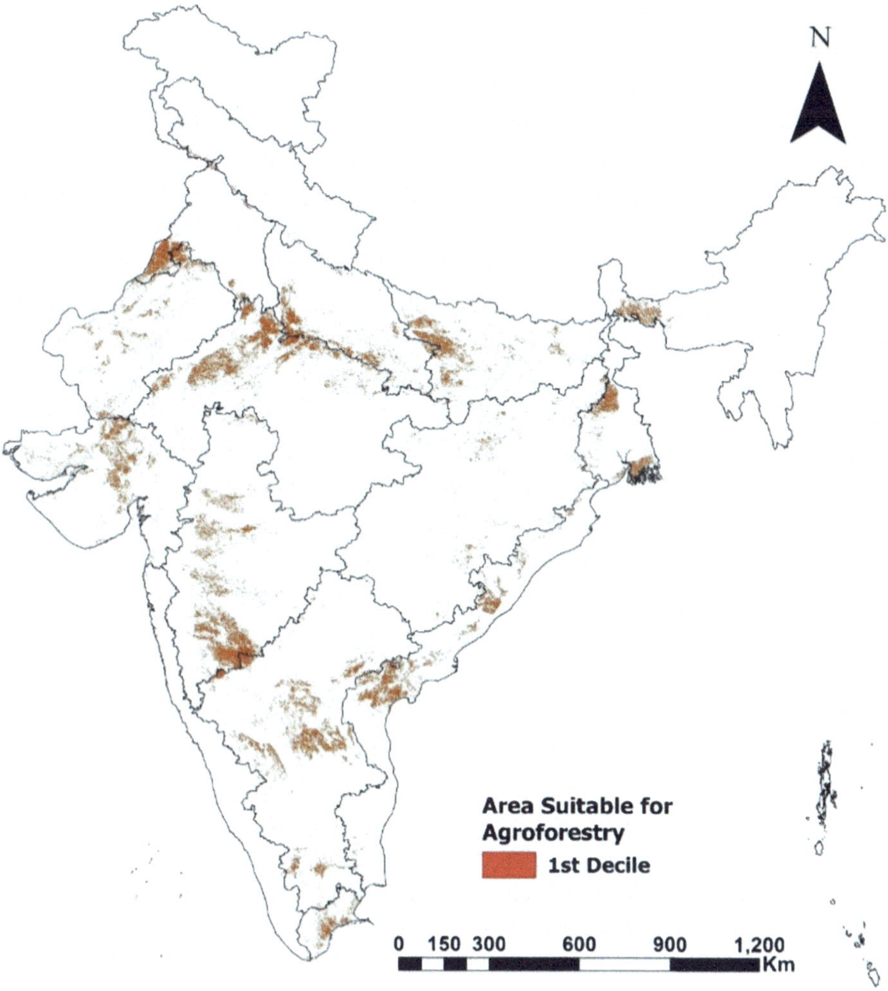

Fig. 15.8 The Map represents the regions with the least SOC stock in the croplands (below ten percentiles in each ACZ); this also serves as the potential zones for Agroforestry to enhance the organic soil stock in the cropland soil. The Thar Desert region of Rajasthan under this category is excluded from consideration for a potential agroforestry zone

6th are considered for SOC enrichment and restoration within cropland-remaining-cropland (CL-CL) land use. We assume a SOC enrichment of 5 t C ha^{-1} to 1 t C ha^{-1} in these decile classes, respectively, from 2020 to 2050. SOC restoration within CL-CL is projected to yield 800–900 million tons of additional CO_2 in soils by 2050, as shown in Table 15.5. Thus, our identified priority areas for Agroforestry and soil carbon enrichment in croplands are projected to yield more than 1300 million tons of additional CO_2 in soils by 2050. The regions of Rajasthan (Zone 14) that were mathematically identified for agro-forestry (first decile soc) are excluded, as it is practically not feasible.

Table 15.5 Distribution of Carbon sequestration potential under different decile classes in Agro-climatic zones by 2050

	Decile classes						
	1st	2nd	3rd	4th	5th	6th	Total
Agro climatic zones	Carbon sequestration estimate in million tons						
ACZ-1	1.03	0.72	0.52	0.39	0.25	0.12	**3.03**
ACZ-2	3.06	1.25	0.83	1.14	0.95	0.44	**7.67**
ACZ-3	2.07	1.10	0.38	0.55	0.57	0.35	**5.03**
ACZ-4	10.34	6.78	4.96	4.32	2.03	1.02	**29.46**
ACZ-5	10.41	5.90	4.74	3.66	2.68	1.31	**28.70**
ACZ-6	8.39	4.66	4.14	2.82	1.88	1.14	**23.04**
ACZ-7	11.02	6.22	5.00	3.77	2.69	1.31	**30.01**
ACZ-8	17.37	10.08	8.13	6.12	3.89	1.99	**47.58**
ACZ-9	22.34	13.10	11.08	8.32	5.10	2.72	**62.66**
ACZ-10	18.45	10.20	8.69	7.68	5.44	2.01	**52.47**
ACZ-11	6.92	4.01	3.39	2.32	1.72	0.77	**19.11**
ACZ-12	3.79	2.88	2.19	1.64	1.13	0.47	**12.09**
ACZ-13	10.64	6.26	5.08	3.56	2.64	1.26	**29.43**
ACZ-14	0.00	2.88	2.39	1.77	1.16	0.58	**8.78**
Total carbon sequestration	125.83	76.05	61.53	48.05	32.12	15.49	**359.07**
Total CO_2 eqv.	**461.79**	**279.10**	**225.81**	**176.33**	**117.89**	**56.85**	**1317.78**

15.4 Discussion

Estimating the current SOC status in India is very important for anticipating soil quality in the future and formulating strategies to reduce the accumulation of gases responsible for global warming, like CO_2, in the atmosphere. It raises the concern about maintaining the quality of soil for a sustainable future, as the population and its demand for resources have increased manifold in the recent past. Several regional scale studies reveal that changing climate extremes reduce the net primary production, eventually decreasing biomass entering the soil (Pan et al. 2014; Tian et al. 2015). The SOC stock is falling globally due to rising atmospheric temperature, overutilization of land, extensive soil fertility mining, soil degradation, inappropriate soil tillage, poor crop management, unsustainable fertilizer use, and enhanced soil erosion (Lal 2016). Our study results reveal that India's current SOC stock shows a negative trend where the value ranges between low and moderate. (Singh et al. 2015) estimated the SOC stock of the Northern Gangetic plain and found that the upper Gangetic plain showed higher levels of SOC and the lower plains a lower level. At the same time, the middle region maintained an optimum value. According to our study, SOC stock in Gangetic Plain is also decreasing, which could be attributed to the advancement of agricultural practices in this zone. The central Indian regions showed moderate to high stocks of SOC. In contrast, Punjab, where higher agricultural production happens, showed a lower SOC stock. Many regions have reported changes in SOC stocks due to changes in land cover (Chatterjee et al. 2018; Pellikka et al. 2018). The conversion of forest land to other land cover patterns also leads to fluctuation in SOC stocks and loss of soil quality (Iqbal and Tiwari 2016). Lan cover changes, unscientific agricultural practices, and extensive cultivation of crops could be the reason for decreasing SOC in Indian croplands.

15.4.1 *Initiative to Restore Soil Organic Carbon*

Soil monitoring is crucial for understanding the nutrient cycle, organic matter sequestration, and other soil dynamics. To monitor soil quality, the Indian government has taken a great initiative called the Soil Health Card, which has an online database for all the essential parameters of the soil in the croplands. Soil Health Card (SHC) is a Government of India scheme supported by the Department of Agriculture & Co-operation under the Ministry of Agriculture and Farmers' Welfare. It is used to monitor the current status of soil health and quality and to estimate the changes in soil health affected by land management. In cycle I (2015–16 to 2016–17), 10.7 crore Soil Health Cards (SHC) were issued, and in cycle II (2017–18 to 2018–19), 11.9 crore SHCs were issued. These health cards contain the details of parameters like pH, EC, organic carbon, available nitrogen, etc., with the tested value and optimum value; a sample Soil Health Card is shown in Fig. 15.7. It is essential for determining fertilizers, and cropping pattern by the farmers. The advantages of this scheme are that it will monitor farmers' soil and advise them on which crop they should cultivate and which one they should skip regularly. The farmers get an accurate evaluation to incorporate management practices for their crops and soil. This scientific way of practice will reduce the degradation of soil to a great extent and help in improving the SOC stock in croplands. The SHC is also important for policymakers and researchers to understand the soil property over a region and the actions needed for sustainable farming practices.

15.4.2 *Promoting Agroforestry and Soil Organic Carbon Restoration*

Another solution for restoring SOC in Indian croplands is through promoting and implementing Agroforestry, which has been recognized in the National Agroforestry Policy of 2014 as the only alternative means to achieve the goal of raising the current forest or tree cover to 33 percent as National Forest Policy (1988) envisions. Forests have a great potential for sequestering carbon (Sharma et al. 2013); thus, introducing forests in agricultural lands in systemic ways will benefit both the environment and the people dependent on agriculture. However, this will require proper planning and aid from the government and knowledge sharing from researchers to effectively plant trees in the areas explored in Fig. 15.8. Further, Table 15.4 defines sequestration potential in each zone. A goal of 4 tons per hectare sequestration is set for the regions with SOC levels below the tenth percentile, which can be achieved by incorporating proper agroforestry methods, and region-specific plantation is adopted. Similarly, by adopting more sustainable cropping patterns, the regions with 20, 30, and 40 percentiles of SOC stock will plausibly achieve the goal of sequestering 3, 2, and 1 ton of carbon per hectare, respectively.

15.5 Conclusions and Future Recommendations

In the current study, we have estimated the status of SOC stock distribution in the Indian croplands and segregated it according to the agroclimatic zones. Our results indicate that SOC stock's mean value (24 t C ha^{-1}) is skewed to the minimum of the range rather than the maximum, indicating a long-term depletion of SOC stocks. However, we could not quantify it further due to the unavailability of data. In our study, we could only quantify the spatial distribution of SOC due to a lack of time series data. Finally, we proposed suitable areas for agro-forestry and proposed sequestration rates for the croplands and is projected that around 1300 million tons of CO_2 can be sequestered in the agricultural land with agro-forestry and sustainable farming methods under a no depletion of SOC stock and constant positive sequestration rates scenario.

We recommend that there is a need for periodic monitoring of soil carbon in croplands to monitor the changes in soil over a while. In this context, the Soil Health Card Scheme of the government of India is very useful, although, as of now, it does not have time-series data; it is expected to collect more samples in the future that can be used to create spatial time-series. However, the Soil health card values are not being updated regularly. Farmlands with low SOC content should be prioritized and targeted for soil carbon improvement-related programs. Similarly, farmlands that have shown considerable improvement in SOC should be highlighted and rewarded as case studies so that other farmers may adopt such successful practices. Conservation and improvement of SOC will help adapt to climate change and go a long way in improving the soil fertility and water holding capacity—thus helping the overall cause of national food security.

References

Agricultural Statistics 2021 (2021) https://eands.dacnet.nic.in/

Batjes NH (2002) A homogenized soil profile data set for global and regional environmental research (WISE, V. 1.1). International Soil Reference and Information Centre, Wageningen, The Netherlands. Available on http://www.isric.org.

Bhattacharyya T, Pal DK, Mandal C, Velayutham M (2000) Organic carbon stock in Indian soils and their geographical distribution. Curr Sci 79(5):655–660

Bhattacharyya T, Pal DK, Easter M, Batjes NH, Milne E, Gajbhiye KS, Chandran P, Ray SK, Mandal C, Paustian K, Williams S, Killian K, Coleman K, Falloon P, Powlson DS (2007) Modelled soil organic carbon stocks and changes in the indo-Gangetic Plains, India from 1980 to 2030. Agric Ecosyst Environ 122(1):84–94

Chatterjee N, Nair PKR, Chakraborty S, Nair VD (2018) Changes in soil carbon stocks across the Forest-Agroforest-Agriculture/Pasture continuum in various agroecological regions: a meta-analysis. Agric Ecosyst Environ 266(July):55–67. https://doi.org/10.1016/j.agee.2018.07.014

Chhabra A, Palria S, Dadhwal VK (2003) Soil organic carbon pool in Indian forests. For Ecol Manag 173(1–3):187–199. https://doi.org/10.1016/S0378-1127(02)00016-6

Dharumarajan S, Kalaiselvi B, Suputhra A, Lalitha M, Vasundhara R, Kumar KSA, Nair KM, Hegde R, Singh SK, Lagacherie P (2021) Digital soil mapping of soil organic carbon stocks in Western Ghats, South India. Geoderma Reg 25:e00387. https://doi.org/10.1016/j.geodrs.2021.e00387

Eliazer Nelson ARL, Ravichandran K, Antony U (2019) The impact of the green revolution on indigenous crops of India. J Ethnic Foods 6(1):1–10. https://doi.org/10.1186/s42779-019-0011-9

Falloon P, Smith P, Szabó J, Smith J, Marshall S, Coleman K (1998) Regional estimates of carbon sequestration potential : linking the Rothamsted carbon model to GIS databases. Biol Fertil Soils 27:236–241

FAO and ITPS (2018) Global Soil Organic Carbon Map (GSOCmap) Technical Report. Rome. 162 pp

Ghosh S, Wilson BR, Mandal B, Ghoshal SK, Growns I (2010) Changes in soil organic carbon pool in three long-term fertility experiments with different cropping systems and inorganic and organic soil amendments in the eastern cereal belt of India. Aust J Soil Res 48(5):413–420. https://doi.org/10.1071/SR09089

Gougoulias C, Clark JM, Shaw LJ (2014) The role of soil microbes in the global carbon cycle: tracking the below-ground microbial processing of plant-derived carbon for manipulating carbon dynamics in agricultural systems. J Sci Food Agric 94(12):2362–2371. https://doi.org/10.1002/jsfa.6577

IBPES (2018) The IPBES assessment report on land degradation and restoration. IBPES

IPCC (2022) Climate change 2022 – impacts, adaptation and vulnerability – summary for policymakers. IPCC

Iqbal S, Tiwari SC (2016) Soil organic carbon pool under different land uses in Achanakmar Amarkantak Biosphere Reserve of Chhattisgarh, India. Curr Sci 110(5):771–773

Jansson C, Wullschleger SD, Kalluri UC, Tuskan GA (2010) Phytosequestration: carbon biosequestration by plants and the prospects of genetic engineering. Bioscience 60(9):685–696

Jobbágy EG, Jackson RB (2000) The vertical distribution of soil organic carbon and its relation to climate and vegetation. Ecol Appl 10(2):423–436. https://doi.org/10.1890/1051-0761(2000)010[0423:TVDOSO]2.0.CO;2

Jones C, McConnell C, Coleman K, Cox P, Falloon P, Jenkinson D, Powlson D (2005) Global climate change and soil carbon stocks; predictions from two contrasting models for the turnover of organic carbon in soil. Glob Chang Biol 11(1):154–166. https://doi.org/10.1111/j.1365-2486.2004.00885.x

Lal R (2004) Soil carbon sequestration in India. Clim Chang 65(3):277–296. https://doi.org/10.1023/B:CLIM.0000038202.46720.37

Lal R (2012) Climate change and soil degradation mitigation by sustainable management of soils and other natural resources. Agric Res 1(3):199–212. https://doi.org/10.1007/s40003-012-0031-9

Lal R (2016) Soil health and carbon management. Food Energy Secur 5(4):212–222. https://doi.org/10.1002/fes3.96

Lal R (2021) Soil organic matter and feeding the future. In: Soil organic matter and feeding the future (Issue August). https://doi.org/10.1201/9781003102762

Minasny B, Malone B, Mcbratney A, Angers D, Arrouays D, Chambers A, Chaplot V, Chen Z-S, Cheng K, Das B, Field D, Gimona A, Hedley CB, Hong S-Y, Mandal B, Marchant B, Martin M, Mcconkey BG, Mulder VL, Winowiecki L (2017) Soil Carbon 4 per mille. Goderma 292:59–86. https://doi.org/10.1016/j.geoderma.2017.01.002

MoEFF G (2019) BUR report final 2019

Pan S, Tian H, Dangal SRS, Zhang C, Yang J, Tao B, Ouyang Z, Wang X, Lu C, Ren W, Banger K, Yang Q, Zhang B, Li X (2014) Complex spatiotemporal responses of global terrestrial primary production to climate change and increasing atmospheric CO_2 in the 21st century. PLoS One 9(11):e112810. https://doi.org/10.1371/journal.pone.0112810

Paris Climate Change Conference (2015) Join the 4 per 1000 initiative: soils for food security and climate change. http://agriculture.gouv.fr/join-40-initiative-soils-food-security-and-climate

Pellikka PKE, Heikinheimo V, Hietanen J, Schäfer E, Siljander M, Heiskanen J (2018) Impact of land cover change on aboveground carbon stocks in Afromontane landscape in Kenya. Appl Geogr 94(September 2017):178–189. https://doi.org/10.1016/j.apgeog.2018.03.017

Scharlemann JPW, Tanner EVJ, Hiederer R, Kapos V (2014) Global soil carbon: understanding and managing the largest terrestrial carbon pool. Carbon Manag 5(1):81–91. https://doi.org/10.4155/cmt.13.77

Sharma J, Chaturvedi RK, Bala G, Ravindranath NH (2013) Challenges in vulnerability assessment of forests under climate change. Carbon Manag 4(4):403–411. https://doi.org/10.4155/cmt.13.35

Singh SK, Singh AK, Sharma BK, Tarafdar JC (2007) Carbon stock and organic carbon dynamics in soils of Rajasthan, India. J Arid Environ 68(3):408–421. https://doi.org/10.1016/j.jaridenv.2006.06.005

Singh VK, Rani M, Dwivedi BS, Singh SK, Gupta VK, Majumdar K, Mishra RP (2015) Soil organic carbon stock variability in the Northern Gangetic Plains of India: interaction between agro-ecological characteristics and cropping systems. Soil Use Manag 31(4):461–473. https://doi.org/10.1111/sum.12217

Stockmann U, Padarian J, McBratney A, Minasny B, de Brogniez D, Montanarella L, Hong SY, Rawlins BG, Field DJ (2015) Global soil organic carbon assessment. Glob Food Sec 6:9–16. https://doi.org/10.1016/j.gfs.2015.07.001

Tian H, Lu C, Yang J, Banger K, Huntzinger DN, Schwalm, Hristopher R, Michalak AM, Cook R, Ciais P, Hayes D, Huang M, Ito A, Jain AK, Lei H, Mao J, Pan S, Post WM, Peng S, Poulter B et al (2015) Global biogeochemical cycles. Glob Biogeochem Cycles 18:1088–1089. https://doi.org/10.1111/1462-2920.13280

Zhou M, Liu C, Wang J, Meng Q, Yuan Y, Ma X, Liu X, Zhu Y, Ding G, Zhang J, Zeng X, Du W (2020) Soil aggregates stability and storage of soil organic carbon respond to cropping systems on Black Soils of Northeast China. Sci Rep 10(1):1–13. https://doi.org/10.1038/s41598-019-57193-1

Zomer RJ, Bossio DA, Sommer R, Verchot LV (2017) Global sequestration potential of increased organic carbon in cropland soils. Sci Rep 7(1):1–8. https://doi.org/10.1038/s41598-017-15794-8

Chapter 16
Challenges and Opportunities for Soil Carbon Sequestration

S. Rakesh, Deepranjan Sarkar, S. Dinesha, Kirttiranjan Baral, Subha Laxmi Mishra, and Pradeep Kumar Dubey

Abstract Climate change is a global concern and soil organic carbon sequestration is a good strategy for mitigation. Carbon neutrality goals set by different countries still face a lot of challenges to minimize the carbon emission pathways. Soil carbon budgeting for progressing toward low-carbon development has multiple benefits besides addressing food security. However, to document the practical progress, we have to overcome sequestration challenges in field conditions. This chapter is a glimpse of the issues and management opportunities in terrestrial carbon sequestration. The adoption of a robust framework to achieve global commitments and climate governance will suppress the high carbon emission processes. We need to put more effort into strengthening the regulatory framework to monitor the changes in soil carbon stocks over time.

S. Rakesh
International Crops Research Institute for the Semi-Arid Tropics, Patancheru, Hyderabad, Telangana, India

D. Sarkar (✉)
Department of Agriculture, Integral Institute of Agricultural Science and Technology, Integral University, Lucknow, Uttar Pradesh, India

International Rice Research Institute, New Delhi, India
e-mail: deep.irri23@gmail.com

S. Dinesha
Department of Forestry, North-Eastern Hill University, Tura, Meghalaya, India

K. Baral
National Academy of Agricultural Research Management, Hyderabad, Telangana, India

S. L. Mishra
ICAR-Indian Institute of Water Management, Bhubaneswar, Odisha, India

P. K. Dubey
Institute of Environment and Sustainable Development, Banaras Hindu University, Varanasi, Uttar Pradesh, India

WRI India, New Delhi, India

S. A. Edrisi et al. (eds.), *Sustainable Plant Nutrition and Soil Carbon Sequestration*, Sustainable Plant Nutrition in a Changing World,
https://doi.org/10.1007/978-3-031-53590-1_16

Keywords CO_2 sequestration · Carbon neutrality · Regulatory framework · Land use change · Management strategies

16.1 Introduction

The act of removing and storing carbon dioxide (CO_2) from the atmosphere in the pedosphere increases the mean resident time and reduces the sink for emissions, the process referred to as "soil carbon sequestration" (Lal et al. 2015). This is done primarily through the breakdown of plant and microbial biomass and the transformation of organic materials into stable forms of carbon. Due to its contribution to lowering the atmospheric concentration of greenhouse gases (GHGs), this mechanism is crucial for preventing climate change. Climate, land use habits, vegetation cover, soil type, and management methods are a few of the variables that have an impact on the global soil carbon sequestration capacity. The overall worldwide soil carbon (C) sequestration capacity is difficult to quantify, although estimations and study findings can provide some insight. In its reports, the Intergovernmental Panel on Climate Change (IPCC) gives estimates of the stocks of soil organic carbon (SOC) around the world. Approximately 1500 gigatons of organic carbon are stored in soils, which is higher than the amount of carbon present in both the atmosphere and terrestrial plants, according to the IPCC's Fifth Assessment Report (2014). In agriculturally farmed soils, the annual rate of soil sequestration ranged from 0.1 to 1.0 tons C ha^{-1} (Paustian et al. 2016). SOC concentrations in the top 30 cm of the soil profile varied on average from 0.30% to 1.05%. The highest active portion of the global terrestrial soil carbon pool of farmland ecosystem, accounting for approximately 10% of SOC stocks (140–170 Pg) in agricultural ecosystems, is found here (Liang et al. 2016; Datta et al. 2017). Global soil carbon sequestration capacity is estimated through studies like the "4 per 1000" project, which aims to increase soil carbon by 0.4% annually to reduce climate change (Minasny et al. 2017). It explores the idea of raising SOC by 0.4% annually (4 per mille) as a method of reducing climate change. This strategy is predicated on the hypothesis that boosting SOC in agricultural soils can aid in carbon sequestration from the atmosphere. On a global scale, however, there are a lot of difficulties in achieving soil carbon sequestration.

16.2 Challenges for Soil Carbon Sequestration

16.2.1 Land Use Change Patterns

Depending on the exact land use transition, land use change can have both beneficial and negative effects on soil carbon sequestration. Forests being converted to cropland results in the breakdown of organic matter and the disruption of the forest ecosystem, logging typically results in a large loss of soil carbon (Guo and Gifford

2002). This may cause the release of CO_2 into the atmosphere. Grassland conversion to cropland results in soil disturbance and increased erosion. Initially, this transformation can release carbon. Urban growth frequently results in soil sealing and compaction, which can reduce the capacity of soils to store carbon. Comparing urban regions to natural ecosystems, soil carbon levels are often lower in urban areas. A study conducted by Yang et al. (2018) in the endorheic Shiyang River basin, a sizable semi-arid to arid region in northwest China estimated the variation in the area's soil carbon store caused by changes in land use between 1973 and 2012. The findings demonstrate that land use change significantly affects the soil carbon budget, with a 3.89 Tg decline in soil carbon storage between 1973 and 2012. Grassland held the most soil carbon (114.34 Mg ha^{-1}), while woods (58.53 Mg ha^{-1}), farmland (26.75 Mg ha^{-1}), and unoccupied land (13.47 Mg ha^{-1}) all stored significantly less carbon. Qin et al. (2016) showed that a particular land shift has a significant impact on the amount and rate of SOC change. Regardless of soil depth or time horizon, conversions to farmland produced an overall SOC gain of 6–14% compared to the baseline SOC level, but conversions from grassland or forest to maize (without residue removal) or poplar resulted in a considerable carbon loss of 9–35%. Land that was changed from grasslands or forests to switchgrass, miscanthus, or willow did not see any noticeable SOC changes. It is crucial to keep in mind that the impacts of land use change on soil carbon sequestration can vary based on regional environmental factors, management techniques, and the exact species of vegetation involved. The negative effects of land use change on soil carbon can be lessened, and soil carbon sequestration can be encouraged, by using sustainable land management techniques and reforestation initiatives.

16.2.2 Loss of Top Fertile Soil

There is a sizable amount of carbon-rich organic matter in topsoil. This organic stuff is lost as a result of topsoil erosion or degradation. In order to sequester carbon, organic matter acts as a source of carbon in the soil (Lal 2004a, b, c). The numerous microbial populations that live in topsoil are in charge of breaking down organic matter and moving carbon throughout the soil. These microbial communities may be disturbed by soil erosion, which will affect their capacity to efficiently break down organic matter and store carbon (Six et al. 2006). The root zone's soil organic carbon threshold is between 1.5% and 2.0%. Land usage, soil management, and farming techniques all have an impact on SOC. Between 0.3 and 1 m deep, more than 50% of the total carbon pool is contained (Lal et al. 2015). All tests revealed an early carbon effect: although soils with high carbon content lost at least 20% during culture, those with very low carbon content typically gained only a small amount of carbon after cultivation. Less than 20% of initial levels, or less than 1.5 kg cm^{-2} within the top 30 cm, are believed to be the average carbon losses from most agricultural soils (Mann 1986). The SOC supply in agroecosystems soils is severely reduced, and the soil quality has declined. Topsoil that has been eroded and the carbon it contains may be carried downstream by erosion or released into the atmosphere as CO_2. The loss of carbon from the environment may be a result of this process.

16.2.3 Soil Degradation

Overgrazing, deforestation, and shifting agriculture methods have all contributed to the degradation of many soils around the world. Rehabilitating these soils to increase carbon storage can be difficult and may involve substantial investments as well as adjustments to how land is managed. As per the estimations, between 25% and 33% of land is directly impacted by land degradation (Gudka et al. 2015), which has a yearly economic impact of about US$300 billion (Nkonya et al. 2016; Thomas et al. 2023). In China, due to the extensive agricultural operations and the region's unique long, gentle slopes, substantial soil erosion and land degradation have occurred. As a result, the top black soil layer's thickness has reduced from 50 to 80 cm in the 1950s to 20–40 cm today (Xie et al. 2019). In fact, degraded soils have a lower potential to store carbon (Sarkar et al. 2020; Rakesh et al. 2023). The processes involved in soil degradation affect the composition, fertility, and biological activity of soils, which can have an adverse effect on their capacity to store carbon. Reduced microbial activity in degraded soils may be caused by elements including pollution, compaction, or chemical contamination. Microbes are essential for the breakdown of organic matter and the soil's carbon cycle. Reduced microbial activity can impede carbon sequestration and the breakdown of organic compounds. Erosion is frequently more likely to occur in degraded soils. Topsoil and the carbon it contains can be carried away from the site by erosion, resulting in a net loss of soil carbon. India pledged at the 14th Conference of Parties to restore 26 million ha of degraded land by 2030 to tackle both land degradation and climate change (Kar et al. 2023).

16.2.4 Climate Change Impact

The sequestration of carbon in soil is already hampered by climate change. Increased soil organic matter decomposition may result in the release of CO_2 into the atmosphere if temperatures rise. The dynamics of soil carbon can also be impacted by changes in precipitation patterns. The burning of fossil fuels and biomass, particularly through the decomposition of organic matter at the surface and at depth in the soil, releases a growing concentration of CO_2 as well as other GHGs, such as methane (CH_4) and nitrogen oxides (NOX), which are the main drivers of the anthropogenic GHG effect. As a result, there is a tremendous increase in the surface temperature of the Earth and loss of soil carbon directly through decomposition and indirectly from the vagaries of climate change such as floods, droughts, cyclones, heatwaves, etc. By 2050 (the target year for achieving carbon neutrality in Europe), it will not be possible to avoid them completely based on current and developing technologies (European Union 2020). Consequently, carbon offsets are needed. These will have to be carbon dioxide removals (CDR) in a world where carbon emissions are neutral. CDRs are included in every IPCC scenario that successfully keeps warming to 1.5 °C above pre-industrial levels (Paul et al. 2023). There is a common misconception that one ton of CO_2 sequestration will equal one ton of emissions; however, research by Zickfeld et al. (2021) suggests that the climatic response may be asymmetric, necessitating a larger sequestration than emissions.

16.2.5 Political, Social, and Economic Factors

How well carbon sequestration schemes in agricultural soils perform will depend on political and economic factors. The issues touch on economics, private property rights, agricultural regulations, institutional and social structures, as well as environments for implementation. Success will ultimately depend on the incentive system created and how carbon sequestration is woven into the overall framework of agriculture policy (Marland et al. 2001). It might be difficult to provide economic incentives for soil carbon sequestration. For farmers and landowners to implement techniques that improve soil carbon storage, there may be a requirement for financial assistance or compensation. Land ownership can also be affected depending on how soil carbon sequestration techniques are used. It is essential to make sure that local communities have the rights and motivations to practice sustainable land management. As part of broader efforts to mitigate climate change, effective development in soil carbon sequestration depends on international cooperation and coordination.

16.2.6 Monitoring and Technical Expertise

It can be challenging and expensive to measure changes in soil carbon stocks over time. For efforts to sequester carbon in soil, developing reliable and affordable monitoring techniques is crucial. Educating land managers on knowledge and best practices is frequently necessary to promote soil carbon sequestration. In this context, education and extension services can be quite important. To fill the gaps in our knowledge and choose the best cost-effective procedures, a thorough evaluation and critical analysis of the many SOC determination techniques used by various research groups in contrasting agroecosystems and management circumstances is needed (Rakesh et al. 2020). Modeling, whole-field sampling (net ecosystem exchange), combined modeling and measurement, etc. are some approaches for calculating the offsets of terrestrial SOC and building a dynamic database (Conant et al. 2011). Repeated soil surveys, long-term field studies, models, and remote sensing are essential elements of national and global monitoring, reporting, and verification

(MRV) platforms (Smith et al. 2020). The open-access database has a key role in strengthening the global MRV system.

16.2.7 Long-Term Commitment and Balancing Multiple Goals

The benefits of soil carbon sequestration may not become fully apparent for many years. Long-term commitment to these initiatives can be difficult to ensure. Land is frequently used for several things, including the production of food, the preservation of biodiversity, and the storage of carbon. A difficult problem is finding a balance between these goals and avoiding trade-offs. The maximum physical potential of soil carbon sequestration and what is feasible in a complicated society differ greatly due to socioeconomic, social, institutional, and political constraints (Amundson and Biardeau 2018). Therefore, conservation agriculture and agroforestry practices are promising options for co-delivery of multiple benefits and overcoming the barriers (Mbow et al. 2014; Corbeels et al. 2019; Dubey et al. 2020). Dinesha et al. (2023) reported mahogany (*Swietenia macrophylla*) based plantations are not only profitable for the farmers of eastern Himalaya (India), but in addition, they also provide substantial improvements in soil fertility and carbon sequestration, which plays a critical role in the long-term quality of the ecosystem. Thus, win-win scenarios can be implemented including biodiversity recovery by proper land use management practices and a positive approach towards conservation and community participation.

16.3 Plant Nutritional Aspects of Carbon Sequestration

Increasing SOC has biological benefits, one of which is nutrient supply. According to Lal et al. (2007), the decline in the SOC stock would have conflicting effects on the quality of the soil. SOC content serves as a stand-in for SOM, which is the primary source of plant nutrients and, consequently, soil fertility (Shi et al. 2016). It preserves soil nutrients, lessens soil erosion, increases the physical agglomeration of soil particles, and safeguards the purity of the water. While cation exchange sites are crucial for nutrient retention, it chemically raises the soil CEC, which increases the retention of positively charged nutrient ions like calcium (Ca), iron (Fe), potassium (K), magnesium (Mg), zinc (Zn), and many others. Biologically, SOC serves as the primary energy source for most of the soil fauna and microorganisms (Elbasiouny et al. 2022). SOM is a vital source of energy, nutrition, and habitat for most soil organisms (Sarkar and Rakshit 2023). It is also necessary for soil biological activity, together with plant and animal waste. SOC and SOM are directly correlated with soil biodiversity. Numerous elements influence C sequestration, including parent material, temperature, rainfall, landscape position, living things, and different management techniques (Conforti et al. 2016; Rakesh et al. 2022b). Decomposition and formation rates of SOM are also important aspects (Rakesh et al. 2021).

Plant nutrition may be obtained through organic matter mineralization, which makes increasing the stock of SOM beneficial for soil fertility. This, however, necessitates the application of regionally appropriate farming techniques that will maximize soil carbon storage by augmenting soil carbon inputs with fixed or declining outputs (Dignac et al. 2017). Conservation tillage (Feng et al. 2020), cropping systems (Crystal-Ornelas et al. 2021), legume-based crop rotation (Nath et al. 2021), cover crops (Dad et al. 2021), integrated nutrient management (Kumari et al. 2020), irrigation management (Chandel et al. 2021), agroforestry systems (Mayer et al. 2021), and grassland/pasture management (Madigan et al. 2021) are some of the management options that have been presented in studies to restore soil carbon representation. Improving soil and crop management techniques to sequester carbon is essential for improving soil quality and the production system (Yadav et al. 2021). Since SOC supports a variety of beneficial processes in the soil ecosystem, increasing the stored SOC may enhance soil quality (i.e., biological, physical, and chemical processes). Soil health may be compromised, and yield potential may not be realized when SOC is less than 1% (Fan et al. 2015; Shahid et al. 2017).

16.3.1 Connection Between Carbon Sequestration and Crucial Soil Nutrients

According to Grover et al. (2015), the main factor influencing carbon sequestration throughout plant development is soil carbon input, which is influenced by nitrogen availability and nutrients needed to sustain atmospheric nitrogen fixation. Artificial fertilizers have an impact on soil health and contain a carbon charge, but they do not contribute to organic matter. Therefore, increasing the ratio of nutrient solubilizers/mobilizers to biological nitrogen fixers may be an economical and sustainable way to improve carbon sequestration in agroecosystems. The soil processes are regulated by the agricultural residues/SOM C:N ratio. Stable forms of SOC, humus, and soil microbial biomass (MB) have average C:N ratios of roughly 10 to 12, 10, and 7, respectively. The C:N ratio of SOM is 12:1, the C:P ratio is 50:1, and the C:S ratio is 70:1, based on weight. Therefore, 83, 20, and 14 kg of N, P, and S, respectively, are needed to sequester one gram of carbon (Singh et al. 2017). Additionally, Richardson et al. (2014), suggested that managing inorganic nutrients (N), phosphorus (P), and sulfur (S) is a crucial viewpoint for methods of constructing soil carbon from agricultural wastes that are high in C. According to some research, one of the main strategies for long-term carbon storage in sediment and marine environments is the biosequestration of carbon through microbial deposition.

16.3.2 Carbon Fractions and Nutrient Dynamics

According to Shahid et al. (2017), active fractions (i.e., microbial biomass carbon, microbial biomass nitrogen, particulate organic carbon and particulate nitrogen, potentially mineralizable C and N, which respond quickly to the change in management practices) can alter the nutrient dynamic due to immobilization mineralization processes, which can better reflect the variations in both soil quality and productivity. The OM coarse fraction-containing particle organic carbon (POC) and particulate organic nitrogen (PON) are in the middle of the active and slow SOC and TN fractions, which fluctuate rapidly due to management decisions (Franzluebbers and Stuedemann 2002; Rakesh et al. 2023). These fractions are good indicators of soil quality and may be more susceptible to these activities than the SOC and TN, which could influence the way that soil operates. As a result, increasing the OC return to soils through conservation agriculture, the use of organic and inorganic mulch to reduce nutrient loss through leaching and volatilization, cover crops, the balanced application of macro- and micronutrients, compost application, rhizobacteria that promote plant growth, biofertilizers, and crop system variation and diversification can all directly increase the SOC storage in agricultural soils (Upadhyay et al. 2019; Mahmud et al. 2021; Sarkar et al. 2021).

16.3.3 Carbon Sequestration and Nutrient Use Efficiency

Benefits to yield were shown when carbon activities such as crop rotation, green manuring, compost addition, and crop residue management were combined with fertilizer application (Fuchs et al. 2008). Carbon-rich organic resources could be added to reduce the usage of mineral fertilizers and occasionally replace the nutrients in them (Lal 2020). When compared to mineral fertilizer treatment, the combined application of chemical fertilizers and compost greatly boosted the grain yield of maize by 23.3–55.3% and the agronomic nutrient use efficiency by 36.5–81.3% (Asaye et al. 2022). In addition to replacing 10–50% of external nitrogen supplies, practices like crop recycling, legume green manuring, and farmyard manure also improve the soil fertility status for wheat agriculture (Panhwar et al. 2019).

16.4 Opportunities for Soil Carbon Sequestration

Soil is the second-largest carbon store on Earth, behind the oceans (38,000 gigatons/ Gt C), and even a small alteration in the soil's organic carbon deposit can have a major impact on atmospheric CO_2. The capacity of global soils to sequester carbon is estimated to be between 0.4 and 1.2 Gt C year^{-1}. Around 750 Pg (picograms) of carbon are found in the atmosphere as CO_2, whereas the upper 100 cm of soil (apart from permafrost) contains approximately 1500 Pg of carbon (1 Pg = 1 Gt = 1015 g)

(2500 Pg of carbon in the top 200 cm) as SOC and 900–1700 Pg as inorganic C. This soil exchanges 60 Pg of carbon with the atmosphere annually (Eswaran et al. 1993; Lal 2010; Meena et al. 2015).

At the global level, soil carbon sequestration offers a number of prospects for reducing climate change and advancing sustainable land management techniques. Through reforestation and agricultural forestry, we can possibly increase soil carbon sequestration which can be directly achieved by planting trees and incorporating them into agricultural activities in agroforestry systems (Edrisi et al. 2022a, b). According to Nair et al. (2010), the agroforestry's capacity for sequestering carbon under different biological and management conditions ranged from 0.29 to 15.21 Mg ha^{-1} $year^{-1}$ in biomass from above-ground plants and 30–300 Mg ha^{-1} $year^{-1}$ in components from below-ground plants down to 1.0 m. Agroforestry has the capacity to sequester 1.1–2.2 Pg C in terrestrial ecosystems over the next 50 years, according to a 2007 IPCC report (Jose 2009). There is a great deal of potential for increasing the rate of soil carbon sequestration by irrigation water application. The improved irrigation produced low to medium moderately high abatement rates of 0.2–3.4 t CO_2 equivalent ha^{-1} $year^{-1}$ (World Bank 2012). Implementing conservation tillage techniques and cover crops can enhance soil health and boost carbon sequestration (Sarkar et al. 2020; Dubey et al. 2022). Lee et al. (1993) estimated that during the next 100 years, switching from conventional tillage to no-tillage on maize and soybean farms in the US corn belt could sequester 3.3×10^6 tons C $year^{-1}$. In a meta-analysis conducted by Poeplau and Don (2015), the change in SOC stock ($R^2 = 0.19$) was monitored for 54 years following the introduction of cover crops into the cropping system. The results showed a yearly change rate of 0.32 ± 0.08 Mg ha^{-1} $year^{-1}$ in a mean 22 cm soil depth. Because forage, cover crops, and green manure replenish a significant amount of carbon and nitrogen in the soil system, the impact of leguminous crop species on SOC sequestration is more noticeable. Crop rotation reduced GHG emissions by 0.7–1.5 t CO_2 equivalent ha^{-1} $year^{-1}$ (World Bank 2012). The intercropping system's SOC accumulation rate ranged from a moderate value of approximately 1.0 Mg C ha^{-1} (Mitran et al. (2018). Soil carbon credits and carbon farming encourage adoption by creating carbon markets and payment plans for farmers who use soil carbon sequestration techniques (Rakesh et al. 2022a). Restoring wetlands and peatlands has the potential to drastically reduce carbon emissions and enhance water quality. Better grazing and pasture management increases carbon sequestration in grasslands. About 3460 Mha, or roughly 31% of the Earth's land surface, are covered by grasslands and grazing fields worldwide (Lal 2004a, b, c). More than 10% of the total carbon storage in all vegetation is shared throughout the ecosystem (Nosberger et al. 2000). Since up to 98% of all carbon is found stored below ground in grassland ecosystems, the soil is the greatest carbon storage body in the terrestrial carbon pool (Jones and Donnelly 2004). The use of biochar and organic soil additives possibly increases soil fertility levels and soil carbon storage. By switching to organic agricultural methods, we can use less synthetic fertilizer while increasing the carbon content of the soil (Dubey et al. 2020; Dubey et al. 2022). The use of synthetic fertilizers, organic manures (such as farmyard manure (FYM), compost, vermicompost, biosolids, and biochar), surface residue retention, and green manuring are examples of fertilizer management techniques in cultivated soils that are effective in increasing SOC accumulation and lowering CO_2 evolution from the soil (Dubey et al. 2021; Edrisi et al. 2022b). Crop production, biomass-C generation, crop residue, and root input in the soil are all improved by adequate availability of nutritional elements from various sources (Kätterer et al. 2011). In Indian soils, Pathak et al. (2011) reported that NPK + FYM treatment documented higher carbon sequestration (0.33 Mg C ha^{-1} $year^{-1}$) and reduced GHG emissions without incurring additional costs than NPK treatment (0.16 Mg C sequestered ha^{-1} $year^{-1}$).

Sustainable land use management options can possibly enhance long-term carbon storage by incorporating soil carbon sequestration into land use planning and regulations. These possibilities show how soil carbon sequestration has the ability to mitigate climate change and promote sustainable land management techniques on a global scale. To take full use of these benefits and successfully solve the problems related to soil carbon sequestration, more research and policy support are required. There should always be a win-win situation when we are dealing with farmers to achieve multiple goals in the field. Food security and water quality/renewability are two of the many benefits of soil carbon sequestration that need robust governmental intervention to encourage farmers and land managers to adopt recommended management practices.

16.5 Legislative Framework for Achieving Global Soil Carbon Sequestration

Legislation to regulate the effects of climate change on soil management can be framed in a variety of ways. However, the institutional and legal components that safeguard the physical processes involved in soil carbon sequestration as well as the creation and upkeep of carbon sinks and reservoirs will determine the legislative ability to accomplish effective soil carbon management. These components ought to be included in policies that control and manage land use practices that degrade land, release soil carbon, and worsen the greenhouse effect. Reforming laws and policies should begin with a solid conceptual foundation, ideally with the overarching social objective of managing and improving sinks. The legislation ought to incorporate aims and objectives that delineate the responsibility for regulating and overseeing soil carbon sequestration, ideally integrated into an environmentally sustainable method of managing soil. The UNFCCC, the Kyoto Protocol, the Marrakech Accords, and other pertinent international treaties, strategies, and policies regarding ecology, biodiversity, conservation of nature, and sustainable land management can all be used as a basis for adopting certain principles (Rolfe 2001). A framework to manage soil carbon sequestration and global commitments has been illustrated in Fig. 16.1. The components that center on the following determine the strategy to enhance carbon sequestration:

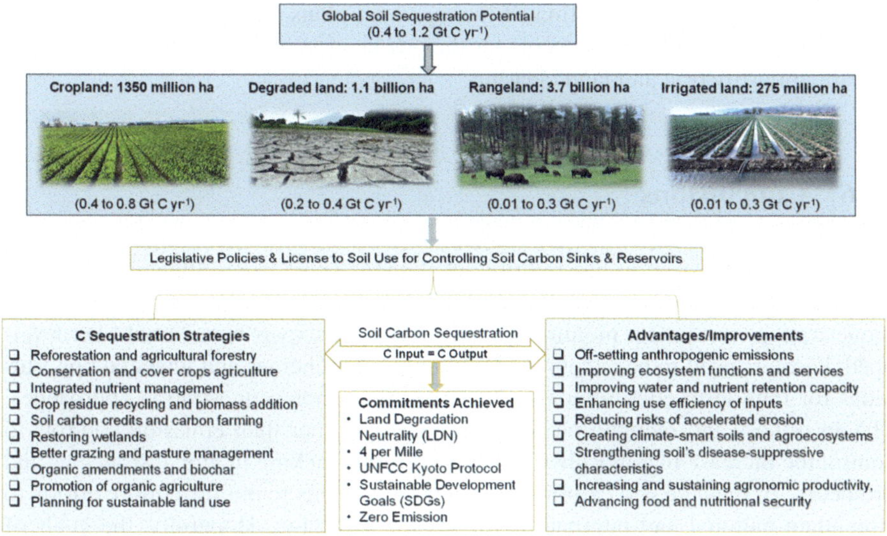

Fig. 16.1 Framework to manage soil carbon sequestration and global commitments. (Source: Lal 2004a, b, c)

- Indicate the emission limitations and the strategy for creating new soil carbon sinks.
- Describe the adjustments that must be made to the land use sectors for agriculture and forestry.
- Methods for creating a national carbon accounting and forecasting system to pinpoint regions where soil carbon management strategies are applicable.
- Encourage the use of techniques to locate both new and current terrestrial sinks.
- Describe the steps involved in locating "hot spots" for soil and land degradation.
- Outline the steps involved in creating awareness and education campaigns about greenhouse land use issues and soil carbon sequestration.
- Describe how to conduct research on how changes in land use affect soil carbon stocks and, in turn, how these changes affect biodiversity conservation overall.
- Granting licenses or permits to regulate soil use; would specify rights to use for things like habitat preservation, soil restoration, sink management, and organic matter content, among other things.
- Prosecution for disregarding guidelines for controlling soil carbon sinks and reservoirs.
- Permit the creation of policies, guidelines, and ecological standards.
- Permit participation in decision-making processes by governments, landowners, land managers, and the community guidelines for creating municipal and state soil management plans.
- Clauses safeguarding biodiversity and significant values associated with nature conservation.

- Clauses requiring the creation of management plans based on sustainable land management standards protocols for the prosecution of grave offenses involving habitat destruction and land clearing.

16.6 Conclusions

Soil carbon sequestration has multiple challenges apart from climate change and soil degradation. For implementing and balancing the multiple goals under carbon sequestration, a constant monitoring and measuring expert team should be developed. Farmers need to be encouraged with multiple schemes and financial opportunities for long-term commitments under various carbon sequestration approaches. The awareness and recognition of using methods other than emission controls to control the increase in atmospheric CO_2 is currently lacking in global strategies and protocols. To manage soil carbon, the community needs to put forth more efforts to strengthen national and international environmental laws. Hopefully, the goals of the UN Decade on Ecosystem Restoration (2021–30) and the Bonn Challenge will be accomplished.

References

Amundson R, Biardeau L (2018) Soil carbon sequestration is an elusive climate mitigation tool. Proc Natl Acad Sci 115(46):11652–11656

Asaye Z, Kim DG, Yimer F, Prost K, Obsa O, Tadesse M, Bruggemann N (2022) Effects of combined application of compost and mineral fertilizer on soil carbon and nutrient content, yield, and agronomic nitrogen use efficiency in maize-potato cropping systems in Southern Ethiopia. Land 11(6):784. https://doi.org/10.3390/land11060784

Chandel S, Datta A, Yadav RK, Dheri GS (2021) Does saline water irrigation influence soil carbon pools and nutrient distribution in soil under seed spices? J Soil Sci Plant Nutr 21:949–966

Conant RT, Ogle SM, Paul EA, Paustian K (2011) Measuring and monitoring soil organic carbon stocks in agricultural lands for climate mitigation. Front Ecol Environ 9(3):169–173

Conforti M, Lucà F, Scarciglia F, Matteucci G, Buttafuoco G (2016) Soil carbon stock in relation to soil properties and landscape position in a forest ecosystem of southern Italy (Calabria region). Catena 144:23–33

Corbeels M, Cardinael R, Naudin K, Guibert H, Torquebiau E (2019) The 4 per 1000 goal and soil carbon storage under agroforestry and conservation agriculture systems in sub-Saharan Africa. Soil Tillage Res 188:16–26

Crystal-Ornelas R, Thapa R, Tully KL (2021) Soil organic carbon is affected by organic amendments, conservation tillage, and cover cropping in organic farming systems: a meta-analysis. Agric Ecosyst Environ 312:107356

Dad JM, Dand SA, Pala NA (2021) The effect of bi-culture cover crops on soil quality, carbon sequestration, and growth characteristics in apple orchards of North Western Himalayas. Agrofor Syst 95:1745–1758

Datta R, Baraniya D, Wang YF, Kelkar A, Moulick A, Meena RS, Yadav GS, Ceccherini MT, Formanek P (2017) Multi-function role as nutrient and scavenger of free radical in soil. Sustain MDPI 9:402. https://doi.org/10.3390/su9081402

Dignac M-F, Derrien D, Barre P, Barot S, Cécillon L, Chenu C, Chevallier T, Freschet GT, Garnier P, Guenet B et al (2017) Increasing soil carbon storage: mechanisms, effects of agricultural practices and proxies. A review. Agron Sustain Dev 37:14. https://doi.org/10.1007/s13593-017-0421-2

Dinesha S, Panda MR, Pradhan D, Rakesh S, Dey AN, Bhat JA, Pandey R (2023) Ecosystem carbon budgeting under *Swietenia macrophylla* King plantation in sub humid foothills of Eastern Himalayans of India. Environ Dev Sustain. https://doi.org/10.1007/s10668-022-02902-6

Dubey PK, Singh GS, Abhilash PC (2020) Adaptive agricultural practices: building resilience under changing climate, Springer brief in environmental science. Springer Nature Switzerland. https://doi.org/10.1007/978-3-030-15519-3. ISBN 978-3-030-15518-6

Dubey PK, Singh A, Chaurasia R, Pandey KK, Bundela AK, Dubey RK, Abhilash PC (2021) Planet friendly agriculture: farming for people and the planet. Curr Res Environ Sustain 3:100041

Dubey PK, Singh A, Merah O, Abhilash PC (2022) Managing agroecosystems for food and nutrition security. Curr Res Environ Sustain 4:100127. https://doi.org/10.1016/j.crsust.2022.100127

Edrisi SA, Dubey PK, Chaturvedi RK, Abhilash PC (2022a) Bioenergy crop production potential and carbon mitigation from marginal and degraded lands of India. Renew Energy 192:300–312

Edrisi SA, Tripathi V, Dubey PK, Abhilash PC (2022b) Carbon sequestration and harnessing biomaterials from terrestrial plantations for mitigating climate change impacts. In: Thakur IS et al (eds) Climate change mitigation, sequestration of green-house gases; biomass, biofuels, biochemicals. Elsevier, pp 299–313. https://doi.org/10.1016/B978-0-12-823500-3.00001-7

Elbasiouny H, El-Ramady H, Elbehiry F, Rajput VD, Minkina T, Mandzhieva S (2022) Plant nutrition under climate change and soil carbon sequestration. Sustain For 14(2):914. https://doi.org/10.3390/su14020914

Eswaran H, Vandenberg E, Reich P (1993) Organic C in soils of the world. Soil Sci Soc Am J 57:192–194

European Union (2020) Long-term low greenhouse gas emission development strategy of the European Union and its member states. In: United Nations framework convention on climate change, vol 6. European Union

Fan H-Z, Chen Q-R, Qin Y-S, Chen K, Tu S-H, Xu M-G, Zhang W-J (2015) Soil carbon sequestration under long-term rice-based cropping systems of purple soil in Southwest China. J Integr Agric 14:2417–2425

Feng Q, An C, Chen Z, Wang Z (2020) Can deep tillage enhance carbon sequestration in soils? A meta-analysis towards GHG mitigation and sustainable agricultural management. Renew Sust Energ Rev 133:110293

Franzluebbers AJ, Stuedemann JA (2002) Particulate and non-particulate fractions of soil organic carbon under pastures in the Southern Piedmont USA. Environ Pollut 116:S53–S62

Fuchs JG, Berner A, Mayer J, Schleiss K, Kupper T (2008) Effects of compost and digestate on environment and plant production–results of two research projects. Orbit 2008, 13–15 Oct. 2008 in Wageningen, The Netherlands

Grover M, Maheswari M, Desai S, Gopinath K, Venkateswarlu B (2015) Elevated CO2: plant associated microorganisms and carbon sequestration. Appl Soil Ecol 95:73–85

Gudka MDJ, Laban P, Metternicht G, Alexander S, Hannam I (2015) Land degradation neutrality: implications and opportunities for conservation. IUCN, Gland

Guo LB, Gifford RM (2002) Soil carbon stocks and land use change: a meta-analysis. Glob Chang Biol 8(4):345–360

IPCC (2014) Climate change 2014: synthesis report. In: Contribution of working groups I, II, and III to the fifth assessment report of the intergovernmental panel on climate change. IPCC

Jones MB, Donnelly A (2004) C sequestration in temperate grassland ecosystems and the influence of management, climate and elevated CO2. New Phytol 164(3):423–439

Jose S (2009) Agroforestry for ecosystem services and environmental benefits: an overview. Agrofor Syst 76:1–10

Kar SK, Singh RM, Patra S, Sankar M, Kumar S, Singh A (2023) Implication of land use shifting on land degradation and restoration potential of conservation agriculture in India's North-West Himalayan region. Geoderma Reg 32:e00616

Kätterer T, Bolinder MA, Andrén O, Kirchmann H, Menichetti L (2011) Roots contribute more to refractory soil organic matter than above-ground crop residues, as revealed by a long-term field experiment. Agric Ecosyst Environ 141:184–192

Kumari R, Kundu M, Das A, Rakshit R, Sahay S, Sengupta S, Ahmad F (2020) Long-term integrated nutrient management improves carbon stock and fruit yield in a subtropical mango (*Mangifera indica* L.) Orchard. J Soil Sci Plant Nutr 20:725–737

Lal R (2004a) Soil C sequestration impacts on global climate change and food security. Science 304:1623–1627

Lal R (2004b) Soil C sequestration in India. Clim Chang 65:277–296

Lal R (2004c) Soil C sequestration to mitigate climate change. Geoderma 123:1–22

Lal R (2010) Enhancing eco-efficiency in agro-ecosystems through soil C sequestration. Crop Sci 50(Suppl 1):S–120

Lal R (2020) Regenerative agriculture for food and climate. J Soil Water Conserv 75:123–124

Lal R, Follett RF, Stewart BA, Kimble JM (2007) Soil carbon sequestration to mitigate climate change and advance food security. Soil Sci 172:943–956

Lal R, Negassa W, Lorenz K (2015) Carbon sequestration in soil. Curr Opin Environ Sustain 15:79–86

Lee JJ, Phillips DL, Liu R (1993) The effect of trends in tillage practices on erosion and C content of soils in the U.S. corn belt. Water Air Soil Pollut 70:389–401

Liang F et al (2016) Three-decade long fertilization-induced soil organic C sequestration depends on edaphic characteristics in six typical croplands. Sci Rep 6:30350. https://doi.org/10.1038/srep30350

Madigan AP, Zimmermann J, Krol DJ, Williams M, Jones MB (2021) Full Inversion Tillage (FIT) during pasture renewal as a potential management strategy for enhanced carbon sequestration and storage in Irish grassland soils. Sci Total Environ 805:150342

Mahmud AA, Upadhyay SK, Srivastava AK, Bhojiya AA (2021) Biofertilizers: a nexus between soil fertility and crop productivity under abiotic stress. Curr Res Environ Sustain 3:100063

Mann LK (1986) Changes in soil carbon storage after cultivation. Soil Sci 142(5):279–288

Marland G, McCarl BA, Schneider U (2001) Soil carbon: policy and economics. Clim Chang 51(1):101–117

Mayer S, Wiesmeier M, Sakamoto E, Hübner R, Cardinael R, Kühnel A, Kögel-Knabner I (2021) Soil organic carbon sequestration in temperate agroforestry systems—a meta-analysis. Agric Ecosyst Environ 323:107689

Mbow C, Smith P, Skole D, Duguma L, Bustamante M (2014) Achieving mitigation and adaptation to climate change through sustainable agroforestry practices in Africa. Curr Opin Environ Sustain 6:8–14

Meena RS, Yadav RS, Meena H, Kumar S, Meena YK, Singh A (2015) Towards the current need to enhance legume productivity and soil sustainability worldwide: a book review. J Clean Prod 104:513–515

Minasny B et al (2017) Soil Carbon 4 per mille. Geoderma 292:59–86

Mitran T, Meena RS, Lal R, Layek J, Kumar S, Datta R (2018) Role of soil phosphorus on legume production. In: Meena et al (eds) Legumes for soil health and sustainable management. Springer, Singapore. https://doi.org/10.1007/978-981-13-0253-4_15

Nair PKR, Nair VD, Kumar BM, Showalter JM (2010) C sequestration in agroforestry systems. Adv Agron 108:237–307

Nath C, Kumar N, Das K, Hazra K, Praharaj C, Singh N (2021) Impact of variable tillage based residue management and legume based cropping for seven years on enzymes activity, soil quality index and crop productivity in rice ecology. Environ Sustain Indic 10:100107

Nkonya E, Mirzabaev A, Von BJ (2016) Economics of land degradation and improvement: an introduction and overview. In: Economics of land degradation and improvement–a global assessment for sustainable development, pp 1–14

Nosberger J, Blum H, Fuhrer J (2000) Crop ecosystem responses to climatic change: productive grasslands. In: Hodges HF (ed) Climate change and global crop productivity. CAB International, Wallingford, pp 271–291

Panhwar QA, Ali A, Naher UA, Memon MY (2019) Fertilizer management strategies for enhancing nutrient use efficiency and sustainable wheat production. In: Chandran S, Unni MR, Thomas S (eds) Organic farming: global perspectives and methods. Woodhead Publishing, pp 17–39

Pathak H, Byjesh K, Chakrabarti B, Aggarwal PK (2011) Potential and cost of carbon sequestration in Indian agriculture: estimates from long-term field experiments. Field Crop Res 120:102–111

Paul C, Bartkowski B, Dönmez C, Don A, Mayer S, Steffens M et al (2023) Carbon farming: are soil carbon certificates a suitable tool for climate change mitigation? J Environ Manag 330:117142

Paustian K, Lehman J, Ogle S, Reay D, Robertson GP, Smith P (2016) Climate-smart soils. Nature 532:49–57

Poeplau C, Don A (2015) C sequestration in agricultural soils via cultivation of cover crops—a meta-analysis. Agric Ecosyst Environ 200:33–41. https://doi.org/10.1016/j.agee.2014.10.024

Qin Z, Dunn JB, Kwon H, Mueller S, Wander MM (2016) Soil carbon sequestration and land use change associated with biofuel production: empirical evidence. GCB Bioenergy 8(1):66–80

Rakesh S, Sarkar D, Shikha SA, Sinha AK, Mukhopadhyay P, Rakshit A (2020) Protocols for determination and evaluation of organic carbon pools in soils developed under contrasting pedogenic processes and subjected to varying management situations. In: Rakshit A, Ghosh S, Chakraborty S, Philip V, Datta A (eds) Soil analysis: recent trends and applications. Springer, Singapore, pp 87–105

Rakesh S, Sarkar D, Sinha AK, Shikha MP, Danish S, Fahad S, Datta R (2021) Carbon mineralization rates and kinetics of surface-applied and incorporated rice and maize residues in Entisol and Inceptisol soil types. Sustain For 13(13):7212. https://doi.org/10.3390/su13137212

Rakesh S, Kamalakar J, Raju B, Jannu N, Pandit VB, Kunta R et al (2022a) Understanding carbon trading: accomplishments and prospects in India and other countries. Clim Change Environ Sustain 10(2):113–122

Rakesh S, Sinha AK, Juttu R, Sarkar D, Jogula K, Reddy SB, Raju B, Danish S, Datta R (2022b) Does the accretion of carbon fractions and their stratification vary widely with soil orders? A case study of an Alfisol and an Entisol of sub-tropical eastern India. Land Degrad Dev 33(12):2039–2049

Rakesh S, Sinha AK, Sarkar D, Roy D, Bodiga D, Sahoo S, Jha PK, Dubey PK, Rakshit A (2023) Active and passive carbon fractions in contrasting cropping systems, tillage practices, and soil types. Land 12:365. https://doi.org/10.3390/land12020365

Richardson EA, Kirkby CA, Banerjee S, Kirkegaard J (2014) The inorganic nutrient cost of building soil carbon. Carbon Manag 5(3):265–268

Rolfe C (2001) Sink solutions. Background analysis for the joint west coast environmental law/ David Suzuki Foundation report. In: Taking credit: Canada and the role of sinks in international climate change negotiations. West Coast Environmental Law, Canada

Sarkar D, Rakshit A (2023) Amalgamation of farmers' bio-priming knowledge in integrated nutrient management for sustainable management of red cabbage soil under Middle Gangetic Plains, India. Environ Manag 72:382–395

Sarkar D, Kar SK, Chattopadhyay A, Shikha RA, Tripathi VK, Dubey PK, Abhilash PC (2020) Low input sustainable agriculture: a viable climate-smart option for boosting food production in a warming world. Ecol Indic 115:106412

Sarkar D, Dubey PK, Chaurasiya R, Sankar A, Shikha CN, Ganguly S, Meena VS, Meena SK, Parewa HP, Rakshit A (2021) Organic interventions conferring stress tolerance and crop quality in agroecosystems during the UN Decade on Ecosystem Restoration. Land Degrad Dev 32(17):4797–4816

Shahid M, Nayak AK, Puree C, Tripathi R, Lal B, Gautam P, Bhattacharyya P, Mohanty S, Kumar A, Panda BB et al (2017) Carbon and nitrogen fractions and stocks under 41 years of chemical and organic fertilization in a sub-humid tropical rice soil. Soil Tillage Res 170:136–146

Shi S, Peng C, Wang M, Zhu Q, Yang G, Yang Y, Xi T, Zhang T (2016) A global meta-analysis of changes in soil carbon, nitrogen, phosphorus and sulfur, and stoichiometric shifts after forestation. Plant Soil 407:323–340

Singh G, Sangwan S, Narwal E, Annapurna K (2017) Microbial proteins and soil carbon sequestration. In: Rakshit A, Abhilash PC, Singh HB, Ghosh S (eds) Adaptive soil management: from theory to practices. Springer, Singapore, pp 131–139

Six J, Frey SD, Thiet RK, Batten KM (2006) Bacterial and fungal contributions to carbon sequestration in agroecosystems. Soil Sci Soc Am J 70:555–569

Smith P, Soussana JF, Angers D, Schipper L, Chenu C, Rasse DP et al (2020) How to measure, report and verify soil carbon change to realize the potential of soil carbon sequestration for atmospheric greenhouse gas removal. Glob Change Biol 26(1):219–241

Thomas A, Bentley L, Feeney C, Lofts S, Robb C, Rowe EC et al (2023) Land degradation neutrality: testing the indicator in a temperate agricultural landscape. J Environ Manag 346:118884

Upadhyay SK, Saxena AK, Singh JS, Singh DP (2019) Impact of Native ST-PGPR (*Bacillus pumilus*; EU927414) on pgp traits, antioxidants activities, wheat plant growth and yield under salinity. Clim Chang Environ Sustain 7(2):157–168

World Bank (2012) C sequestration in agricultural soils, Report No. 67395-GLB. World Bank

Xie Y, Lin H, Ye Y, Ren X (2019) Changes in soil erosion in cropland in northeastern China over the past 300 years. Catena 176:410–418

Yadav GS, Das A, Babu S, Mohapatra KP, Lal R, Rajkhowa D (2021) Potential of conservation tillage and altered land configuration to improve soil properties, carbon sequestration and productivity of maize based cropping system in eastern Himalayas, India. Int Soil Water Conserv Res 9:279–290

Yang S, Sheng D, Adamowski J, Gong Y, Zhang J, Cao J (2018) Effect of land use change on soil carbon storage over the last 40 years in the Shi Yang River Basin, China. Land 7(1):11

Zickfeld K, Azevedo D, Mathesius S, Matthews HD (2021) Asymmetry in the climate–carbon cycle response to positive and negative CO2 emissions. Nat Clim Chang 11(7):613–617

Chapter 17
Global and Regional Sustainability Challenges for Plant and Soil Nutrient Management

Vivek Verma and Sheikh Adil Edrisi

Abstract The global population is expected to reach 9.5 billion people by 2050. It is projected to cross 11 billion by the end of the century. Consequently, there is a need to expand crop production by 1.5 folds to satisfy the projected demand. Therefore, a major emphasis is ironically made on increasing the quantity of agricultural produce through different mismanaged agricultural practices. These practices are intensified in developing and underdeveloped nations facing socio-economic issues to feed their ever-increasing population. Therefore, there is an urgent requirement for smart, semi-sustainable nutrient management to improve soil and plant productivity, focusing on their resilience and endurance. Moreover, plant nutrient management often connects with food and nutrition security, mitigating pollution and climate change, limiting biodiversity loss, and other associated socio-economic issues. This management refers to improving the soil's overall vital nutrient resources and their transformation for improving plant vigor. However, the fortification of nutrients also does not meet the sustainability scenario standards. Furthermore, this prospective practice also seeks to be overhauled under the appropriate socio-economic framework to offer wider benefits to the end-users. Therefore, there are multiple bottlenecks in various adopted practices under sustainable plant nutrition management at the regional and global levels. Hence, examining the role of plant and soil nutrient management and its plausible limitation in providing agro-ecosystem services is essentially required to develop a holistic framework. Therefore, the present chapter discusses the diverse efforts in managing the plant and soil nutrients and highlights some tipping challenges at the regional and global levels, providing a window of opportunity for further exploratory research.

Keywords Nutrient management · Sustainability · Soil nutrients · Agroecosystems · Agricultural practices

---

V. Verma · S. A. Edrisi (✉)
Thapar School of Liberal Arts and Sciences, Thapar Institute of Engineering and Technology, Patiala, Punjab, India
e-mail: sheikhadil.edrisi@thapar.edu

S. A. Edrisi et al. (eds.), *Sustainable Plant Nutrition and Soil Carbon Sequestration*, Sustainable Plant Nutrition in a Changing World,
https://doi.org/10.1007/978-3-031-53590-1_17

17.1 Introduction

In the past 60 years, worldwide agricultural production has increased 2.2% annually (Fuglie 2018). The enhanced nutrient and pollutant discharge from the domestic and industrial sectors under the expanding global economy is crossing the maximum threshold of the carrying capacity of the ecosystem (Goulding et al. 2008; McMichael and Powles 1999). Furthermore, global variations in land productivity have increased from 2.1% to 4.6% for fats and from 2.1% to 2.9% for calories per year (Tuninetti et al. 2020). Sustainable intensification and expansion are considered critical factors in the development of agricultural production. As a result, both have contributed to a significant rise in fertilizer consumption. The 'Industrial Revolution' has altered the global carbon cycle since pre-industrial times. In contrast, the 'Green Revolution' during the 1960s has contributed to enhancing food production at the cost of other ecosystem services (IPCC 2021). Globally, mineral leaching, nutrient deficiencies, and toxicities are estimated to be present in around 60% of the cultivated soil, limiting the growth of planted crops and produce (Cakmak 2002). Providing nutrients for growing plants at the appropriate time and an adequate level is necessary to maintain soil fertility (Johnston and Bruulsema 2014; Withers et al. 2018).

Unfavorable soil conditions frequently limit nutrient delivery and the development of plants as they influence the soil parameters like the soil pH, redox state, etc., ultimately affecting the phyto-availability of nutrient-rich minerals within the soil (White and Greenwood 2013). A fundamental difficulty associated with plant nourishment is understanding the phyto-available stock of nutrients (Tandy et al. 2012; Mason et al. 2013). Furthermore, the soil must have apposite depth (topography) to withstand the development of root growth and ensure anchorage and adequate nutrient absorption (White et al. 2013a, b). However, the ability of soil to supply nutrients is also influenced by climatic and anthropogenic factors. Therefore, climate change may also have an increased effect on soil and plant nutrition (Fanzo and Downs 2021). There is a burgeoning challenge for the farmers and associated stakeholders to maintain the ample nutrient status in the soil and plants during crop production (Abhilash et al. 2016a, b). To this, various practices have been adopted to check the leaching of beneficial nutrients, ranging from the agronomic practices in the field to the technological and biotechnological advances from the scientific laboratories. However, the practices need to be updated, amended, and cautiously adopted under the site-specific scenarios for their successful ventures in soil and plant nutrition management.

Since many current agricultural practices are becoming eco-friendly, there is still a need to amplify the efficacies of these practices for attaining sustainable agriculture and multiple United Nations Sustainable Development Goals (UN-SDGs). Therefore, reviewing these practices at different steps and levels is imperative to achieve optimum yield and promote the best practices for further development (Priydarshini et al. 2021; Dubey et al. 2021). The application of fertilizers must be augmented to minimize the costs and protect ecosystems. In contrast, developing nations usually face technological limitations, population explosion, non-alignment of various policies, and hidden challenges. Therefore, the present chapter is aimed to discuss the diverse practices adopted in managing plant and soil nutrients. It also tried to highlight some burgeoning hidden challenges from the regional and global platforms that could provide an opportunity for further exploratory research.

17.2 Advances in Strategies for Plant and Soil nutrition Management

In the twentieth century, the science of plant nutrition was significantly advanced by the enhanced understanding of the physiological and biochemical processes involved in maintaining plant growth that enhanced plant productivity and soil quality. Sustainable food production by maintaining the viability of the soil system and plant nutrition could be ensured by holistic approaches through systematic science of practice. Additionally, nutrient management aims to frame decisions under the action plan for different agroecosystems by utilizing information about soil and plants' nutrient status. This strategy further facilitates optimizing different agricultural practices being adopted in the field. Generally, these practices are agronomic practices, soil amendments, microbial interventions, and biotechnological advances. Though numerous other practices are also considered under plant and soil nutrient management, for the sake of the current discussion, the present chapter tried to restrict its discussion to the most diverse and common interventions, as mentioned below.

17.2.1 Agronomic Practices

Agronomic practices can significantly reduce the consumption of synthetic chemical fertilizers, freshwater for irrigation, and chemical pesticides in agriculture. Subsequently, the yield and quality of crops will remain the same (Young et al. 2021). Achieving food security and combating malnutrition have been considered under the two UN-SDGs, i.e., No poverty (Goal no. 1) and Zero hunger (Goal no. 2), respectively. These goals are considered to be attained with the help of bio-based agronomic practices for primary production (Chojnacka et al. 2020). These practices also have a more positive impact on ecological functions and economic sustainability. Therefore, these methods can protect natural resources and lower the relative economic input in agricultural crop production. Among the most significant sustainable agronomic practices, intercropping, organic farming, and reduce-to-no-tillage practices have enhanced soil structure and health and improved soil biodiversity (Morugán-Coronado et al. 2020).

With the intercropping system, crops can be spatiotemporally intensified to meet regional future demands. Multiple advantages of intercropping include improved yield, food security, production sustainability, and better ecosystem services. Under intercropping, two or more crop species can be grown simultaneously in a crop cycle, interacting with one another in agroecosystems. In this system, legumes serve a variety of roles, including biological N fixation, soil quality enhancement, enhanced yields, particularly protein production, and greater functional diversity (Maitra et al. 2021). However, organic farming enhances local species richness and abundance by 34% and 50%, with plants and bees benefiting the most (Tuck et al. 2013; Bengtsson et al. 2005; Smith et al. 2020).

Furthermore, organic farming forbids synthetic pesticides, fertilizers, and Genetically Modified Organisms (GMOs) to protect the environment and maintain soil fertility and biodiversity (Seufert and Ramankutty 2017; Niggli 2015). Interestingly, numerous other agronomic practices, such as weed control techniques, sowing frequency, pretreatment of seeds, irrigation, and plant variety, influence soil fertility and management strategies (Prasad and Prasad 1998). It has been observed that the amount of intercepted photosynthetically active radiation and leaf area index considerably increase with increasing planting density, encouraging plant development and agricultural productivity (Zhang et al. 2021).

According to Ramirez et al. (2022), evidence of improved yield, profitability, water productivity, and nutrient-use efficiency likely raise the SOC levels in the recent development of agronomic practices. It strongly suggested the practice of potato zero-tillage and mulching (PZTM), a highly promising practice for smallholder farmers in the Patna district of Bihar, India. It was observed that there was a 12.8% increment in SOC under PZTM compared to the control. It greatly aided in mitigating the adverse effects of agro-climatic changes that many nations are presently experiencing (Sarangi et al. 2020). Furthermore, implementing cultivars with high yields, superior product qualities, and increased resistance to biotic and abiotic stresses should also be considered a viable option to improve plant performance and productivity (Rai et al. 2020).

Improved soil management will eventually increase crop yields when all the recommended, timely-tested agronomic practices are applied appropriately to their optimal capacity for different crops at regular intervals (Ghosh and Singh 1994). Various biotechnological research has examined interactions between plants and microbes to promote sustainable development and protect the environment. As a result, plant species significantly influence the variety of the plant-microbiome interaction (Berg et al. 2016). An environmentally beneficial substitute for chemical fertilization is the inoculation of plant growth-promoting microbes (PGPM) utilized for crop production such as *Azospirillum, Bacillus, Burkholderia, Enterobacter, Flavobacterium, Pseudomonas, Rhizobium, Frankia, Klebsiella, Clostridium, Trichoderma, Beauveria, Serratia and Streptomyces* (Abhilash et al. 2016a, b; Oosten et al. 2017; Gouda et al. 2018).

Seeds, roots, and soil are mostly infected with microbial inoculants. PGPM increases plant development by increasing nutrient availability, controlling phytohormones, and improving plant tolerance to biotic and abiotic challenges (Abhilash et al. 2016a, b; Asghari et al. 2020; Etesami 2020). Further, using transgenic and designer plants provides an effective mechanism for nutrient uptake. As a result, it also helps in plant-microbe responses to climate change and their function in reducing soil stress, which is advantageous for the long-term health of agroecosystems and maintaining food security (Sankar Ganesh et al. 2017).

However, there are also a few burgeoning challenges in such promising practices, particularly at regional levels, during their application in the field. Most importantly, familiarization with the concerned practices needs to be prioritized during implementation and naturalization. It is important because most unsuccessful projects have highlighted the issue of the lack of knowledge of concerted practices or technologies. Furthermore, the societal acceptance and awareness from the different stakeholders, such as the stringently associated partners (like farmers and others), are also a few of the crucial bottlenecks. Local farmers should have fundamental knowledge and understanding of crop plantations. Farmers must get the appropriate education and training, as depicted in Fig. 17.1, to enable them to comprehend the mechanisms governing nutrient cycling and the variables influencing nutrient losses.

Moreover, taking feedback and conducting surveys at regular intervals can help transform a robust framework for plant and soil nutrition management. Continuous

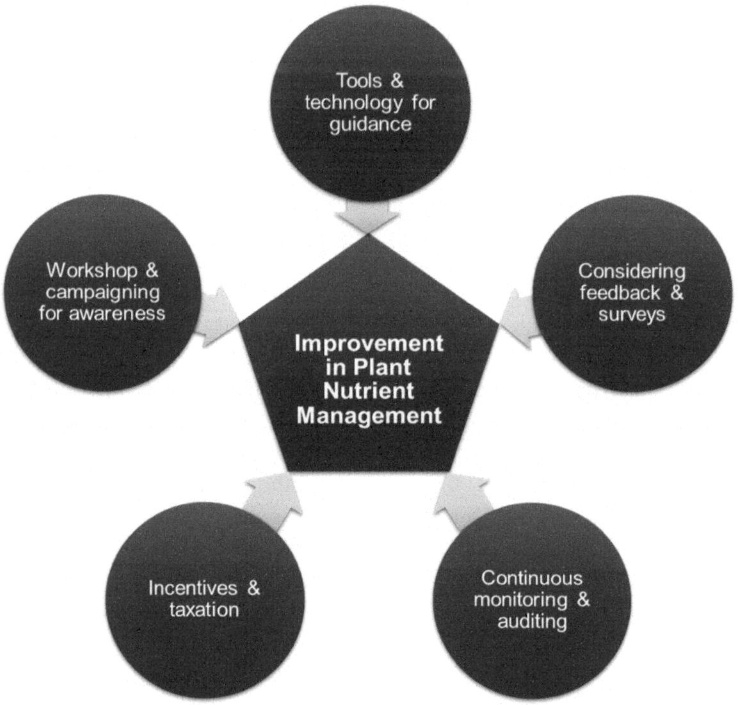

Fig. 17.1 Conceptual design for managing plant nutrition through monitoring and quality control. (Source: Modified from Oenema and Pietrzak 2002)

monitoring and auditing could also ascertain if the policy ventures are going in the right direction. It will also facilitate the identification of chances for augmented incentivization and taxation (Fig. 17.1).

17.2.2 Soil Amendment

Soil amendment is an approach to enhance soil quality, plant growth, disease resistance, and carbon sequestration. It also helps optimize soil-air-water interactions, varying soil bulk density, and strengthening soil structure by stabilizing soil aggregates. Various soil amendments have different impacts on the soil, as shown in Table 17.1. According to Chan et al. (2008), the use of poultry litter biochar as soil amendments in New South Wales, Australia, resulted in significant changes in the dry matter yield of radish crops even at the lower rate of application, i.e., 10 t ha^{-1}. These modifications may significantly affect the nutrient content and availability and the biochar's other qualities as a soil supplement.

Moreover, reducing soil-borne diseases, particularly those stimulated by worms that parasitize plants is a popular practice. Although amendments are known to release harmful substances directly or indirectly during their breakdown in soil, they may also enhance the populations of nematode-avoidant microbes in the soil. For example, applying the amendments along with the neem extracts significantly decreased root galling in tomato cultivation in Japan (Oka et al. 2007). It also revealed that it increases the persistence of ammonium concentrations in the soil by preventing the nitrification of the ammonium released from the amendments. The increased nematicidal activity of the ammonia-releasing amendments appeared to be caused by worms being exposed to ammonia for extended periods due to nitrification suppression by the neem extracts (Oka et al. 2007).

Organic amendment refers to various added soil additives that serve as an energy source for heterotrophic bacteria, fungi, and invertebrates that are good for plant health (Sankar Ganesh et al. 2017). They simultaneously supplement deficient soil fertility and lower agricultural waste. Traditional organic fertilizers applied over an extended period increase microbial diversity and biomass, the composition of which varies depending on the type of amendment (Saeid and Chojnacka 2019). For instance, manure has been utilized for centuries due to its abundance in N, P, and organic matter (Edrisi et al. 2021). It can change the soil microbiome and its physical and chemical properties to boost agricultural output (Qian et al. 2016). As per Adeniyan et al. (2011), applying several kinds of organic manures in maize crops located in the coastal area of Epe, Nigeria, results in lowered acidity levels in soils, e.g., in acidic and nutrient-depleted soil, cow dung treatment produced the most outstanding pH levels of 6.37 and 6.50, respectively. In contrast, NPK fertilizer had the lowest pH levels of 5.28 and 5.74 for both soils. Additionally, applying various kinds of organic manures improved soil organic C, total N, accessible P, exchangeable K, and cation exchange capacity (CEC) in both soils more effectively than NPK fertilizer.

Table 17.1 Various soil amendments and their impacts on the soil systems

S. No.	Amendments in soil	Impact on soil	Significant changes	Country	Source
1.	Poultry litter-based biochar	Increased fertility and carbon sequestration	Increase from 42% at 10 t ha^{-1} to 96% at 50 t ha^{-1}	Australia	Chan et al. (2008)
2.	Composted waste from animals and organic matter	Enhanced fertility and increase in mineral content	Increase in N (189.3 kg ha^{-1}), P (54.8 kg ha^{-1}), K (193.8 kg ha^{-1}), Zn (1.83 mg kg^{-1}), soil carbon stock (12.5 t ha $year^{-1}$), carbon sequestration (3.0 t ha $year^{-1}$), microbial carbon biomass (1741.2 μg g^{-1}), marketable yield (7.9 t ha^{-1}), crop biomass (7.0 t ha^{-1}), litterfall (0.54 t ha^{-1})	India	Parmar et al. (2016)
3.	*Azadirachta indica* oil cakes and leaves	Nematode suppression	A 0.025 g kg^{-1} ammonium sulfate (AS concentration) combined with a neem extract (Azatin) reduced galling index and suppressed nematode eggs	Japan	Oka et al. (2007)
4.	Fly-ash based material	Increased field capacity and changes in pH	For example, the pH of 1:5 $CaCl_2$ extracts ranged from 3.8 to 9.9	Australia	Pathan et al. (2003)
5.	*Brassica juncea*	Heavy metal immobilization	–	Spain	Lwin et al. (2018)
6.	Effluent from olive mills	Increase aggregate stability, resistance to water, and decrease in infiltration rate.	Stability increased from 18% (control) to 31 and 38%, water resistance from 1.8 kg cm^{-2} (control) to 3.5 and 4.5 kg cm^{-2}, and water repellency from <5 s (control) to 120 and 261 s	Jordan	Albalasmeh et al. (2019)

(continued)

Table 17.1 (continued)

S. No.	Amendments in soil	Impact on soil	Significant changes	Country	Source
7.	Flour made from hoofs and horns	Support lowering the usage of nitrogen fertilizer	–	Brazil	da Silva et al. (2019)
8.	Mulch	Decrease in N leaching caused by $N\text{-}NO_3^-$ immobilization	Reduction in soil erosion (> 97%)	Germany	Doring et al. (2005)
9.	Composted cassava peelings, poultry manure, rabbit and cane rat droppings, cow dung	Increase in soil pH and organic matter N, P, K.	Increased pH from 5.28 to 6.37 in acidic soil and 5.74–6.50 in nutrient-depleted soil	Nigeria	Adeniyan et al. (2011)
10.	Charred(ashes) rice husks	Increase in pH, the concentration of anions and silica	Increased P-availability up to 120% and plant growth from 17% to 197%	Indonesia	Utami et al. (2012)

Interestingly, one more technique is widely utilized to eliminate contaminants from the soil or reduce their toxicity by applying plants, such as phytoremediation. Plants are a broad set of creatures cultivated and used in various situations where changes occur due to location, seasonal shifts, and daily, weekly, monthly, and yearly climate fluctuations. The nutrient content of soils in both agricultural systems and natural environments can be very low, adequate, or excessive. Two different types of plant development and metabolism can be considered to create a framework for understanding how plants adjust to various soil nutrients. One mode is when there is an abundance of mineral resources and nutrient-sufficient conditions for plant growth. Plants may find it relatively easy to access nutrients and water in soils with abundant sunlight, and development occurs at optimum rates.

On the other hand, when nutrients are limited, growth is delayed, and plants modify certain aspects of their acquisition and morphology to optimize and get restricted resources. Therefore, phytoremediation assisted with microbial and soil amendments could be one of the promising integrated efforts during the production of edible crops for sustainable plant and nutrient management (Abhilash et al. 2016b). Based on their evolutionary habitat and soil mobility, plant development reacts differently to nutrient restrictions (Hodge 2004). Although the reaction of plants to the deprivation of critical mineral elements is well established, there is limited understanding regarding plant perception and signal changes regarding nutrient availability (Schachtman and Ryoung 2007). Therefore, there are existing challenges in building a holistic understanding of the role of the nexus between the plant-soil-microbe interactions, which could pave the way for future research. However, the application of microbial technology has been crucial to sustainability and agricultural production.

17.2.3 Microbial Intervention and Biotechnological Advances

Biotechnology and microbiology currently coexist to train and selectively breed animals, grow crops, and use microbes to create products like cheese, yogurt, bread, beer, and wine (Gupta et al. 2018). Meanwhile, agriculturalists in ancient times used primitive biotechnology to create higher-quality species of plants and animals by cross-pollination or crossbreeding (Lammerts van Bueren et al. 2018). The techniques used in microbial biotechnology significantly widen the gene pool and enable their integration into plants. Biotechnology can shorten the time needed to produce new plant varieties with its assistance. A good example of the introduction of genes into plant cells is transgenic plants (Padikasan et al. 2018). Transgenic plants can thrive under unfavorable conditions (Raman 2017). In addition, with the increase in their habitat range from cold, heat, and drought, they can withstand high moisture and salt concentrations. *Bacillus thuringiensis* (Bt) production is now induced in plants through genetic modification. This bacterium is naturally found in soil and can create a specific protein that aids in killing some insects. After being consumed by the insects, this protein impairs the operation of their digestive system (Castagnola and Jurat-Fuentes 2012; Abbas 2018). This bacterium also has the benefit of not being dangerous to humans, fish, birds, or precious insects. However, pesticide inefficiency has several causes, including the photolysis of active chemicals by sunshine and insecticide runoff by rainwater. Therefore, Bt works well under the plant system to counter the extrinsic stresses induced by insects (Van Huis et al. 2013; Abbas 2018).

Using pesticides over a long period will render the soil unsuitable for agriculture in the future, even though current agricultural practices unquestionably increase crop yield. Microbial technology has been recognized as the most appropriate alternative to maintain soil sustainability without affecting soil ecology or climatic factors (Singh et al. 2010; Sengupta and Gunri 2015; Schafer and Adams 2015). The rhizosphere of the soil is a vast repository for microbial variety. Numerous metabolic processes carried out by microbes are necessary for their survival. They can directly or indirectly help the biosphere using nutrient recycling, environmental detoxification, soil wellness enhancement, wastewater treatment, etc. There are still many unknown useful soil microorganisms, and little is known about their ecological functions. Therefore, extensive tests of microbial activity are the first stages in creating new technologies that effectively use microorganisms to achieve sustainability in agriculture.

Meanwhile, only plants can transport and store minerals. Determining how plants mobilize minerals from the soil may be possible for food crops with higher mineral content. Additionally, plants that can store various minerals could be used in bioremediation, cleaning up soil with noxious mineral contamination. Since not all soils contain enough minerals to support crop growth, fertilization is required. Therefore, it is cost intensive and can have a harmful impact on the environment. Ironically, even barren soils contain few minerals to support crop growth. However, those minerals are not available to plants for uptake. Microbial intervention (Fig. 17.2) may help increase productivity while maintaining sustainability in agriculture.

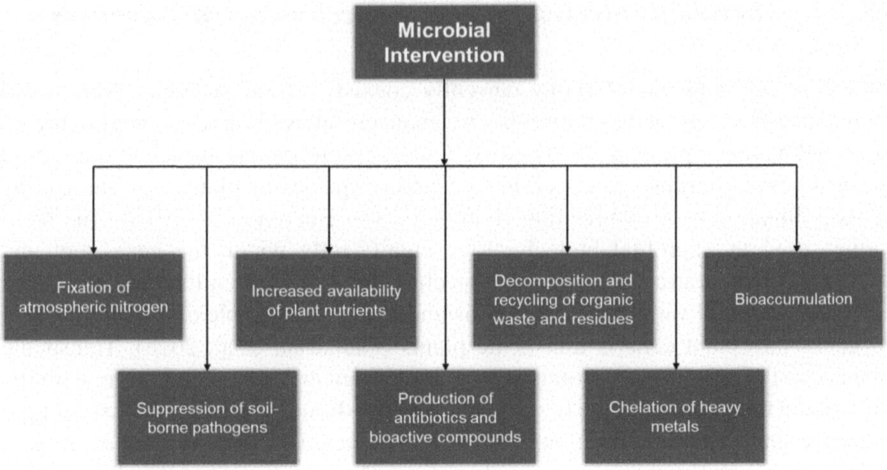

Fig. 17.2 Diverse benefits of microbial interventions in soil. (Source: Adapted from Brierley 1985; Ehrlich 1990)

It includes fixing atmospheric nitrogen, enhancing the plant nutrients in terms of their availability, decomposing and recycling organic waste and debris, bioaccumulation or microbial leaching of inorganics, suppressing soil-based pathogens, biodegrading contaminants like insecticides, and producing antibiotics and other bioactive substances (Brierley 1985; Ehrlich 1990). Interestingly, the bio-priming technique (seed coating) is also used to overcome farmers' significant agricultural challenges in the Varanasi district of Uttar Pradesh state, India. It includes the physiological and biological aspects of crop growth, disease, or pest control. Hence, a combination of three compatible bio-agents, viz., *Trichoderma harzianum*, *Pseudomonas fluorescens*, and *Bacillus subtilis*, were applied alone and in combination with a 75% recommended dose of fertilizer (RDF). As a result, microbial intervention increases the soil fertility of the red cabbage farming systems (Sarkar and Rakshit 2023).

Furthermore, fortified crops become a great source of nutrition for malnourished children in developing countries or places with a severe food crisis. Potato is one example of such a fortified crop. This genetically modified potato is produced and consumed frequently in India. It offers around 30–50% more protein than a typical potato, hence called 'Protato.' It also contains many essential amino acids, including lysine and methionine.

Moreover, in nations where potatoes are a key staple meal, such as China, Germany, Netherlands, and France, this 'Protato' could be an up-and-coming food source (Rajasree and Pugalendhi 2021). There are immense prospects for exploring cutting biotechnological developments for managing plant and soil nutrition. These biotechnological advancements are the omics technologies comprising the exploitation of functional and active genes in the microbial population for pollutant degradation and nutrient enrichment (Tripathi et al. 2017). However, there is apprehension among the field experts, particularly during the implementation of the biotechnological tools, because wide-scale and cautious testing is utmost required during translational research. Therefore, such sustainability challenges are obvious and usually untested in the field but still require attention.

17.3 Sustainability Challenges and Upbringing of United Nations Sustainable Development Goals (UN-SDGs)

According to Vollset et al. (2020), the world's population is predicted to reach 9.5 billion by 2050, so feeding the burgeoning global population needs enhanced agricultural production. Due to mismanaged agricultural practices and nutrient management strategies, soil health is deteriorating. Therefore, it limits food security, which is one of the major challenges being highlighted at the global level (Fig. 17.3).

These mismanaged practices are usually associated with the limited knowledge of the diverse biological processes that fundamentally optimize nutrient cycling. Furthermore, increased dependencies on non-renewable resources and lower reliance on renewables also lower the systems' sustainability, as Edrisi et al. (2022) reported. In addition, smallholder farmers in tropical countries also face the issue of land constraints regarding the availability of areas, which makes them unable to improvise new big machinery and technologies. Moreover, there is growing pressure to increase the population, which impels the governance for intensifying

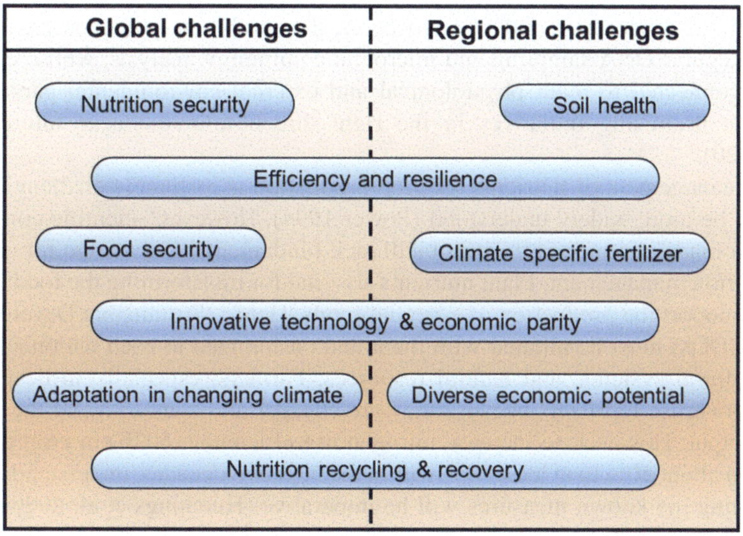

Fig. 17.3 Global and regional challenges of plant-soil nutrition management. (Adapted from Dobermann et al. 2022)

agriculture with proper training to the local farmers, leading to the failure of innovative technologies (Fig. 17.3) (Kumwenda et al. 1996).

Hence, there is an urgent need to develop different integrated and transdisciplinary approaches for enhancing and managing plant nutrition, especially while restoring marginal and degraded lands (Edrisi and Abhilash 2021). It will eventually help cope with numerous local, national, and international issues associated with nutrients in the dietary chain system. Crop production intensively relies on applying different fertilizers for the prospects of nutrition management. However, applying chemical fertilizers enhances the quantity of produce; still, there are shortcomings regarding overall sustainable and nutritional growth. There are various challenges associated with them due to their nutritional aspect. For instance, trade in food and feed globally causes disparity on a regional level (Grote et al. 2005).

As discussed in previous sections, intercropping has advantages, but there are also certain challenges compared to mono-cropping. These are caused mainly by inter-species competition for limited resources, the manifestation of allelopathic effects, and difficulty in managing crops in combination from an agronomic perspective (Cenpukdee and Fukai 1992; Maitra et al. 2021). Moreover, there are also significant drawbacks to organic farming's benefits for biodiversity, including lower yields, false beliefs about pesticide use, taxon-specific advantages, and commercial intensification of productivity. Furthermore, it is misleading that organic farms mostly avoid pesticides. Under organic labeling, pesticides are permitted as long as they are made from natural rather than synthetic substances (Seufert and Ramankutty 2017).

Interestingly, plant-microbe interactions aim to create healthy microbiomes to comprehend how plants and microbiomes interact. However, there are technical difficulties with real-time phenotypic/physiological plant characterization and monitoring of the intricate spatial and biochemical microbiome activities (Arif et al. 2020). Regular DNA sampling and microbial community analysis, whose changes can be connected to plant physiological and external environmental parameters, might be promising initiatives in the right direction in the near future (Arif et al. 2020).

The management of nutrients for crop production is extremely challenging and needs to be more widely understood (Power 1994). However, scientific communities have made enormous efforts but still lack fundamental knowledge for enhancing nutrition management. Plant nutrients are vital for transforming the food system as they support the production of essential foods. Hence, Sustainable Development Goals (SDGs) must be aligned with them and customized to each nation's unique food, cultural systems, and natural resources. Farmers, especially in developing nations, overuse fertilizers because they are cheaper and want to avoid the risk of lower output. However, to increase nitrogen use efficiency (NUE) in crop production from about 50% to at least 70% within the next two decades, understanding and overcoming the known measures will be imperative (Hutchings et al. 2020). Plant nutrients are at the core of the food system transformation because they drive both primary food production and many of the externalities caused by it.

Crop yields can be increased by sustainable intensification. However, the ratio of nutrient input to output may fall, and there is a chance that nutrient surpluses will become unsustainable. A prominent example is India, where several fertilizer price control and subsidy programs have significantly increased fertilizer consumption since the 1970s. Since then, N and P fertilizer application on the cropland in India has escalated to more than doubled. However, its use efficiency has been lowered to around 30–40% and is still at that lower level (Dobermann et al. 2022).

However, the hidden challenges of plant and soil nutrition management are also raised due to the lack of field experts with the necessary training, in-depth understanding of soil systems, and the capacity to communicate effectively with other stakeholders. The UN-SDG numbers 2, 6, 13, and 15 directly focus on the uplifting of the soil and allied systems in which soil serves a larger role. In contrast, other UN-SDGs are indirectly associated and primarily linked to socio-economics of the system (Ussiri and Lal 2018). Therefore, it is necessary to focus on the critical attributes and guidelines that direct attaining the UN-SDGs mentioned above. These attributes could be the systematic framing of action plans under sustainable intensification in agricultural lands.

Furthermore, building cohort and strategic approaches are required to address some of the challenges discussed in the previous sections. Moreover, the absence of a thorough disciplinary framework to assist those working as soil scientists or in other capacities that directly relate to achieving the SDGs (Brevik et al. 2020; Collins 2008; Diochon et al. 2016; Field et al. 2017; Havlin et al. 2010), has been a major concern. Agricultural production must be increased in areas with high demand for food and where the rural and urban populations are expected to grow rapidly in future years. This condition currently exists in sub-Saharan Africa and various other regions of Latin America, Asia, and the Pacific (UNDESA 2013).

17.4 Conclusions and Future Recommendations

Nutrient management must be considered a specialized activity to achieve economic and environmental targets. The targets must be explicit, flexible, measurable, controllable, and set at the appropriate temporal and spatial dimensions. Specialized activities may vary and be challenging in practice depending on the type of agroecosystem and other external factors. Hence, management demands skill, decisiveness, precision, appropriate equipment, and in-depth knowledge of the nutrition cycle. The correct economic incentives, farmer education, and training are needed to focus policies and activities. The use of GMOs is a development of microbial technology and is crucial to the sustainability and productivity of agriculture. Modern microbial and omics technology is more sophisticated and applicable. Hence, combining it with other basic sciences leads to further cutting-edge research and nutrient management improvement, provided wider and intensive testing is accomplished at the field levels.

Furthermore, policies supporting nutrient additions should emphasize achieving food security goals and concentrating more on the environmental effects of agriculture. Adopting such notions in ecological and agronomic practices can help foster soil and nutrient management. Using nutrients more efficiently and recycling organic nutrient sources are a few strategies to protect the environment. They can also boost farming's profitability and help address global food and nutrition security issues.

Acknowledgments Authors are thankful to the Director, Dean, and Head TSLAS, TIET, Patiala, India for providing institutional support and encouragement. SAE is grateful to SERB for providing Start-up Research Grant (file no. SRG/2021/000281) and TIET for Seed Funding (ref. no. TU/DORSP/1596). VV is grateful to SERB for providing Junior Research Fellowship under the project.

References

Abbas MST (2018) Genetically engineered (modified) crops (*Bacillus thuringiensis*) and the world controversy on their safety. Egypt J Biol Pest Control 28(1):1–12

Abhilash PC, Dubey RK, Tripathi V, Gupta VK, Singh HB (2016a) Plant growth-promoting microorganisms for environmental sustainability. Trends Biotechnol 34:847–850. https://doi.org/10.1016/j.tibtech.2016.05.005

Abhilash PC, Tripathi V, Edrisi SA, Dubey RK, Bakshi M, Dubey PK, Ebbs SD (2016b) Sustainability of crop production from polluted lands. Energ Ecol Environ 1:54–56

Adeniyan ON, Ojo AO, Akinbode OA, Adediran JA (2011) Comparative study of different organic manures and NPK fertilizer for improvement of soil chemical properties and dry matter yield of maize in two different soils. J Soil Sci Environ Manag 2(1):9–13. https://doi.org/10.5897/JSSEM.9000022

Albalasmeh AA, Alajlouni MA, Ghariabeh MA, Rusan MJ (2019) Short-term effects of olive mill wastewater land spreading on soil physical and hydraulic properties. Water Air Soil Pollut 230:208. https://doi.org/10.1007/s11270-019-4243-5

Arif I, Batool M, Schenk PM (2020) Plant microbiome engineering: expected benefits for improved crop growth and resilience. Trends Biotechnol 38(12):1385–1396

Asghari B, Khademian R, Sedaghati B (2020) Plant growth promoting rhizobacteria (PGPR) confer drought resistance and stimulate biosynthesis of secondary metabolites in pennyroyal (*Mentha pulegium* L.) under water shortage condition. Sci Hortic 263:1–10. https://doi.org/10.1016/j.scienta.2019.109132

Bengtsson J, Ahnström J, Weibull A-C (2005) The effects of organic agriculture on biodiversity and abundance: a meta-analysis. J Appl Ecol 42:261–269. https://doi.org/10.1111/j.1365-2664.2005.01005.x

Berg G, Rybakova D, Grube M, Köberl M (2016) The plant microbiome explored: implications for experimental botany. J Exp Bot 67(4):995–1002

Brevik EC, Dolliver H, Edinger-Marshall S, Itkin D, Johnson-Maynard J, Liles G, Mbila M, Moorberg C, Sanchez-de Leon Y, Steffan JJ, Ulery A, Vaughan K (2020) Soil Sci Soc Am J. https://doi.org/10.1002/saj2.20140

Brierley JA (1985) Use of microorganisms for mining metals. In: Halvorson HO, Pramer D, Rogul M (eds) Engineered organisims in the environment: scientifc issues. ASM Press, Washington, pp 141–146

Cakmak I (2002) Plant nutrition research: priorities to meet human needs for food in sustainable ways. Plant Soil 247:3–24

Castagnola AS, Jurat-Fuentes JL (2012) Bt crops: past and future. In: Bacillus thuringiensis biotechnology. Springer, Dordrecht, pp 283–304

Cenpukdee U, Fukai S (1992) Cassava/legume intercropping with contrasting cassava cultivars. 1. Competition between component crops under three intercropping conditions. Field Crop Res 29:113–133

Chan KY, Van Zwieten L, Meszaros I, Downie A, Joseph S (2008) Using poultry litter biochars as soil amendments. Soil Res 46(5):437–444

Chojnacka K, Moustakas K, Witek-Krowiak A (2020) Bio-based fertilizers: a practical approach towards circular economy. Bioresour Technol 295:122223

Collins ME (2008) Where have all the soils students gone? J Nat Resour Life Sci Educ 37:117–124. https://doi.org/10.2134/jnrlse2008.371117x

da Silva RR, Leite RC, Carneiro JSS, de Freitas GA, dos Santos ACM, dos Santos AC, Kuyumjian LA (2019) Application of slaughterhouse residues as nitrogen source replacing commercial fertilizers on mombasa grass (*Megathyrsus maximus*). Aust J Crop Sci 13(02):294–299

Diochon A, Basiliko N, Krzic M, Yates TT, Olson E, Masse J, Amiro B, Kumaragamage D, Yang XM (2016) Profiling undergraduate soil science education in Canada: status and projected trends. Can J Soil Sci 97:122–132

Dobermann A, Bruulsema T, Cakmak I, Gerard B, Majumdar K, McLaughlin M, Reidsma P, Vanlauwe B, Wollenberg L, Zhang F, Zhang X (2022) Responsible plant nutrition: a new paradigm to support food system transformation. Glob Food Secur 33:100636. https://doi.org/10.1016/j.gfs.2022.100636. ISSN 2211-9124

Doring TF, Brandt M, Heß J, Finckh MR, Saucke H (2005) Effects of straw mulch on soil nitrate dynamics, weeds, yield and soil erosion in organically grown potatoes. Field Crop Res 94:238–249. https://doi.org/10.1016/j.fcr.2005.01.006

Dubey PK, Singh A, Chaurasia R, Pandey KK (2021) Fostering Nature-Based Solutions for Ecorestoration and Attaining United Nations Sustainable Development Goals. J Clean Pro 298:126680

Edrisi SA, Abhilash PC (2021) Need of transdisciplinary research for accelerating land restoration during the UN Decade on Ecosystem Restoration. Restor Ecol 29(8):e13531

Edrisi SA, Tripathi V, Chaturvedi RK, Dubey DK, Patel G, Abhilash PC (2021) Saline soil reclamation index as an efficient tool for assessing restoration progress of saline land. Land Degrad Dev 32(1):123–138

Edrisi SA, Sahiba SA, Chen B, Abhilash PC (2022) Emergy-based sustainability analysis of bioenergy production from marginal and degraded lands of India. Ecol Model 466:109903

Ehrlich HL (1990) Geo-microbiology, 2nd edn. Dekker, New York, p 646

Etesami H (2020) Plant–microbe interactions in plants and stress tolerance. In: Plant life under changing environment. Academic, pp 355–396. https://doi.org/10.1016/B978-0-12-818204-8.00018-7

Fanzo JC, Downs SM (2021) Climate change and nutrition-associated diseases. Nat Rev Dis Primers 7:90. https://doi.org/10.1038/s41572-021-00329-3

Field DJ, Yates D, Koppi AJ, McBratney AB, Jarrett L (2017) Framing a modern context of soil science learning and teaching. Geoderma 289:117–123

Fuglie KO (2018) Is agricultural productivity slowing? Glob Food Sec 17:73–83

Ganesh K, Sundaramoorthy P, Nagarajan M, Xavier R (2017) Role of organic amendments in sustainable agriculture. In: Sustainable agriculture towards food security, pp 111–124. https://doi.org/10.1007/978-981-10-6647-4_7

Ghosh PK, Singh NP (1994) Soil-nitrogen status under summer legumes-maize (Zea mays) sequence. Indian J Agric Sci 64(12):856–857

Gouda S, Kerry RG, Das G, Paramithiotis S, Shin H, Patra JK (2018) Revitalization of plant growth promoting rhizobacteria for sustainable development in agriculture. Microbiol Res 206:131–140. https://doi.org/10.1016/j.micres.2017.08.016

Goulding K, Jarvis S, Whitmore A (2008) Optimizing nutrient management for farm systems. Philos Trans R Soc Lond B Biol Sci 363(1491):667–680. https://doi.org/10.1098/rstb.2007.2177. PMID: 17652069; PMCID: PMC2610177

Grote U, Craswell E, Vlek P (2005) Nutrient flows in international trade: ecology and policy issues. Environ Sci Pol 8:439–451

Gupta VK, Treichel H, Shapaval V, de Oliveira LA, Tuohy MG, Wiley J (eds) (2018) Microbial functional foods and nutraceuticals. Wiley. ISBN:9781119048961

Havlin J, Balster N, Chapman S, Ferris D, Thompson T, Smith T (2010) Trends in soil science education and employment. Soil Sci Soc Am J 74:1429–1432

Hodge A (2004) The plastic plant: root responses to heterogenous supplies of nutrients. New Phytol 162:9–24

Hutchings NJ, Sørensen P, Cordovil CM, Leip A, Amon B (2020) Measures to increase the nitrogen use efficiency of European agricultural production. Global Food Secur 26:Article 100381

IPCC (2021) Climate change 2021: the physical science basis. In: Masson-Delmotte V, Zhai P, Pirani A, Connors SL, Péan C, Berger S, Caud N, Chen Y, Goldfarb L, Gomis MI, Huang M, Leitzell K, Lonnoy E, Matthews JBR, Maycock TK, Waterfield T, Yelekçi O, Yu R, Zhou B (eds) Contribution of working group I to the sixth assessment report of the intergovernmental panel on climate change. Cambridge University Press, Cambridge/New York. https://doi.org/10.1017/9781009157896

Johnston AM, Bruulsema TW (2014) 4R nutrient stewardship for improved nutrient use efficiency. Procedia Eng 83:365–370

Kumwenda JDT, Waddington SR, Snapp SS, Jones RB, Blackie MJ (1996) Soil fertility management research for the maize cropping systems of smallholders in southern Africa: a review. Natural Resource Group Paper 96-02

Lammerts van Bueren ET, Struik PC, Eekeren NV, Nuijten E (2018) Towards resilience through systems-based plant breeding. A review. Agron Sustain Dev 38(42):1–21

Lwin CS, Seo BH, Kim HU, Owens G, Kim KR (2018) Application of soil amendments to contaminated soils for heavy metal immobilization and improved soil quality.A critical review. Soil Sci Plant Nutr 64(2):156–167

Maitra S, Hossain A, Brestic M, Skalicky M, Ondrisik P, Gitari H, Sairam M (2021) Intercropping—a low input agricultural strategy for food and environmental security. Agronomy 11(2):343

Mason SD, McLaughlin MJ, Johnston C, McNeill A (2013) Soil test measures of available P (Colwell, resin and DGT) compared with plant P uptake using isotope dilution. Plant Soil 373:711–722

McMichael AJ, Powles JW (1999) Human numbers, environment, sustainability, and health. BMJ 319(7215):977–980. https://doi.org/10.1136/bmj.319.7215.977. PMID: 10514165; PMCID: PMC1116806

Morugán-Coronado A, Linares C, Gómez-López MD, Faz Á, Zornoza R (2020) The impact of intercropping, tillage and fertilizer type on soil and crop yield in fruit orchards under Mediterranean conditions: a meta-analysis of field studies. Agric Syst 178:102736

Niggli U (2015) Sustainability of organic food production: challenges and innovations. Proc Nutr Soc 74:83–88

Oenema O, Pietrzak S (2002) Nutrient management in food production: achieving agronomic and environmental targets. AMBIO J Hum Environ 31(2):159–168. https://doi.org/10.1579/0044-7447-31.2.159

Oka Y, Tkachi N, Shuker S, Yerumiyahu U (2007) Enhanced nematicidal activity of organic and inorganic ammonia-releasing amendments using neem extracts. J Nematol 39:9–16

Oosten MJV, Pepe O, Pascale S, Silletti S, Maggio A (2017) The role of biostimulants and bioeffectors as alleviators of abiotic stress in crop plants. Chem Biol Technol Agric 4:5. https://doi.org/10.1186/s40538-017-0089-5

Padikasan IA, Chinnannan K, Kumar S, Subramaniyan G (2018) Agricultural biotechnology: engineering plants for improved productivity and quality. Omics Technol Bio-Eng 2:87–104. https://doi.org/10.1016/B978-0-12-815870-8.00006-1

Parmar DK, Thakur DR, Jamwal RS (2016) Effect of long term organic manure application on soil properties, carbon sequestration, soil – plant carbon stock and productivity under two vegetable production systems in Himachal Pradesh. J Environ Biol 37:333–339

Pathan SM, Aylmore LAG, Colmer TD (2003) Properties of several fly ash materials in relation to use as soil amendments. J Environ Qual 32(2):687–693

Power JF (1994) Understanding the nutrient cycling process. J Soil Water Conserv 49(Suppl):16–23

Prasad UK, Prasad TN (1998) Effect of irrigation and nitrogen on growth, yield and water use of summer sunflower. In: Proceedings of extended summaries of first international agronomy congress, November 23–27, New Delhi, pp 88–89

Priydarshini P, Dubey PK, Singh A, Chaurasia R, Abhilash PC (2021) The Dasgupta review: resetting the stage for a new paradigm. Front Ecol Environ. https://doi.org/10.1002/fee.2451

Qian M, Wu H, Wang J, Zhang H, Zhang Z, Zhang Y, Lin H, Ma J (2016) Occurrence of trace elements and antibiotics in manure-based fertilizers from the Zhejiang Province of China. Sci Total Environ 559:174–181. https://doi.org/10.1016/j.scitotenv.2016.03.123

Rai PK, Singh M, Anand K, Saurabh S, Kaur T, Kour D, Yadav AN, Kumar M (2020) Role and potential applications of plant growth-promoting Rhizobacteria for sustainable agriculture. Elsevier, Amsterdam; ISBN 9780128205266

Rajasree V, Pugalendhi L (2021) Breeding vegetables for nutritional security. In: Jovandaric LPE-MZ (ed) Veganism–a fashion trend or food as a medicine. IntechOpen, Rijeka. p Ch. 7

Raman R (2017) The impact of genetically modified (GM) crops in modern agriculture: a review. GM Crop Food 8(4):195–208

Ramirez DA, Silva-Díaz C, Ninanya J, Carbajal M, Rinza J, Kakraliya SK, Kreuze J (2022) Potato zero-tillage and mulching is promising in achieving agronomic gain in Asia. Agronomy 12(7):1494

Saeid A, Chojnacka K (2019) Fertilizers: need for new strategies. In: Chandran S et al (eds) Organic farming, pp 91–116

Sarangi SK, Maji B, Sharma PC, Digar S, Mahanta KK, Burman D, Mandal UK, Mandal S, Mainuddin M (2020) Zero-tilled-Paddy straw mulched potato (*Solanum tuberosum*) cultivation in the coastal saline soils reduce soil salinity, increase yield and profitability. Proceedings 36:147

Sarkar D, Rakshit A (2023) Amalgamation of farmers' bio-priming knowledge in integrated nutrient Management for Sustainable Management of red cabbage soil under middle Gangetic Plains, India. Environ Manag 72:382–395

Schachtman PD, Ryoung S (2007) Nutrient sensing and signaling: NPKS. Annu Rev Plant Biol 58:47–69

Schafer T, Adams T (2015) The importance of microbiology in sustainable agriculture. In: Lugtenberg B (ed) Principles of plant-microbe interactions. Springer, Cham. https://doi.org/10.1007/978-3-319-08575-3_2

Sengupta A, Gunri SK (2015) Microbial intervention in agriculture: an overview. Afr J Microbiol Res 9(18):1215–1226

Seufert V, Ramankutty N (2017) Many shades of gray – the context-dependent performance of organic agriculture. Sci Adv 3:e1602638

Singh B, Singh BP, Cowie AL (2010) Characterisation and evaluation of biochars for their application as a soil amendment. Soil Res 48(7):516–525

Smith OM, Cohen AL, Reganold JP, Jones MS, Orpet RJ, Taylor JM, Crowder DW (2020) Landscape context affects the sustainability of organic farming systems. Proc Natl Acad Sci 117(6):2870–2878

Tandy S, Mundus S, Yngvesson J, de Bang TC, Lombi E, Schjoerring JK, Husted S (2012) The use of DGT for prediction of plant available copper, zinc and phosphorus in agricultural soils. Plant Soil 346:167–180

Tripathi V, Edrisi SA, Chen B, Gupta VK, Vilu R, Gathergood N, Abhilash PC (2017) Biotechnological advances for restoring degraded land for sustainable development. Trends Biotechnol 35(9):847–859

Tuck S, Winqvist C, Mota F, Ahnström J, Turnbull L, Bengtsson J (2013) Land-use intensity and the effects of organic farming on biodiversity: a hierarchical meta-analysis. J Appl Ecol 51. https://doi.org/10.1111/1365-2664.12219

Tuninetti M, Ridolfi L, Laio F (2020) Ever-increasing agricultural land and water productivity: a global multi-crop analysis. Environ Res Lett 15:Article 0940a2

UNDESA (2013) World population prospects: the 2012 revision, medium variant. United Nations Department of Economic and Social Affair. Available at: http://esa.un.org/unpd/wpp/unpp/panel_population.htm

Ussiri DAN, Lal R (2018) The soil-livestock-climate nexus. In: Lal R, Stewart BA (eds) Soil and climate. CRC Press, Boca Raton, p 44

Utami SR, Kurniawan S, Situmorang B, Rositasari ND (2012) Increasing P-availability and P-uptake using sugarcane filter cake and rice husk ash to improve Chinesse cabbage (*Brassica* Sp) growth in Andisol, EastJava. J Agric Sci 4(10):153–160. https://doi.org/10.5539/jas.v4n10p153

Van Huis A, Van Itterbeeck J, Klunder H, Mertens E, Halloran A, Muir G et al (2013) Edible insects: future prospects for food and feed security, No. 171. Food and Agriculture Organization of the United Nations. ISBN 978-92-5-107595-1

Vollset SE, Goren C, Yuan W, Cao J, Smith AE, Hsiao T, Bisignano C, Azhar GS, Castro E, Chalek J, Dolgert AJ, Frank T, Fukutaki K, Hay SI, Lozano R, Mokdad AH, Nandakumar V, Pierce M, Pletcher M, Robalik T, Steuben KM, Wunrow HY, Zlavog BS, Murray CJL (2020) Fertility, mortality, migration, and population scenarios for 195 countries and territories from 2017 to 2100: a forecasting analysis for the Global Burden of Disease Study. Lancet. https://doi.org/10.1016/S0140-6736(20)30677-2

White PJ, Greenwood DJ (2013) Properties and management of cationic elements for crop growth. In: Gregory PJ, Nortcliff S (eds) Soil conditions and plant growth. Blackwell publishing, Oxford, pp 160–194. ISBN 9781405197700

White PJ, George TS, Dupuy LX, Karley AJ, Valentine TA, Wiesel L, Wishart J (2013a) Root traits for infertile soils. Front Plant Sci 4:193. https://doi.org/10.3389/fpls.2013.00193

White PJ, George TS, Gregory PJ, Bengough AG, Hallett PD, McKenzie BM (2013b) Matching roots to their environment. Ann Bot 112:207–222

Withers PJA, Doody DG, Sylvester-Bradley R (2018) Achieving sustainable phosphorus use in food systems through circularisation. Sustain For 10:1804

Young MD, Ros GH, de Vries W (2021) Impacts of agronomic measures on crop, soil, and environmental indicators: a review and synthesis of meta-analysis. Agric Ecosyst Environ 319:107551

Zhang Y, Xu Z, Li J, Wang R (2021) Optimum planting density improves resource use efficiency and yield stability of rainfed maize in semiarid climate. Front Plant Sci 12:752606

Index

S. A. Edrisi et al. (eds.), *Sustainable Plant Nutrition and Soil Carbon Sequestration*, Sustainable Plant Nutrition in a Changing World,
https://doi.org/10.1007/978-3-031-53590-1

The manufacturer's authorised representative in the EU is Springer Nature Customer Service Centre GmbH, Europaplatz 3, 69115 Heidelberg, Germany. If you have any concerns regarding our products, please contact ProductSafety@springernature.com

Printed and bound by CPI Group (UK) Ltd, Croydon, CR0 4YY
01/10/2026
02225906-0014